ISO/IEC 导则 第1部分
ISO 技术工作程序合集

(第10版,2019)

ISO/IEC 著

国家标准化管理委员会 译
国家市场监督管理总局标准创新管理司

中国标准出版社

北 京

International Organization for Standardization
CP 401 • Ch. de Blandonnet 8
CH-1214 Vernier, Geneva
Phone: +41 22 749 01 11
Fax: +41 22 749 09 47
Email: copyright@iso.org
Website: www.iso.org
Published in Switzerland
© ISO/IEC 2019

International Electrotechnical Commission
CP 131 • 3, rue de Varembé
CH-1211 Geneva 20
Phone: + 41 22 919 02 11
Fax: + 41 22 919 03 00
Email: inmail@iec.org
Website: www.iec.org

All rights reserved. Unless otherwise specified, or required in the context of its implementation, no part of this publication may be reproduced or utilized otherwise in any form or by any means, electronic or mechanical, including photocopying, or posting on the internet or an intranet, without prior written permission. Permission can be requested from either ISO at the address below or ISO's member body in the country of the requester.

北京市版权局著作权合同登记号：图字 01-2019-0118

图书在版编目（CIP）数据

ISO/IEC 导则.第1部分.ISO技术工作程序合集：第10版：2019/ 国家标准化管理委员会，国家市场监督管理总局标准创新管理司译．—北京：中国标准出版社，2021.4
书名原文：ISO/IEC Directives，Part 1 Consolidated ISO Supplement —Procedures specific to ISO
ISBN 978-7-5066-9496-4

Ⅰ.①I… Ⅱ.①国… ②国… Ⅲ.①电子工业—质量管理体系—国际标准—2019 Ⅳ.① TM-65

中国版本图书馆 CIP 数据核字（2019）第 228350 号

中国标准出版社　出版发行
北京市朝阳区和平里西街甲 2 号（100029）
北京市西城区三里河北街 16 号（100045）
网址：www.spc.net.cn
总编室：（010）68533533　发行中心：（010）51780238
读者服务部：（010）68523946
中国标准出版社秦皇岛印刷厂印刷
各地新华书店经销
*
开本 787×1092　1/16　印张 23　字数 401 千字
2021 年 4 月第一版　2021 年 4 月第一次印刷
*
定价 89.00 元

**如有印装差错　由本社发行中心调换
版权专有　侵权必究
举报电话：（010）68510107**

审定和翻译委员会

主任委员 田世宏

副主任委员 崔 钢

主 审 李玉冰　　肖 寒　　白德美

审稿人员（按姓氏音序排序）

　　　　　　高丽稳　　黄 立　　李东方
　　　　　　李鹏程　　刘永东　　田 武
　　　　　　余云涛　　余 化　　张 亮

翻译人员（按姓氏音序排序）

　　　　　　蔡利花　　丁 宁　　陆小伟

译者的话

国际标准是全球治理体系和经贸合作发展的重要技术基础。随着我国对外开放程度的不断扩大，经济快速发展，科技水平持续进步，对我国标准化工作者参与国际标准化活动的能力提出了更高的要求，迫切需要我国专家和相关人员熟悉和掌握国际标准制修订工作程序和规则。为此，国家标准化管理委员会组织翻译了《ISO/IEC 导则》的重要出版物。

《ISO/IEC 导则》是国际标准化组织（ISO）和国际电工委员会（IEC）技术工作的重要文件，是 ISO/IEC 各成员体参与国际标准化技术活动的指导性文件。随着国际标准化事业的不断发展，ISO/IEC 在认真分析和总结实践经验的基础上，对技术工作程序不断进行修改和完善。2019 年版《ISO/IEC 导则》分三本书出版：第 1 本为《ISO/IEC 导则　第 1 部分　ISO 技术工作程序合集（第 10 版，2019）》；第 2 本为《ISO/IEC 导则　IEC 技术工作程序补充部分（第 13 版，2019）》；第 3 本为《ISO/IEC 导则　第 1 部分　JTC 1 技术工作程序合集（第 15 版，2019）》。

《ISO/IEC 导则　第 1 部分　ISO 技术工作程序合集（第 10 版，2019）》由《ISO/IEC 导则》的第 1 部分和 ISO 专用程序组成。书中提供了完整的 ISO 技术工作程序，供 ISO 委员会遵守，主要涉及通过技术委员会及其附属机构的活动制定和维护国际标准。

《ISO/IEC 导则　IEC 技术工作程序补充部分（第 13 版，2019）》，包括了对 ISO/IEC 导则的修改和补充内容，提供了 IEC 委员会应遵循的完整程序规则，经 IEC 标准化管理局批准，已在 IEC 内部实施。

《ISO/IEC 导则 第 1 部分 JTC 1 技术工作程序合集（第 15 版，2019）》包含了《ISO/IEC 导则》的第 1 部分和联合技术委员会（JTC 1）补充部分给出的程序规则，纳入了自 2018 年第 14 版出版以来 ISO 和 IEC 一致同意修改的所有内容。带修订模式（灰色底纹）的第 15 版可用于查阅修改的细节，英文版本中已删除的内容未作翻译，为方便读者查阅，保留了英文原文（含删除线）。

上述三本书中文版的出版，有助于我国标准化工作者不断跟踪和研究导则的变化内容，进而更加实质性地参与 ISO/IEC 国际标准化技术工作。为方便读者对照参考，书中附有相应的英文版。同时，希望通过本书普及国际标准化组织的相关知识，加强我国专家国际标准化能力建设，不断提升我国参与国际标准化活动的整体水平。

在本书编译过程中，我们得到了国内外标准化专家和标准化出版机构专家的大力支持和帮助，在此谨向他们表示衷心的感谢。

由于时间仓促，书中疏漏在所难免，恳请广大读者和标准化同行提出宝贵意见。

译 者

2020 年 9 月

目 录

0 引言（《ISO/IEC 导则 第 1 部分 ISO 技术工作程序合集》） 1

前言 3

1 技术工作的组织构架和职责 7

    1.1 技术管理局的任务 7

    1.2 技术管理局咨询组 7

    1.3 联合技术工作 8

    1.4 首席执行官的职责 8

    1.5 技术委员会的设立 9

    1.6 分委员会的建立 11

    1.7 参加技术委员会和分委员会工作 11

    1.8 技术委员会和分委员会主席 13

    1.9 技术委员会和分委员会秘书处 15

    1.10 项目委员会 18

    1.11 编辑委员会 18

    1.12 工作组 18

    1.13 委员会中具有咨询职能的小组 21

    1.14 临时工作组 21

    1.15 技术委员会间的联络 22

    1.16 ISO 与 IEC 之间的联络 22

    1.17 与其他组织的联络 22

2 国际标准的制定 26

    2.1 项目管理方法 26

    2.2 预研阶段 31

2.3 提案阶段 31

2.4 准备阶段 33

2.5 委员会阶段 34

2.6 征询意见阶段 36

2.7 批准阶段 38

2.8 出版阶段 39

2.9 可交付使用的成果的维护 39

2.10 更正和修正 42

2.11 维护机构 43

2.12 注册机构 43

2.13 版权 43

2.14 专利项目的引用（见附录 I）... 44

3 其他可交付使用的文件的制定 45

3.1 技术规范 45

3.2 可公开提供的规范（PAS）... 45

3.3 技术报告 46

4 会议 46

4.1 概述 46

4.2 召开会议的程序 47

4.3 会议语言 48

4.4 会议的取消 48

4.5 文件分发 48

4.6 远程参加委员会会议 48

5 申诉 49

5.1 概述 49

5.2 对分委员会决定的申诉 49

5.3 对技术委员会决定的申诉 50

5.4 对技术管理局决定的申诉 50

5.5 申诉期间工作的运行 50
附录 A（规范性） 指南 51
附录 B（规范性） ISO/IEC 联络和分工程序 54
附录 C（规范性） 对制定标准的提案的论证 58
附录 D（规范性） 秘书处资源和秘书资格 63
附录 E（规范性） 语言使用通用政策 65
附录 F（规范性） 项目制定的选择方案 69
附录 G（规范性） 维护机构 72
附录 H（规范性） 注册机构 73
附录 I（规范性） ITU-T/ITU-R/ISO/IEC 共用专利政策实施指南 74
附录 J（规范性） 技术委员会和分委员会的制定范围 90
附录 K（规范性） 项目委员会 92
附录 L（规范性） 管理体系标准提案 94
附录 M（规范性） 制定行业专用管理标准和管理体系标准
（MSS）的政策 117

以下附录仅适用于 ISO 的附录

附录 SA（规范性） ISO 行为准则 122
附录 SB（规范性） 文件分发 123
附录 SC（规范性） 战略业务计划 125
附录 SD（规范性） 项目阶段的矩阵表（见表 SD.1）....................................... 127
附录 SE（规范性） 文件编号 130
附录 SF（规范性） 承办会议 131
附录 SG（规范性） ISO 标准的第二（及后续）语言文本 136
附录 SH（规范性） 图形符号的标准化程序 138
附录 SI（规范性） 国际研讨会协议（IWA）的制定程序 143
附录 SJ（规范性） 表格 149
附录 SK（规范性） 委员会及工作组会议文件邮寄截止日期 150
附录 SL（当前空白——预留给新增附录）... 152
附录 SM（规范性） ISO 技术工作及发布文件的全球相关性 153
附录 SN（规范性） 注册机构（RA）政策 156
附录 SO（规范性） 与公共政策举措相关或支持性作用的 ISO、
IEC 标准的制定原则 162

附录 SP（当前空白——预留给新增附录） 164
附录 SQ（规范性） 技术工作领导人员的选拔标准 165
附录 SR（规范性） 限制可交付使用文件目的或用途的声明 169
附录 SS（规范性） 选择性执行委员会草案（CD）阶段——委员会
指南 172
附录 ST（规范性） 结对政策 174

参考文献 179

ISO/IEC Directives,Part 1 Consolidated ISO Supplement—
Procedures specific to ISO 181

0 引言（《ISO/IEC 导则　第 1 部分　ISO 技术工作程序合集》）

0.1 关于《ISO/IEC 导则　第 1 部分　ISO 技术工作程序合集》

《ISO/IEC 导则》规定了制定国际标准和其他出版物应遵循的基本程序。《ISO/IEC 导则　第 1 部分　ISO 技术工作程序合集》是《ISO/IEC 导则　第 1 部分》和《ISO 专用程序》组成的合集。

《ISO/IEC 导则　第 1 部分》以及《ISO/IEC 导则　第 1 部分　ISO 技术工作程序合集》规定了 ISO 委员会应遵循的整套程序规则。此外，可能还需参照其他文件。特别是，对于按照 ISO 与欧洲标准委员会（CEN）之间的技术合作协议（《维也纳协议》）与 CEN 合作的委员会，应参照"主席和秘书处关于实施《维也纳协议》的指南"。

还应注意：ISO/IEC JTC 1 "信息技术委员会"已同时采用了《ISO/IEC 导则　第 1 部分》和《JTC 1 技术工作程序合集》。

0.2 《ISO/IEC 导则　ISO 技术工作程序合集》的结构

《ISO/IEC 导则　第 1 部分　ISO 技术工作程序合集》的章节结构与《ISO/IEC 导则　第 1 部分》保持一致。为了使用方便，在《ISO/IEC 导则》的条款下，用带阴影字体突出标注 ISO 专用条款。如果某一特定条款未用灰色底纹突出标注，则意味着此条款非 ISO 专用条款。

《ISO/IEC 导则　第 1 部分　ISO 技术工作程序合集》中以"S"开头的表格和附录均为 ISO 专用，不按照《ISO/IEC 导则》的顺序排列。如果 ISO 以"S"开头的附录中引用《ISO/IEC 导则　第 1 部分》附录的内容，将明确加以说明。

对 ISO 补充部分前一版修改的条款如下：前言、1.7、1.8.1、1.8.3、1.9.2、1.9.3、1.9.4、1.12.1、1.12.6、2.1.6、2.9.1、3.2.2、附录 SD、附录 SH、附录 SI、附录 SK、附录 SQ、附录 SR、附录 ST 和参考文献。第 10 版的修订内容可在 www.iso.org/directives 查阅。

前言

《ISO/IEC 导则》分为两部分出版：

—— 第 1 部分：技术工作程序（以下简称《ISO/IEC 导则 第 1 部分》）

—— 第 2 部分：ISO 和 IEC 文件的结构和起草原则与规则（以下简称《ISO/IEC 导则 第 2 部分》）

另外，国际标准化组织（ISO）、国际电工委员会（IEC）和 ISO/IEC 第一联合技术委员会（JTC 1）还分别出版了对第 1 部分的独立补充部分，包括各自的专用程序。

本部分阐述了 ISO 和 IEC 在开展技术工作时应遵循的程序：主要涉及通过技术委员会及其附属机构的活动制定和维护国际标准。

ISO、IEC 和 ISO/IEC JTC 1 还在各自网站（http：//www.iso.org/directives、http：//www.iec.ch/ members_experts/refdocs/ 和 http：//www.jtc1.org）上向那些关注技术文件制定的人员提供其他指南和工具。

本文件为第 15 版，纳入了自 2018 年第 14 版出版以来 ISO 技术管理局和 IEC 标准化管理局一致同意修改的所有内容。《ISO/IEC 导则》的专用程序，分别以《ISO 技术工作程序》（亦称《ISO 技术工作程序合集》）《IEC 技术工作程序补充部分》或 ISO/IEC《JTC 1 技术工作程序》（亦称《JTC 1 技术工作程序合集》）的形式给出。这些补充部分应与本文件结合使用。

对前一版本修改的条款如下：前言、1.7.3、1.12.2、1.13.2、1.13.6、1.17.2、3.2.1、3.2.2、3.2.4、附录 B、附录 D、附录 E、附录 F、附录 I 和附录 M。带修订模式的第 15 版可用于查阅修改的细节。考虑到国际标准具有成本效益和适时性的特点，ISO 和 IEC 制定本程序，并且得到广泛认可和普遍应用。为了实现这些目标，本程序的制定基于以下原则。

a） 现行技术和项目管理

在这些程序框架内，通过应用现行技术（如信息技术工具）和项目管理方法，提高工作速度，方便专家和秘书处完成任务。

b） 协商一致

协商一致原则旨在解决实质性的反对意见，这是制定能获得许可并被广泛采用的国际标准的一条基本原则和必要条件。虽然加速开展技术工作是必要的，但在标准进入批准阶段之前要为讨论、协商和解决重大技术争议问题留出足够的时间。

关于"协商一致"原则的详细信息，见 2.5.6。

c) 纪律

国家成员体必须保证遵守时间要求的纪律，以避免过长和不确定的"停滞时间"。同样，为了避免重复讨论，国家成员体有责任保证其技术观点是在国家层面考虑所有相关利益方的基础上提出的，并且应在技术工作初期明确提出观点而非在最后（批准）阶段提出。国家成员体还必须认识到不要把实质性意见提交会议上讨论，因其他会议代表没有机会就该意见提前在本国进行充分磋商，因而无法迅速达成协商一致。

d) 成本效益

本程序考虑了运行总成本。"总成本"的概念包括各国家成员体的直接支出、日内瓦办公机构的支出（主要由国家成员体交纳的会费资助）、差旅费、国家和国际工作组以及委员会专家的工时费。

尽可能提供参加远程会议的机会。

e) 表决和决定的通用原则

委员会秘书处应确保委员会无论是在全体会议上还是以通信方式作出的决定都应有记录，在ISO以委员会决议的形式体现。

对于采用通信方式的投票或在委员会会议期间进行的投票，除《ISO/IEC导则》另有规定，需P成员简单多数投票同意方可通过。

委员会领导层应确保，会议前以书面形式提交的投票在会上予以考虑。

在文件投票表决阶段（NP、CD或其他后续阶段），禁止在会议期间对其进行正式讨论或通过委员会正式渠道分发国家成员体意见。

在所有投票中，弃权不计算在内。

ISO不允许委托投票。

f) ISO国际标准的全球相关性

ISO的目标和期望是，其每一项国际标准都能代表全球共识，且满足全球市场需求。为实现这一目标，需采取特别措施，尤其要确保ISO的技术工作兼顾发展中国家的需求。如其中一项措施是在《ISO/IEC导则 第1部分 ISO技术工作程序合集》中给出了关于"结对关系"的具体规定为提升能力，两个ISO成员体之间建立伙伴关系（见1.7、1.8.3、1.9.2、1.9.3、1.9.4和附录ST）。

虽然该等规定限定于技术工作，但"结对关系"可发生在多个层面，特别是协助来自发展中国家的被结对方开展其标准化、合格评定和IT基础设施建设，其目标是被结对方在开展其活动方面实现自给自足。

g) 委员会专用程序

《ISO/IEC导则 第1部分 ISO技术工作程序合集》是基于多年来多个委员会和

领域的标准制定经验制定的。因此，委员会没有必要制定委员会专用程序（CSP）（包括委员会事实程序的决议）。如果委员会认为《ISO/IEC 导则》（或《ISO/IEC 导则 第 1 部分 ISO 技术工作程序合集》）不能充分解决某一具体问题，宜向 ISO 技术管理局提交修订/扩大《ISO/IEC 导则》（或《ISO/IEC 导则 第 1 部分 ISO 技术工作程序合集》）的申请，而不是制定委员会专用程序。如果委员会认为《ISO/IEC 导则》（或《ISO/IEC 导则 第 1 部分 ISO 技术工作程序合集》）完全无法解决某一具体操作问题，可制定 CSP，包括制定该程序的明确理由，并提交 ISO 技术管理局审批，ISO 技术管理局将全力作出快速地回应。

h) 标准的可持续性

在制定标准时，鼓励委员会参考《ISO 指南》"标准指南 82"），在标准中关注可持续性，见http：//isotc.iso.org/livelink/livelink?func=ll&objId=8389141&objAction=browse&viewType=1。

i) 本文件使用的术语

注 1：为方便起见，本文件在适用处均采用了下表中的术语，表示 ISO 和 IEC 中相似或相同的概念。

本文件术语	ISO	IEC
国家成员体 National Body	成员体 Member Body（MB）	国家委员会 National Committee（NC）
技术管理局 Technical Management Board（TMB）	ISO 技术管理局 Technical Management Board（ISO/TMB）	标准化管理局 Standardization Management Board（SMB）
首席执行官 Chief Executive Officer（CEO）	秘书长 Secretary-General	秘书长 General Secretary
CEO 办公室 Office of the CEO	中央秘书处 Central Secretariat（CS）	中央办公室 Central Office（CO）
理事局 Council Board	理事会 Council	理事局 Council Board（CB）
咨询组 Advisory Group	技术咨询组 Technical Advisory Group（TAG）	咨询委员会 Advisory Committee
秘书（委员会或分委员会） Secretary（of a Committee or Subcommittee）	委员会主管 Committee Manager	秘书 Secretary
关于其他概念，请参照《ISO/IEC 指南 2》。		

注2：此外，本文件中还使用了下列缩略语：

JTAB	联合技术咨询局
JCG	联合协调小组
JPC	联合项目委员会
JTC	联合技术委员会
JWG	联合工作组
TC	技术委员会
SC	分委员会
PC	项目委员会
WG	工作组
PWI	预研工作项目
NP	新工作项目提案
WD	工作草案
CD	委员会草案
DIS	国际标准草案（ISO）
CDV	投票用委员会草案（IEC）
FDIS	最终国际标准草案
PAS	可公开提供的规范
TS	技术规范
TR	技术报告

1 技术工作的组织构架和职责

1.1 技术管理局的任务

各组织的技术管理局负责技术工作的全面管理,具体如下:

a) 设立技术委员会。

b) 任命技术委员会主席。

c) 指派或重新指派技术委员会秘书处和分委员会秘书处(在某些情况下)。

d) 批准技术委员会的名称、范围和工作计划。

e) 批准由技术委员会提出的设立和解散分委员会的决定。

f) 如有必要,安排技术工作特定项目的优先顺序。

g) 协调技术工作,包括对多个技术委员会相关主题的标准制定进行指派、分工或协调。为协助此项工作,技术管理局可在相关领域设立专家咨询小组,就其基础、行业及跨行业协调工作提供咨询,并对相关的计划及新工作需求提出建议。

h) 在 CEO 办公室的协助下,监控技术工作的进展情况,并采取适当的行动。

i) 审查对新技术领域的工作的需求并进行规划。

j) 维护 ISO/IEC 导则和其他技术工作规则。

k) 考虑国家成员体提出的原则性事项及对新工作项目提案、委员会草案、征询意见草案或最终国际标准草案决定的申诉。

注1:对新工作项目提案、委员会草案、征询意见草案及最终国际标准草案的术语解释见第 2 章。

注2:关于 ISO 技术管理局的任务和职责的详细信息,见 ISO/TMB(技术管理局)的职权范围——https://www.iso.org/committee/4882545.html;关于 IEC,见 http://www.iec.ch/dyn/www/f?p=103:47:0::::FSP_ORG_ID,FSP_LANG_ID:3228,25。

1.2 技术管理局咨询组

1.2.1 就 1.1 g) 而言,具有咨询职能的小组可由:

a) 其中一个技术管理局设立;

b) 两个技术管理局联合设立。

注:在 IEC 中,这类小组被称为咨询委员会。

1.2.2 成立这类小组的提案中应包括拟成立小组的职责和组成建议,切记为了保证其有效运作,在确保充分代表相关方的同时,要尽可能限制小组的规模。例如,可以决定其成员仅由相关技术委员会的主席和秘书组成。无论如何,技术管理局都应决定拟实施的准则,并指定小组的成员。

这类小组对其职责、组成或工作方法(适当时)提出的任何变更建议都应提交技术管理局审批。

1.2.3 指派给这类小组的任务可包括对出版物（特别是国际标准、技术规范、可公开提供的规范及技术报告）的起草或协调提出建议，但不应包括起草这些文件，除非由技术管理局特别授权。

1.2.4 任何准备出版的文件都应按照附录 A 的程序制定。

1.2.5 咨询组的工作结果应以建议的形式提交到技术管理局。这种建议可包括为起草出版物成立工作组（见 1.12）或联合工作组（见 1.12.6）的提案。如有这类工作组，应在相关的技术委员会内开展工作。

1.2.6 咨询组内部文件只应分发给其成员，并抄送给 CEO 办公室。

1.2.7 规定任务一旦完成或如果随后决定其工作可由正常的联络机构完成，咨询组应解散（见 1.16）。

1.3 联合技术工作

1.3.1 联合技术咨询局（JTAB）

联合技术咨询局的任务是避免或消除 ISO 和 IEC 技术工作中可能或实际重复的工作，并在其中一个组织认为需要作出联合计划时采取措施。联合技术咨询局只处理现行程序下较低层面不能解决的问题（见附录 B）。这些问题除了技术工作外，可能还涉及策划和程序等内容。

联合技术咨询局的决定要通知 ISO 和 IEC 两个组织立即实施，这种决定至少在 3 年内不得申诉。

1.3.2 联合技术委员会（JTC）和联合项目委员会（JPC）

1.3.2.1 根据 ISO 技术管理局和 IEC 标准化管理局的共同决定或联合技术咨询局的决定可设立联合技术委员会和联合项目委员会。

1.3.2.2 应由两个组织协商同意，由其中一个组织承担联合项目委员会的管理责任。按照一个成员/国家一票的原则参与其工作。

若在同一国家的两个国家成员体都选择参加一个联合项目委员会的工作，应确定一个成员体承担该委员会的管理责任。承担管理责任的国家成员体有责任协调在其国家的活动，包括文件分发、评论和投票。

此外，还要遵循项目委员会的正常程序（见 1.10）。

1.4 首席执行官的职责

各组织的首席执行官主要负责在其组织内部实施《ISO/IEC 导则》和其他技术工作规则。为此，CEO 办公室负责各技术委员会、理事局和技术管理局间的衔接。

未经 ISO 或 IEC 首席执行官、ISO/IEC 联合技术咨询局或技术管理局的授权，不得制定与本文件不一致的程序。

1.5 技术委员会的设立

1.5.1 技术委员会由技术管理局设立和解散。

1.5.2 经与相关技术委员会协商后，技术管理局可将现有的分委员会转换成新的技术委员会。

1.5.3 针对新技术活动领域设立新技术委员会的提案可由下列组织提出：

—— 国家成员体；

—— 技术委员会或分委员会；

—— 项目委员会；

—— 政策委员会；

—— 技术管理局；

—— 首席执行官；

—— 有相关组织支持的负责管理认证体系的机构；

—— 具有国家成员体资格的其他国际组织。

1.5.4 提案应使用规定的表格，可从 www.iso.org/forms 和 http：//www.iec.ch/standardsdev/resources/docpreparation/forms_templates/ 下载电子版（通常为 word 文档），提案的内容包括：

a） 提案方；

b） 提案的主题；

c） 提案的范围和初始工作计划；

d） 提案的论证；

e） 如果适用，在其他机构开展的类似工作的调查报告；

f） 认为有必要与其他机构的联络。

有关新工作提案的详细内容见附录 C。该表格应提交给 CEO 办公室。

1.5.5 CEO 办公室应确保提案是按 ISO 和 IEC 的要求（见附录 C）正确编制的，并应提供足够信息以支持国家成员体作出明确决定。CEO 办公室还应评估提案与现行工作的关系，可以征询有关方面意见，包括技术管理局或主持现行有关工作的委员会。必要时，可设立一个临时工作组对该提案进行审查。

CEO 办公室审查后，可以决定将提案退给提案方，以便在分发投票前作进一步修改。在这种情况下，提案方应按建议进行修改或提供不修改的理由。如提案方不进行修改并且要求对原提交的提案分发投票，技术管理局将决定采取适当的措施。这可能包括停止该提案直到对提案进行修改或对收到的建议进行投票。

在所有情况下，CEO 办公室都可以将评议意见和建议填入提案表格中。

有关提案论证的详细内容，见附录 C。

积极鼓励提案方在编制提案过程中与其他国家成员体进行非正式沟通。

在某些情况下，ISO/TMB 认为可考虑非正式的意见征询。

技术委员会由 ISO/TMB 临时设立。在技术委员会首次会议后的 18 个月内，临时设立的技术委员会需要编制一份战略业务计划，供 ISO/TMB 审查（见附录 SC）。技术委员会在战略工作计划审查通过时，由 ISO/TMB 正式设立。这并不妨碍在这 18 个月期间启动标准化项目。

1.5.6　CEO 办公室应将提案分发给各组织（ISO 或 IEC）的所有国家成员体，询问他们是否：

　　a）　支持设立新的技术委员会，并提供一份陈述理由的说明（"理由陈述"）；

　　b）　计划积极参与新技术委员会的工作（见 1.7.1）。

该提案还应提交给另一组织（IEC 或 ISO）征询意见并达成协议（见附录 B）。

应在提案分发后的 12 周内用适当的表格对提案作出答复。若国家成员体未提供上述 1.5.6 a）提及的说明，其赞成票或反对票均不予以登记和考虑。

对提案的答复意见表格已替换为电子投票系统，不使用电子投票系统的答复意见将不计算在内。

1.5.7　技术管理局对答复进行评估，并且决定设立一个新的技术委员会，基于：

　　1）　参加投票表决的国家成员体中 2/3 多数赞成此项提案；

　　2）　至少有 5 个投赞成票的国家成员体表达了其积极参加活动的意向。

并指派秘书处（见 1.9.1），或者根据上述相同的验收标准，把工作指派给现有技术委员会。

1.5.8　应按其成立的顺序对技术委员会进行编号，如果某个技术委员会解散，其编号不得指派给另一个技术委员会。

1.5.9　在作出设立新的技术委员会决定之后应尽快安排必要的联络（见 1.15～1.17）。

1.5.10　新技术委员会在成立之后应就其名称和范围尽快形成一致意见，最好采用通信方式。

技术委员会的名称和范围需经 P 成员表决，获得 2/3 多数赞成票方可通过。

范围是明确界定技术委员会工作界限的声明。

规定技术委员会的范围应以"……的标准化"或"……领域的标准化"句子开始，措辞应尽可能简明。

有关业务范围的建议见附录 J。

达成一致意见的名称和范围应由首席执行官提交技术管理局批准。

1.5.11 技术管理局或技术委员会可以提出修改技术委员会名称和范围的建议，修改的措辞应由技术委员会确定并由技术管理局批准。

1.5.12 "待命"——当技术委员会或分技术委员会在工作方案中没有任何任务，但仍保留其名称、范围和秘书处以便在需要分配新任务时予以重新激活，则称该技术委员会或分技术委员会处于"待命"状态。

由负责相关委员会提案的技术管理委员会决定，将该委员会置于"待命"状态或予以重新激活。

1.6 分委员会的建立

1.6.1 设立和解散分委员会由上级技术委员会参加投票的 P 成员的 2/3 多数决定，并需技术管理局批准。只有当某一国家成员体表示愿意承担秘书处工作的条件下方可设立分委员会。

1.6.2 分委员会设立时，应该包括至少 5 个表示愿意积极参加（见 1.7.1）分委员会工作的上级技术委员会成员。

1.6.3 一个技术委员会的分委员会应按顺序设立编号，如果某个分委员会已经解散，其编号不得分配给另一个分委员会，除非解散分委员会是技术委员会进行全面调整工作的一部分。

1.6.4 分委员会的名称及范围应由上级技术委员会决定，并应包括在已确定的上级技术委员会范围内。

1.6.5 上级技术委员会秘书处应采用适当表格将其设立分委员会的决定通知 CEO 办公室，CEO 办公室再将表格提交技术管理局批准。

1.6.6 设立新分委员会的决定得到批准后，应尽快安排与其他机构建立必要的联络关系（见 1.15 ~ 1.17）。

1.7 参加技术委员会和分委员会工作

由于发展中国家成员体时常缺乏足够的资源，无法参加对其国民经济产生重要作用的全部委员会。因此，发展中国家成员体被邀请与更有经验的 P 成员达成 P 成员结对协议，根据结对协议，结对主导方的 P 成员将确保其被结对 P 成员的意见传达给对应的 ISO 委员会且由其纳入考虑。被结对 P 成员也应具有 P 成员身份（见注）且在中央秘书处注册为被结对 P 成员。

注：相关成员体自行决定最有效的实施结对办法。如该结对 P 成员赞助被结对 P 成员的专家参加委员会会议或作为工作组专家，或该结对 P 成员就特定议程项目/文件征求被结对 P 成员的意见，并确保被结对 P 成员的意见以书面形式提交至委员会秘书处。

所有结对协议的细节应通知有关委员会秘书处和主席，并通知委员会成员和 CEO 办公室，以确保最大程度的透明度。

在任何一个委员会中，结对主导方的 P 成员只能与一个 P 成员建立结对关系。

被结对成员应以通信的方式对委员会所有问题自行投票。

有关结对成员的更多信息，见附录 ST《结对政策》。

根据《ISO 章程和议事规则》，通信成员和注册成员不具备 P 成员资格，ISO 的通信成员可注册为委员会观察员，但无权提出意见。在 ISO 内，选择成为委员会 P 成员的国家成员体拥有额外的义务，按照该委员会的职责对所有的系统复审投票进行表决。

1.7.1 所有的国家成员体都有权参加技术委员会和分委员会工作。

为了获得最高的工作效率，维护必要的工作纪律，各国家成员体都应针对每个技术委员会和分委员会的情况向 CEO 办公室明确表示它是否打算：

—— 积极参加其工作，有义务对技术委员会或分委员会内正式提交投票的所有问题、新工作项目提案、征询意见草案和最终国际标准草案进行投票以及对会议作出贡献（P 成员）；

—— 或者，以观察员身份跟进工作，因此接收委员会文件并有权提出评议意见和参加会议（O 成员）。

ISO 的 P 成员有义务对所有的复审标准进行投票。

国家成员体可以选择既不作为某一技术委员会的 P 成员，也不作为 O 成员。在这种情况下，该国家成员体对该技术委员会工作既无以上规定的权利，也不承担上述义务。但所有国家成员体无论其在技术委员会或分委员会中的身份如何，都有权对国际标准的征询意见草案（见 2.6）和最终草案（见 2.7）进行投票。

国家成员体有责任组织国内各相关方，考虑其利益，有效、及时地提交国家层面的意见。

1.7.2 分委员会的成员资格对任何国家成员体开放，无论他们在上级技术委员会中的身份如何。

在分委员会成立之际，应为技术委员会成员提供机会明确他们成为分委员会 P 成员或 O 成员的意向。

技术委员会成员不意味着自动成为分委员会成员，因此国家成员体应正式告知其成为分委员会身份的意向。

1.7.3 国家成员体可在任何时间开始、结束或改变其在技术委员会或分委员会中的成员身份。对于 IEC，须通知 CEO 办公室或相关委员会秘书处；对于 ISO，须按照 1.7.4 和 1.7.5 的要求在全球目录系统中直接输入。

1.7.4 如果某技术委员会或分委员会的 P 成员出现以下情况，该技术委员会或分委员会秘书处应通知 CEO 办公室：

—— 长期不积极，并且连续两次既未直接参加，也未以通信方式为会议作出贡献，未任命专家参加技术工作；

—— 在IEC，对技术委员会或分委员会内正式提交投票的问题没有投票（见1.7.1）；

—— 在ISO，在超过一个日历年内对技术委员会或分委员会内正式提交进行委员会内部投票（CIB）的20%以上（并至少2个）的问题没有投票（见1.7.1）。

一旦收到这种通知，首席执行官应提醒国家成员体有义务积极参加技术委员会和分委员会活动。如果该国家成员体对此提示没有作出令人满意的回复，而且连续出现不符合P成员要求的行为，该国家成员体的身份应自动降为O成员。这样改变身份的国家成员体在12个月之后可以向首席执行官表明其希望重新成为该技术委员会P成员的愿望，在这种情况下，可以批准其成为P成员。

注：本条款不适用于指南的制定。

1.7.5 如果某技术委员会或分委员会的P成员没有参加对相关委员会起草的征询意见草案和最终国际标准草案投票表决，或者没有对本委员会负责的可交付使用的文件进行系统复审投票，首席执行官应提醒该国家成员体履行投票义务。如果对该提醒没有作出满意的回应，则该国家成员体应自动降为O成员。这样改变身份的国家成员体在12个月后可以向首席执行官表明其重新成为该委员会P成员的愿望，在这种情况下，可以批准其为P成员。

注：本条款不适用于指南的制定。

1.8 技术委员会和分委员会主席

1.8.1 任命

技术委员会主席应由本技术委员会秘书处提名，由技术管理局批准，任期最长为6年或酌情缩短。允许延长任期，累计最长为9年。

分委员会主席应由该分委员会秘书处提名，并由相关技术委员会批准，任期最长为6年或酌情缩短。允许延长任期，累计最长为9年。批准任命和延期的准则是相关技术委员会P成员中2/3多数赞成。

技术委员会或分委员会秘书处可在现有主席任期结束前一年内提交新主席的提名。提前一年任命的主席应指定为该委员会的"候任主席"。这一做法旨在为候任主席担任委员会主席之前提供学习机会。

如果两个国家委员会（NSB）同意共享一个ISO委员会的领导权，他们可共同指定一名主席，并共同商定支持主席工作的方式。在这种情况下，鼓励承担委员会秘书处工作的国家委员会从另一个国家委员会推举主席。

为确保透明，国家委员会应将任何合作关系告知委员会成员。

主席候选人应具备附录 SQ《技术工作负责人的选拔标准》列出的能力和特点（见 SQ.3.1）。

1.8.2 职责

技术委员会主席负责全面管理其技术委员会的工作，其中包括各分委员会和工作组的工作。技术委员会或分委员会主席应：

a) 仅以国际身份工作，放弃他/她的本国观点，因此，他/她不能在其委员会内同时作为国家成员体的代表。

b) 指导技术委员会或分委员会秘书履行其职责。

c) 召集会议，以对委员会草案达成一致意见（见 2.5）。

d) 保证在会议上充分归纳所表达的各种观点，以便所有与会者都能理解。

e) 保证在会议上明确阐述所有决定，并由秘书提供书面文件，以便会议期间确认。

f) 在征询意见阶段作出适当的决定（见 2.6）。

g) 就技术委员会的重要事项通过技术委员会秘书处向技术管理局提供咨询意见。为此，他/她应通过分委员会秘书处得到任意一个分委员会主席的报告。

h) 确保技术管理局的政策和战略决定在委员会内的实施。

i) 确保制定和持续维护战略业务计划，战略业务计划涵盖技术委员会和所有向技术委员会报告工作的组织，包括所有分委员会的活动。

j) 确保委员会战略业务计划得到正确一贯的实施并应用于技术委员会或分委员会的工作计划中。

k) 在对委员会的决定有投诉的情况下给予协助。

如果主席因不可预见的原因不能出席会议，与会代表可选举本次会议主席。

分委员会主席应按要求出席上级委员会的会议，并可参加讨论，但无投票权。在特殊情况下，如果主席不能出席，应委派秘书（或 ISO 和 IEC 中的一名代表）代表分委员会。如果分委员会代表不能出席，则应提交书面报告。

1.8.3 联合主席

鼓励各委员会制定主席层面的结对协议，其中一个 ISO 成员体作为结对主导方，另一个 ISO 成员体作为被结对方（一个委员会仅限一名联合主席）。联合主席应是相应委员会的 P 成员（即无论是否通过结对协议成为 P 成员）。主席和联合主席的委任和任期制度一致。职责的划分应经双方协商决定，并写入提名主席和联合主席的两成员体的结对协议，并通知委员会成员和 CEO 办公室。

关于结对的更多信息，见附录 ST《结对政策》。

1.9 技术委员会和分委员会秘书处

1.9.1 指派

技术委员会秘书处工作应由技术管理局指派给国家成员体承担。

分委员会秘书处应由上级技术委员会指派给国家成员体承担。但是，如有两个或多个国家成员体提出申请承担同一分委员会秘书处工作，技术管理局应对分委员会秘书处的指派作出决定。

两个国家委员会（NSB）可就 ISO 委员会秘书处的工作进行合作，在战略思考上相互支持，分担工作的资源负担。但只有其中一方将视为委员会正式秘书处，且该等协议不得视为"联合秘书处"。

为确保透明，相关国家标准机构应将合作关系通知委员会成员。

无论是技术委员会还是分委员会，其秘书处应指派给符合下列条件的国家成员体，即该国家成员体应：

a) 表明其积极参与技术委员会或分委员会工作的意愿；

b) 同意履行秘书处的职责，保证有足够的资源用于支持秘书处工作（见 D.2）。

一旦将技术委员会或分委员会秘书处指派给某一国家成员体，该国家成员体则应任命一位合格人员担任秘书（见 D.1 和 D.3）。

秘书应具备附录 SQ《技术工作负责人的选拔准则》列出的能力和特点（见 SQ.3.2）。

TC/SC 秘书处一般每隔 5 年应接受 ISO 技术管理局的再次确认。

如果在 TC/SC 秘书处应接受复审的某一特定年份，TC/SC 工作产出数据显示该委员会遇有困难，这将提请 ISO 技术管理局注意是否就再次进行确认征询作出决定。秘书处未获悉接受详细检查的，其安排情况将自动再次确认。

如果秘书长或委员会 P 成员请求，也可随时启动再次确认调查。该请求应附以书面理由，由 ISO 技术管理局审议并决定是否开展再次确认调查。

调查应在委员会 P 成员中展开，以确定 P 成员是否对秘书处具备足够的可用资源和对其工作表现是否感到满意。作出否定答复的任何 P 成员应表明是否愿意承担委员会秘书处的工作。

ISO 技术管理局应对 TC 秘书处进行调查，而 TC 秘书处应对 SC 秘书处进行调查。如果，同一成员体同时承担 TC 秘书处和 SC 秘书处，应由 ISO 技术管理局开展调查。

如果没有否定答复，秘书处的安排情况应进行再次确认。TC 和 SC 秘书处的所有否定答复均应提交给 ISO 技术管理局裁决。

1.9.2 职责

已被指派承担秘书处工作的国家成员体应确保对相应的技术委员会或分委员会提

供技术或行政服务。

秘书处负责监控、报告并确保工作积极推进，应竭力早日圆满完成工作。这些任务应尽可能以通信方式进行。

秘书处有责任确保 ISO/IEC 导则和技术管理局的决定得以执行。

秘书处仅以国际身份工作，放弃其本国观点。

秘书处应确保按时完成下列各项工作：

a） 工作文件

   1） 准备委员会草案，安排草案的分发，并处理收到的意见。

   2） 准备征询意见草案和作为最终国家标准草案分发或作为国际标准出版的文本。

   3） 必要时，在能够并愿意负责语言版本的其他国家成员体的协助下，确保英语和法语文本的等效（另见 1.11 和对《ISO/IEC 导则》的独立补充部分）。

b） 项目管理

   1） 协助确定每个项目的优先次序和目标日期。

   2） 向 CEO 办公室通报所有工作组和维护小组召集人和项目负责人的姓名等。

   3） 积极提出出版备选的可交付使用的文件，或由于严重超时和/或缺乏足够的支持取消项目的建议。

c） 会议（见第 4 章）

   1） 确立会议议事日程，并安排分发。

   2） 安排分发会议议事日程中列出的所有文件，包括工作组报告，并指明会议期间需要讨论的所有其他文件（见 E.5）。

   3） 起草会议的决定（亦称"决议"），确保批准工作组建议的决定，包含被赞同的具体要素：

   —— 以书面形式提供决定，供会议期间确认（见 E.5）；

   —— 在会议结束后 48 小时内将决定公布在委员会的电子文件夹中。

   4） 编写会议记录，在会议后 4 周内分发传阅。

   5） 编写报告提交技术管理局（TC 秘书处），IEC 会后 4 周内提交，或提交至上级委员会（SC 秘书处）。

d） 建议

   向主席、项目负责人和召集人就项目进展相关的程序提供咨询建议。

在任何情况下，各秘书处应与其技术委员会或分委员会的主席保持密切联系。

技术委员会秘书处应就其活动，包括其分委员会和工作组的活动，与 CEO 办公室和技术委员会成员保持密切联系。

分委员会秘书处应与上级技术委员会秘书处保持密切联系，并在必要时与CEO办公室保持密切联系。委员会还应就其活动，包括其工作组的活动，与分委员会成员保持联系。

技术委员会或分委员会秘书处应与CEO办公室一起更新委员会成员身份的记录。CEO办公室也应保留其工作组成员名单。

鼓励成员体在秘书处和被结对秘书处之间建立结对协议，每个委员会只能有一个被结对秘书处。牵头伙伴将承担秘书处和被结对伙伴秘书处的工作。被结对秘书处应来自相应委员会的P成员（直接或通过一个结对协议）。秘书处和被结对秘书处的指派规则是一致的，秘书和被结对秘书的指派规则也是一致的。职责的划分应由相应成员体协商决定，并记录在结对协议中，同时通知委员会成员和CEO办公室。

关于结对的更多信息，见附录ST《结对政策》。

1.9.3 技术委员会秘书处的变更

如果某一国家成员体希望放弃技术委员会秘书处工作，该国家成员体应立即通知CEO办公室，至少提前12个月发出通知。由技术管理局作出将该秘书处移交另一国家成员体的决定。

如果某一技术委员会秘书处始终未能履行本程序规定的职责，首席执行官或国家成员体可将此事提交技术管理局，技术管理局可对秘书处的指派情况进行评价，以便将秘书处工作移交另一个国家成员体。

若技术委员会秘书处结对协议的结对主导方希望放弃秘书处工作，需执行相应的程序，见附录ST《结对政策》ST.5.7"终止结对协议"。

1.9.4 分委员会秘书处的变更

如果某一国家成员体希望放弃分委员会秘书处工作，该国家成员体应立即通知上级技术委员会秘书处，至少提前12个月发出通知。

如果某一分技术委员会的秘书处始终未能履行本程序中规定的职责，首席执行官或国家成员体可以将此事提交上级技术委员会处理，可以通过P成员的多数票同意决定重新指派分委员会秘书处。

无论是上述哪种情况。技术委员会秘书处都应发出通知征询分委员会其他P成员承担秘书处的意愿。

如有两个或多个国家成员体愿意承担同一分委员会秘书处工作，或是由于技术委员会结构的原因，该秘书处的重新指派与技术委员会秘书处的重新指派密切相关，则由技术管理局对分委员会秘书处的重新指派作出决定。如果只有一个国家成员体愿意承担秘书处工作，则由上级技术委员会自行指派。

若分委员会秘书处的结对协议结对主导方希望放弃秘书处工作，需执行相应的程序，见附录 ST《结对政策》ST.5.7 "终止结对协议"。

1.10 项目委员会

项目委员会由技术管理局设立，制定不属于现有技术委员会范围内的标准。

注：此类标准有一个参考编号但可以分为多个部分。

附录 K 中规定了项目委员会的程序。

希望转化为技术委员会的项目委员会应遵循设立新技术委员会的程序（见 1.5）。

1.11 编辑委员会

建议技术委员会成立一个或多个编辑委员会，以便更新和编辑委员会草案、征询意见草案和最终国际标准草案，确保这些文件符合《ISO/IEC 导则 第 2 部分》的有关要求（见 2.6.6）。

这类委员会至少应包括：

—— 一位以英语为母语并具有一定法语知识的技术专家；

—— 一位以法语为母语并具有一定英语知识的技术专家；

—— 项目负责人（见 2.1.8）。

项目负责人和 / 或秘书可直接负责一种相关语言的文本。

为更新和编辑已经以通信方式通过的草案，以便进一步处理，若相关技术委员会和分委员会秘书处提出要求，编辑委员会应举行会议。

编辑委员会应配备处理和提供电子文本的手段（见 2.6.6）。

1.12 工作组

1.12.1 技术委员会或分委员会可根据委员会决定为完成专项任务设立工作组（见 2.4）。工作组应通过上级技术委员会指定的召集人向上级技术委员会或分委员会报告工作。

工作组召集人应由委员会任命，最长任期为 3 年，在任期期满后上级委员会的下一次全会上终止其任期。这种任命应经相关国家成员体（或联络组织）确认。召集人可被再次任命，最长任期为 3 年。任期次数不限。

召集人的变更均由委员会负责，而不是由国家成员体（或联络组织）负责。

需要时秘书处可支持召集人工作。

工作组秘书的提名应经其国家确认。

鼓励委员会在召集人和被结对召集人层面达成协议，一个工作组限定一名被结对召集人。结对主导方（召集人）和被结对方（联合召集人）由双方协商决定。召集人和被结对召集人应来自相应委员会的 P 成员，召集人和被结对召集人的任命规则一致。

注意，被结对召集人只可能通过结对协议产生。

联合召集人只可能在联合工作组（JWG）内，见 1.12.6。

关于结对的更多信息见附录 ST《结对政策》。

工作组是由 P 成员、上级委员会的 A 类联络组织和 C 类联络组织分别指定的限定数量的专家组成，旨在承担指派给工作组的特定任务。专家以个人身份工作，不作为指派他们的 P 成员或 A 类联络组织（见 1.17）的官方代表，那些由 C 类联络组织（见 1.17）指派的专家除外。但建议他们与指派他们的 P 成员或组织保持密切联系，以便将工作组的工作进展情况和各种意见尽量在早期阶段通报给 P 成员或组织。

建议对工作组的规模予以合理限制。为此，技术委员会或分委员会可决定每个 P 成员和联络组织指派专家的最大数。

一旦作出成立工作组的决定，应正式通知 P 成员及 A 类联络组织和 C 类联络组织，以便任命专家。应按其成立的顺序对工作组编号。

当委员会在其会议上决定成立工作组时，应立即指定召集人或代理召集人，召集人或代理召集人应在 12 周内安排召开第 1 次工作组会议。委员会会议结束后应立即将上述信息通报给技术委员会 P 成员、A 类联络组织和 C 类联络组织，并邀请他们在 6 周内任命专家。可酌情将其他项目指派给现有的工作组。

1.12.2 工作组的组成在 ISO 全球目录系统（GD）或 IEC 专家管理系统（EMS）中界定。未在 ISO GD 或 IEC EMS 中注册的工作组的专家不得参加其工作。召集人可邀请相关人员旁听会议。

如果利益相关方类别缺失，召集人应通知委员会秘书处，秘书处应再次召集专家。

1.12.3 持续不积极的专家是指没有通过参加工作组会议或以通信方式为会议作出贡献的专家。CEO 办公室可以根据技术委员会或分委员会秘书的请求，征求 P 成员意见将其从工作组中清除。

如果工作组专家未达到最少专家数量（依据 2.3.5 规定），召集人应通知委员会秘书处，秘书处应再次召集专家。

如果后续召集专家仍未达到最少专家数量，召集人应咨询委员会秘书处，以决定能否继续或是否应该继续该项目。如果在达不到最少专家数的情况下决定继续进行，委员会秘书处应请求获得 ISO 技术管理局的许可。

1.12.4 其任务一旦完成（通常是在上一项目征询意见阶段结束时）（见 2.6），工作组应由委员会决定解散，项目负责人继续承担顾问工作，直至出版阶段完成（见 2.8）。

1.12.5 工作组的内部文件及其报告应依据 ISO/IEC 导则独立补充部分中规定的程序发行。

1.12.6 在特殊情况下可设立联合工作组（JWG），承担多个 ISO 和/或 IEC 技术委员会或分委员会相关的专项任务。收到联合工作组请求的委员会应及时答复。

注：有关 ISO 和 IEC 委员会的联合工作组的具体规则，除下列规定外还见附录 B。

设立联合工作组的决定应附有涉及下列事项的委员会双方协议：

—— 负责管理该项目的委员会/组织；

—— 联合工作组的召集人，应由其中一个委员会的 P 成员提名，可选择由其他委员会任命共同召集人；

—— 联合工作组的成员资格（成员资格可对希望参加的所有 P 成员、A 类和 C 类联络组织开放，如果同意，可以限定每个委员会派出同等数量的代表）。

负责管理该项目的委员会/组织应：

—— 在其工作计划中记录该项目；

—— 在构成 JWG 的所有委员会中召集专家；

—— 负责处理评议意见（通常反馈给联合工作组），并确保该项目在其各个阶段的评议意见和投票得到适当的汇总和处理（见 2.5、2.6 和 2.7），并向委员会领导提供所有评议意见；

—— 根据 2.5、2.6 和 2.7 规定的程序准备用于委员会阶段、征询意见阶段和批准阶段的草案；

—— 向其他委员会秘书处发送所有相关文件（会议纪要、工作草案、委员会草案、咨询及审批阶段），在其各自委员会中传阅和/或采取行动。

—— 负责该出版物的维护。

批准准则依据牵头管理的委员会使用的导则。如果牵头委员会是 JTC 1 委员会，JTC 1 补充部分合集也适用。

对于提案阶段（NP）：

—— 对于 ISO/ISO JWGs，只需要进行 NP 投票：如果已在一个委员会中启动或批准一个 NP，将不能在另一个 TC 中再次投票，但 ISO/IEC JWG（每个组织一个）应进行两个 NP 投票。

—— 在较后阶段建立联合工作组，在这种情况下将由相关技术委员会确定其牵头管理。

—— 一旦同意设立联合工作组，负责牵头管理的委员会将其牵头事宜和参加其工作的委员会分别通告 ISO CEO 办公室和 IEC 中央办公室。

—— 向其他技术委员会发出征集参加联合工作组的专家的通知。

对于准备阶段（WD）：

—— 联合工作组与其他工作组的功能一样：需要协商一致进入 CD。

对于委员会阶段（CD）：

—— 分发 CD，供各委员会审阅和评论；

—— 按《ISO/IEC 导则 第 1 部分》的规定，CD 最终要求所有委员会的协商一致。

对于 DIS 和 FDIS 投票：

—— 要求国家成员体征求所有相关的国家对口委员会意见以确定立场。在征求意见函的封面页有提请 NSB 注意的说明。

—— 对于 ISO/IEC 联合工作组（JWG），进行两次 DIS/FDIS 投票，每个组织各进行一次。

"前言"中明确参与制定可交付使用的文件的所有委员会。

1.13 委员会中具有咨询职能的小组

1.13.1 技术委员会或分委员会可成立具有咨询职能的小组，帮助委员会主席和秘书完成与协调、策划及指导委员会工作或其他具有咨询特性的具体任务。

1.13.2 成立咨询组的提案中应包括关于咨询组的组成建议和职责范围，包括成员资格条件。为保证其有效工作，在尽可能限制其规模的同时要切记充分代表利益相关方的要求。咨询组成员应为委员会官员和其他个人，由国家成员体和/或相关 A 类联络组织提名。上级委员会应在咨询组成立和提名前批准其最终组成和职责范围。

对于主席的咨询组，应审议平等参与条款。

为了让发展中国家的国家成员体更多地参与 ISO 委员会的管理，强烈建议作出特别规定，为发展中国家的代表在委员会建立的任何顾问组中分配席位。这些代表应由相应委员会的 P 成员体提名（直接或通过结对协议）。

1.13.3 指派给这类小组的任务可包括提出有关起草或协调出版物（特别是国际标准、技术规范、可公开提供的规范及技术报告）的提案，但不包括这些文件的起草。

1.13.4 咨询组应以建议的形式将其成果提交设立该小组的机构。建议中可包括为准备出版物设立工作组（见 1.12）或设立联合工作组（见 1.12.6）的提案。

1.13.5 咨询组内部文件仅发给本组成员，并抄送给相关委员会秘书处及 CEO 办公室。

1.13.6 专项任务一旦完成并得到上级委员会批准，咨询组即宣布解散。

1.14 临时工作组

技术委员会或分委员会可成立临时工作组，研究某一特定问题。该小组应在本次会议或最迟在下一次会议上向其上级委员会报告其所研究的问题。

临时工作组成员应从出席上级委员会会议的代表中选出，如有必要，可由委员会指派的专家补充。上级委员会还应指定一名召集人。

临时工作组应在提交报告的会议上自动解散。

1.15 技术委员会间的联络

关于图形符号标准化的协调问题，见附录 SH。

1.15.1 每个组织内部工作在相关领域的技术委员会和/或分委员会间应建立并保持联络关系。需要时，还要与负责标准化基础工作（如术语和图形符号）的技术委员会建立联络关系，联络中还应包括基础文件（包括新工作项目提案和工作草案）的交换。

委员会可作出决议，即决定建立内部联络。委员会不能拒绝收到的内部联络请求，且收到请求的委员会无需通过确认接受决议。

1.15.2 保持上述这种联络关系是相关技术委员会秘书处的职责。技术委员会秘书处可将这项任务指派给分委员会秘书处。

1.15.3 技术委员会或分委员会可指定一名或若干名联络组织代表跟进另一个与其建立联络关系的技术委员会或其所属的一个或多个分委员会的工作，应将这些联络组织代表的有关信息通知相关委员会秘书处，该秘书处应将所有相关的文件传递给联络组织代表及其所属的技术委员会或分委员会秘书处。被任命的联络组织代表应向其所属秘书处提交工作进展报告。

1.15.4 这些联络组织代表应有权参加指派跟进其工作的技术委员会或分委员会会议，但他们不应有投票权。他们可以参加会议讨论，包括技术委员会权限内的相关事宜和基于从所属委员会收集的反馈信息提交书面意见。他们还可参加该技术委员会或分委员会的工作组会议，但是只能就自己权限内的相关事宜发表自己委员会的观点，不能参加其工作组的活动。

1.16 ISO 与 IEC 之间的联络

1.16.1 安排 ISO 和 IEC 技术委员会和分委员会之间适当的联络十分重要。ISO 和 IEC 技术委员会和分委员会之间建立联络关系的通信渠道是通过 CEO 办公室。就任一组织研究的新课题而言，每当某一组织考虑一个新的或修订的工作计划并且可能与另一个组织相关时，首席执行官就力求使两个组织间达成协议，从而使此项工作不出现重叠或重复等现象（见附录 B）。

1.16.2 由 ISO 或 IEC 指定的联络组织代表应有权参加其跟进的另一个组织的技术委员会或分委员会的讨论，可以提交书面建议，但他们不应有投票权。

1.17 与其他组织的联络

1.17.1 适用于所有联络类别的通用要求

为确保有效性，联络应双向运行并有合适的互惠安排。

在工作初期阶段就应考虑对联络的需求。

无论版权归联络组织所有，还是归其他方所有，联络组织应遵循依据《ISO/IEC 导则》的有关版权的政策（见 2.13）。应将版权政策声明提供给联络组织并要求联络组织就其可接受性提出明确说明。联络组织无权就提交文件收取费用。

联络组织应按需要自愿参与 ISO 或 IEC 的技术工作。

联络组织应在其限定的相关技术和工业领域内具有足够的代表资格。

联络组织应遵循 ISO/IEC 程序，包括知识产权（IPR）（见 2.13）。

联络组织应接受 2.14 有关专利权的要求。

技术委员会和分委员会应对他们所有联络协议定期进行复审，至少每两年复审一次或在委员会的每次会议上进行复审。

1.17.2 联络类别

1.17.2.1 在技术委员会 / 分委员会层面（A 类和 B 类联络）

在技术委员会 / 分委员会层面的联络类别如下：

—— A 类：在技术委员会或分委员会解决问题中，对技术委员会或分委员会工作作出有效贡献的组织。这类组织收到所有相关文件，并被邀请参加会议，他们可以提名专家参加 WG（见 1.12.1）。

—— B 类：已表明希望了解技术委员会或分委员会工作的组织。这类组织可收到技术委员会或分委员会的工作报告。

注：B 类联络供政府间组织使用。

A 类和 B 类联络的建立程序：

—— 希望建立 A 类和 B 类联络的组织应向 CEO 办公室递交申请，并向技术委员会或分委员会官员和 IEC CO 技术官员和 ISO CS 技术项目主管递交申请副本，申请包括下列内容：

　　—— 组织为非盈利组织；

　　—— 组织将向全球或较大范围区域成员开放；

　　—— 组织的活动和成员资格能够说明其具备起草国际标准的能力和专业技能，并且在推动标准在相关技术委员会或分委员会领域实施具备权威性（只与 A 类联络相关）；

　　—— 主要联系人姓名。

注：在提交申请之前，组织将一直与技术委员会或分委员会官员保持联系，在这种情况下，技术委员会或分委员会官员应确保组织了解 1.17.1 规定的版权义务，并且同意 ISO/IEC 程序，包括知识产权和专利权。

—— CEO办公室将对满足合格标准予以确认，然后与申请组织总部所在地的国家成员体进行商议。
—— 如果申请组织总部所在地国家成员体反对，申请事宜将移交至技术管理委员会予以裁决。
—— 如果申请组织总部所在地国家成员体未反对，申请事宜将提交至技术委员会或分委员会秘书，并要求分发投票。
—— 批准A类和B类联络须P成员的2/3多数赞成。

1.17.2.2 在工作组层面（C类联络）

在工作组层面的联络类别为：

C类：积极参与工作组活动并作出技术贡献的组织，包括制造商协会、商业协会、产业联盟、用户群以及行业协会和科学学会。联络组织应是具有个人、公司或国家成员资格的多国组织（在其目标和标准制定活动中），性质上可以是长期的或临时的。

1.17.3 资格

1.17.3.1 在技术委员会/分委员会层面（A类和B类）

某一组织申请作为ISO技术委员会/分委员会的联络组织时，CEO办公室将与该组织所在的国家成员体进行核查。如果该国家成员体认为该联络申请不符合资格准则，应将该问题提交TMB确定资格。

CEO办公室还要确保该组织符合下列资格准则：

—— 不以营利为目的；
—— 是合法实体，CEO办公室将要求其提供其章程；
—— 其成员资格对全球或广泛区域成员开放；
—— 通过其活动和成员资格证明其有能力和专门知识参与国际标准的制定或有权推进国际标准的实施；
—— 有利益相关方参与和达成协调一致的决策过程，以便形成利益相关方提供的建议（见ISO联络组织指南：凝聚利益相关方并达成协商一致，http://www.iso.org/iso/guidance_liaison-organizations.pdf）。

1.17.3.2 在工作组层面（C类联络）

当某组织申请与某一工作组建立联络关系，CEO办公室将与该组织所在的国家成员体进行核查，并且要确保该组织符合下列资格准则：

—— 不以营利为目的；
—— 通过其活动和成员资格证明其有能力和专业知识参与国际标准的制定或有

权利推进国际标准的实施；

—— 有利益相关方参与和达成协调一致的决策过程，以便形成利益相关方提供的建议（见 ISO 联络组织指南：凝聚利益相关方并达成协商一致，http：//www.iso.org/iso/guidance_liaison-organizations.pdf）。

1.17.4 受理（A、B 和 C 类联络）

建立 A、B 和 C 类联络组织的协议需要参加投票 P 成员的 2/3 多数赞成。

鼓励委员会在一项工作项目研制的初期就寻求所有相关方的参与。如果 C 类联络组织在某一特定工作项目研制阶段晚期提交申请，相关的 P 成员将考虑该组织加入后的价值。

1.17.5 权利与义务

1.17.5.1 在技术委员会/分委员会层面（A 和 B 类联络）

技术委员会和分委员会如有可能，应寻求具有联络身份的组织对每份相关文件的充分且正式的支持。应同样对待来自联络组织的任何评议意见和来自国家成员体的评议意见。联络组织未提供充分支持不应认为就是反对。若认为这些不支持（异议）是反对，委员会可参考 2.5.6 寻求进一步指导。

1.17.5.2 在工作组层面（C 类联络）

C 联络组织有权作为正式成员参加工作组、维护组或项目组（见 1.12.1），但不能作为项目负责人或召集人。C 类联络组织专家是任命他们的组织的正式代表。他们只可应邀参加委员会全体会议。如果他们受委员会邀请参会，只可作为观察员。

1.17.6 项目委员会转换为技术委员会或分委员会时带转联络关系

若项目委员会转化为技术委员会或分委员会，新的技术委员会或分委员会应通过一项决议确认带转 A 类和 B 类联络组织。这一决议的批准要求参加投票 P 成员的 2/3 多数赞成。联络类别见表 1。

表 1 联络类别

类别	A	B	C
目的	为委员会的工作作出有效的贡献	了解委员会的工作情况	为工作组起草标准作出技术贡献
资格	——非盈利 ——合法实体 ——会员制（全球或广泛区域） ——相关能力和专门知识 ——利益相关方参与和达成协商一致的决策过程 （详见 1.17.3.1）	仅政府间组织 ——非盈利 ——合法实体 ——会员制（全球或广泛区域） ——相关能力和专门知识 ——利益相关方参与和达成协商一致的决策过程 （详见 1.17.3.1）	——非盈利 ——相关能力和专门知识 ——利益相关方参与和达成协商一致的决策过程 （详见 1.17.3.1）

表 1（续）

类别	A	B	C
层面	TC/SC	TC/SC	工作组
参与	参加 TC/SC 会议，获取文件，可任命专家参加工作组，这些专家可担任召集人或项目负责人	仅了解工作（获取文件）	作为工作组成员充分参与（但不能 成为召集人或项目负责人）
权利与义务	没有投票权，但可以发表意见（其评论与成员体的评论相同对待）。可以提出新的工作项目（见2.3.2）	没有投票权，但可以发表意见（评论与成员体的评论相同对待）。不能提出新的工作项目	如果委员会明确邀请，专家可出席委员会会议，但只能以观察员身份出席。不能提出新的工作项目

2 国际标准的制定

2.1 项目管理方法

2.1.1 概述

技术委员会或分委员会的主要职责是制定并维护国际标准，同时，也要积极鼓励技术委员会和分委员会考虑出版第 3 章中所描述的可交付中间出版物。

应按照下面描述的项目管理方法制定国际标准。

2.1.2 战略业务计划

附录 SC 描述了战略工作计划的目标及其制定和审批流程。

每个技术委员会都应为其专业领域活动制定战略业务计划：

a）考虑制定工作计划的商业环境；

b）指明工作计划中正在拓宽的领域、已经完成的领域、接近完成或稳步发展的领域，以及没有进展宜撤销的领域（见 2.1.9）；

c）评价需要修订的工作（见 ISO/IEC 导则独立补充部分）；

d）提出对新需求的展望。

战略业务计划应由技术委员会正式商定，并纳入到技术委员会报告中以供技术管理局定期复审和批准。

注：在 ISO 内，TMB 仅审批首个战略工作计划，各委员会负责确保其常规更新和修订。

2.1.3 项目阶段

2.1.3.1 表 2 说明了技术工作项目各阶段的顺序，并列出了与项目各阶段有关的文件名称。有关技术规范、技术报告和可公开提供的规范的制定在第 3 章中予以描述。

表 2 项目阶段及有关文件

项目阶段	有关文件	
	名称	缩写
预研阶段	预研工作项目 [a]	PWI
提案阶段	新工作项目提案 [a]	NP
准备阶段	工作草案 [a]	WD
委员会阶段	委员会草案 [a]	CD
征询意见阶段	征询意见草案 [b]	ISO/DIS IEC/CDV
批准阶段	最终国际标准草案 [c]	FDIS
出版阶段	国际标准	ISO、IEC 或 ISO/IEC

[a] 如附录 F 所述,这些阶段可省略。
[b] ISO 为国际标准草案,IEC 为投票用委员会草案。
[c] 可省略(见 2.6.4)。

2.1.3.2 F.1 用图示说明了国际标准出版前的步骤。

2.1.3.3 ISO/IEC 导则 ISO 补充部分和 IEC 补充部分给出了项目阶段矩阵图,并用数字标出相关的子阶段。

为了便于监控项目开发,ISO 采用了系统的项目管理方法,将项目划分为阶段和子阶段。

项目管理体系与详细的项目跟踪系统关联,即分级协调阶段代码系统《ISO 导则 69:1999 协调阶段代码系统(第二版)——原理和使用指南》。附录 SD 给出了该项目跟踪系统的矩阵图,并用数字标出了相关子阶段。在特定阶段完成相应节点规定的活动和决定且按时通知 ISO 中央秘书处的项目,将被录入 ISO 中央秘书处的数据库。

2.1.4 项目描述和接受

项目是推进完成一个新的、修正或修订国际标准的任何工作。一个项目可细分成若干子项目(见 2.1.5.4)。

只有根据相关程序(见 2.3 中关于新工作项目提案及 ISO/IEC 导则独立补充部分中关于现行国际标准的复审和维护)接受了某个项目的提案,此项目方可开展工作。

2.1.5 工作计划

2.1.5.1 技术委员会或分委员会的工作计划包括所有指派给该技术委员会和分委员会的项目,其中包括对已出版标准的维护。

2.1.5.2 在制定工作计划过程中,每个技术委员会或分委员会都应考虑行业规划的需求,以及外部技术委员会对国际标准的需求,即其他技术委员会、技术管理局咨询组、政策级委员会及 ISO 和 IEC 以外的组织对国际标准的需求（见 2.1.2）。

2.1.5.3 项目应在技术委员会通过的范围内，项目的选择应依据 ISO 和 IEC 的政策目标和资源进行仔细推敲（见附录 C）。

2.1.5.4 对工作计划中的每个项目都应给出编号（见《ISO/IEC 导则　IEC 技术工作程序》：有关 IEC 的文件编号），并应将这个编号保留在工作计划中直至该项目完成，或直至同意撤销该项目为止。如果技术委员会或分委员会认为有必要把一个项目分成若干子项目，则可给每个子项目一个分编号。子项目应被完全涵盖在原项目的范围之内，否则应提交一个新工作项目提案。

2.1.5.5 工作计划应注明每个项目酌情分配的分委员会和 / 或工作组。

2.1.5.6 新技术委员会一致同意的工作计划应提交技术管理局批准。

2.1.6　目标日期

对于工作计划中的每个项目，技术委员会或分委员会应确定完成下列每个阶段的目标日期：

—— 完成工作草案初稿（如果新工作项目提案方只提供工作文件大纲，见 2.3）；

—— 分发委员会草案初稿；

—— 分发征询意见草案；

—— 分发最终国际标准草案（经 CEO 办公室同意）；

—— 出版国际标准（经 CEO 办公室同意）。

注：委员会可依照附录 SS 决定省略 CD 阶段。根据 2.6.4，如果不载入技术更改，则应省略 FDIS 阶段。

这些目标日期应适应尽可能短的制定时间，以便快速制定国际标准，并应向 CEO 办公室报告，由 CEO 办公室向各国家成员体发送相关信息。关于目标日期的确定，请见 ISO/IEC 导则的独立补充部分。

在确定目标日期时，应考虑项目之间的关系。应优先考虑作为其他国际标准实施基础的国际标准制定项目。对于技术管理局认可的对国际贸易有重要影响的项目应给予最优先考虑。

技术管理局还可指示相关技术委员会或分委员会秘书处向 CEO 办公室提交最新的有效草案作为技术规范出版（见 3.1）。

应对所有的目标日期经常进行复审，必要时应进行修改，并在工作计划表中清楚标明。修改后的目标日期应通知技术管理局。对于工作计划表中 5 年以上尚未达到批准阶段的所有工作项目，技术管理局将予以撤销（见 2.7）。

2.1.6.1　概述

当新提案获批（无论是新的可交付使用文件，还是对现行可交付使用文件的修订），将结论提交至 ISO 中央秘书处时，委员会秘书处应同时表明选定标准的制定轨迹〔所有目标日期从项目获批的日期开始计算，获批工作项目（AWI），阶段10.99〕。

> 注：制定轨迹的各阶段截止日期应根据具体问题而设立。
> **SDT 18 标准制定轨迹——18 个月发布；**
> **SDT 24 标准制定轨迹——24 个月发布；**
> **SDT 36 标准制定轨迹——36 个月发布；**
> **SDT 48 标准制定轨迹——48 个月发布。**
>
> 注：如果采用 18 个月制定轨迹的项目在项目登记 13 个月内成功完成 DIS 的投票，则应适用 ISO/CS 规定的"直接发布程序"。该程序将发布时间减少约 1/3。

要求委员会秘书处在项目计划期间进行风险评估，为提前识别潜在问题和合理性设置目标日期。委员会秘书处应持续审查目标日期，并应确保这些目标日期持续得到评估，且在各委员会会议上进行确认或修订。这些审查也应努力确认各项目仍与市场相关，如果发现不再需要这些项目或者可能需要太长时间完成，从而导致市场参与者采用替代方案时，这些项目应被取消或转化为其他可交付使用文件（见 2.1.6.2）。

> 注：在标准制定期间，花费在循环试验上的时间不应计算在整体制定时间内。循环试验期间，可根据秘书处向 ISO/CS 申请暂停标准制定轨迹。

2.1.6.2 自动取消（和恢复）项目

如果超出 DIS（阶段 40.00）或发布（阶段 60.60）的目标日期，委员会应在 6 个月内作出以下决定：

a) 处于准备阶段或委员会阶段的项目：提交 DIS——如果技术内容是可接受的或成熟的；处于征询意见阶段的项目：提交第二个 DIS 或 FDIS——如果技术内容是可接受的或成熟的。

b) 发布一个 TS——如果技术内容是可接受的，但作为未来的一项国际标准可能是不完全成熟的。

c) 发布一个 PAS——如果技术内容是可接受的但作为未来的一项国际标准或一个 TS，可能是不完全成熟的。

d) 发布一个 TR——如果技术内容作为一个 TS 或未来的一项国际标准不被认可接受，资料丰富且被认为是公众利益。

e) 向 ISO/TMB 提交延期请求——如果未达成共识，但利益相关方有强烈的兴趣继续，委员会可授予项目总周期最多 9 个月的延期，但建议发布期间提供可交付使用文件（如 PAS 或 TS）。

f) 取消工作项目——如果委员会没有解决方案。

如果在 6 个月结束时未采取以上任何措施，ISO 中央秘书处应自动取消该项目。该项目只能由 ISO 技术管理局批准恢复。

2.1.7 项目管理

技术委员会或分委员会秘书处负责管理该技术委员会或分委员会工作计划中的所有项目，包括对已达成一致同意的目标日期的进展情况进行监控。

如果没有在目标日期内完成工作（见 2.1.6），并对此项工作没有给予充分的支持（即不再符合 2.3.5 中规定的新工作接受要求），负责管理的委员会应撤销这个工作项目。

2.1.8 项目负责人

对于每个项目的起草，技术委员会或分委员会应在考虑新工作项目提案方提名（见 2.3.4）的基础上指定项目负责人（工作组召集人、指派的专家，如果适当时，也可包括秘书）。应确定项目负责人能够获得开展起草工作的适宜资源。项目负责人应仅从国际立场出发，放弃其国家的观点。当对提案阶段到出版阶段（见 2.5 ~ 2.8）产生的技术问题提出要求时，项目负责人应作好顾问角色的准备。

秘书处应将项目负责人的姓名、地址及相关项目的编号报送 CEO 办公室。

2.1.9 进度控制

分委员会和工作组应定期向技术委员会报告项目的进展情况（见《ISO/IEC 导则》的 ISO 和 IEC 技术工作程序），他们的秘书处之间的会议将有助于控制进度。

CEO 办公室负责监控所有工作的进度，并定期向技术管理局汇报有关情况。为此，CEO 办公室应收到在《ISO/IEC 导则》的 ISO 和 IEC 各自技术工作程序中指明的文件副本。

为支持中央秘书处来监控所有工作进度并定期向 ISO 技术管理局汇报，委员会秘书处应确保每次文件分发后 ISO 中央秘书处收到通知。

2.1.10 保存记录的职责

对于委员会工作和国际标准及其他 ISO 可交付使用文件出版背景资料的记录保存，在委员会秘书处和中央秘书处之间进行职责划分。当秘书处职责从一个成员体转换到另一个成员体时，维护这些记录尤为重要。同样重要的是，在国际标准和其他 ISO 可交付使用文件的准备阶段，若出版物的技术内容来源引起纠纷，可随时检索关键决定和重要的相关通信记录。

委员会秘书处应建立和维护本委员会所有的官方交易记录，特别是经审批的会议纪要和决议的副本。工作文件的副本、投票结果等应至少保留至所涉出版物已修订或完成其下一次系统性复审，但在任何情况下均应至少保留至相关国际标准或其他可交付使用文件出版后 5 年。

ISO 中央秘书处应保留所有国际标准和其他可交付使用文件的参考副本，包括已撤销版本，并应保留成员体对这些出版物的最新投票记录。DIS 和 FDIS 的副本，包括相关的投票报告和最终证据应至少保留至所涉出版物已修订或完成其下一次系统性复审,但在任何情况下至少保留至出版后 5 年。

2.2 预研阶段

2.2.1 通过其 P 成员的简单多数表决,技术委员会或分委员会可将尚不完全成熟、不能进入下一阶段且不能确定目标日期的预研工作项目(例如涉及新兴技术的项目)纳入工作计划中。

这类项目中可包括战略业务计划中列出的那些项目,特别是在 2.1.2 d)"对新需求的展望"中所列的项目。

2.2.2 所有预研工作项目都应列入工作计划中。

2.2.3 对于所有的预研工作项目,有关委员会都应定期复审。委员会应对每个预研工作项目的市场相关性和所需要的资源进行评价。

任何预研工作项目若在 TC/SC 规定的截至期满时(针对 IEC)、在 3 年内还未进入提案阶段(针对 ISO),将自动从工作计划中删除。

2.2.4 本阶段可用于预研新工作项目提案(见 2.3)并制定初始草案。

2.2.5 所有这类项目在进入准备阶段前,都应经过 2.3 中规定的程序批准。

2.3 提案阶段

关于编制管理体系可交付使用文件的提案情况见附录 L。

2.3.1 NP 涉及下列内容:

—— 新标准;

—— 现行标准的新增部分;

—— 技术规范(见 3.1)或公开可用规范(见 3.2)。

以下情况不要求 NP 阶段(2.3):

—— 现有标准、TS 或 PAS(如果在其 6 年使用期限内)的修订或修正;

—— TS 或 PAS 转化为国际标准(IS)。

然而,委员会应通过包含以下要素的决议:

1)目标日期。

2)确认不扩大范围。

3)召集人或项目负责人。同时委员会应发起召集专家。

对于 TS 或 PAS 转化为 IS,需要 2/3 同意。

如果修订或修正导致范围扩大,则适用 2.3[应发起 NP 投票且要求表 4[1)]]。

2.3.2 现有技术委员会或分委员会范围内新工作项目提案可由下列相关组织提出:

---

1) ISO 的第 4 号表格,新工作项目提案表——译者注。

—— 国家成员体；

—— 技术委员会或分委员会秘书处；

—— 另一个技术委员会或分委员会；

—— A 类联络组织；

—— 技术管理局或其中某个咨询组；

—— 首席执行官。

2.3.3 如果涉及 ISO 和 IEC 两个组织的技术委员会，首席执行官应安排必要的协调（见附录 B）。

2.3.4 每个新工作项目提案均应使用适当表格提交，且经过充分论证并恰当编制成文件（见附录 C）。

新工作项目提案方应：

—— 尽量提供工作草案初稿供讨论，或至少应提供该工作草案大纲；

—— 提名一名项目负责人；

—— 提交适当的表格之前，应与委员会讨论提案，以决定适当的制定周期（根据市场需求）和起草工作计划，包括关键节点和建议的第一次会议日期。

对于属于现有委员会范围内的建议，该表应提交给 CEO 办公室或相关的委员会秘书处。

CEO 办公室或相关委员会主席和秘书处应确保该提案按照 ISO 和 IEC 要求（见附录 C）正常制定，并且提供充分信息支持国家成员体作出合理决定。

CEO 办公室或相关委员会主席和秘书处还应评定提案与现行工作的关系，他们可以咨询利益相关方，包括技术管理局或开展现行工作的委员会。必要时，可以成立临时工作组审查提案。对提案的任何评审均不应超过 2 周时间。

在所有情况下，CEO 办公室或相关委员会主席和秘书处都可在提案表中加入评议意见和建议。

项目委员会的新工作项目提案见附录 K。

已完成的表格的副本应分发给技术委员会或分委员会成员，供 P 成员进行书面投票和供 O 成员和联络组织成员参考。

应在表格中标明建议出版的日期。

可以用通信方式对一项新工作项目提案作出决定。

应在 12 周内反馈投票结果。

委员会可以根据具体情况通过决议来决定对新工作项目投票期限缩短至 8 周。

填写投票表格时，国家成员体应提供论证其反对票决定的陈述（见论证说明）。如果未提供这种说明，该国家成员体投的反对票将不予登记和考虑。

2.3.5 满足以下要求，项目即获得许可：

a) 参加投票的技术委员会或分委员会 P 成员的 2/3 多数票批准工作项目，统计时不包括弃权票；

b) 在 16 个或以下 P 成员的委员会至少有 4 个 P 成员，在 17 个或以上的 P 成员的委员会至少有 5 个 P 成员，承诺积极参与项目制定工作，也就是在准备阶段通过提名技术专家和对工作草案提出评议意见作出有效贡献。在统计时，只考虑已经批准将该工作项目［见 2.3.5a)］纳入其工作计划的 P 成员。如果一个国家成员体承诺积极参与，但在附有赞成票的表中未指定专家，那么在确定本次投票是否满足批准准则时不予以登记和考虑。

如果成员体对 NP 所投反对票没有提出明确的理由陈述，则委员会秘书处应联系该成员体，要求其在两周内作出解释。

如果成员体未在 2 周内作出回应，该投票结果将不计算在内。

秘书处不应对理由作出价值评判，如有疑问应向成员体提出。

如果成员体未在表格中指定专家，则将在产生投票结果后 2 周内指定专家。如果超过期限未指定，该成员体的参与将不计算在内，从而影响［2.3.5b)］的审批要求。

各委员会可提高对指定专家的最低要求。

在可以证明只有极少数 P 成员具有产业和/或技术知识的前提下，技术委员会可以请求技术管理局允许开始时可少于 4 位或 5 位提名专家。

2.3.6 某个新工作项目提案一旦获得许可，它将以适当的顺序作为一个新项目纳入相关技术委员会或分委员会的工作计划中。在适当表格中应注明一致同意的目标日期（见 2.1.6）。

投票结束后 4 周内向 ISO 中央秘书处［用表 6[2)]］和 IEC 中央办公室［用表 RVN[2)]］报告表决结果。

2.3.7 工作项目被纳入工作计划后，提案阶段即告结束。

2.4 准备阶段

2.4.1 准备阶段包括：依据《ISO/IEC 导则 第 2 部分》的要求准备 WD。

2.4.2 如果一个新工作项目获得许可，项目负责人应与 P 成员在批准阶段提名的专家一起工作［见 2.3.5 a)］。

2.4.3 秘书处可在会议上或以通信方式向技术委员会或分委员会提出成立工作组的建

---

2) ISO 的第 6 号表格：Result of voting on new work item proposal（新提案投票结果报告表）——译者注。

议。通常情况下,工作组的召集人担任项目负责人职务。

这样的工作组应由技术委员会或分委员会负责成立,技术委员会或分委员会应确定工作组的任务,并确定向技术委员会或分委员会提交草案的目标日期(见1.12)。工作组召集人应保证所开展的工作在已投票的工作项目的范围内。

2.4.4 在答复设立工作组的提议时,表示同意积极参与[见2.3.5 a)]的每个P成员应确认其技术专家。其他P成员或A类、C类联络组织也可提名专家。

2.4.5 项目负责人负责项目的制定工作,通常负责召集和主持工作组所有会议。项目负责人可邀请工作组的一名成员担任秘书。

2.4.6 为避免项目制定在后期拖延,应尽一切努力准备英语、法语两种语言文本。

如果要制定三种语言(英语、法语、俄语)的标准,应包括俄语文本在内。

2.4.7 关于本阶段的时间限制见2.1.6。

2.4.8 当工作草案作为CD初稿分发给技术委员会或分委员会成员时,CEO办公室负责登记,准备阶段即告结束。委员会还可决定将最终工作草案作为PAS(见3.2)出版,以满足市场的特定需求。

如果委员会选择省略 CD 阶段(见 2.5),待征询意见草案(DIS)可供传阅后准备阶段结束。

2.5 委员会阶段

2.5.1 委员会阶段是考虑国家成员体意见的主要阶段,旨在在技术内容上达成一致。因此国家成员体应认真研究委员会草案文本并在本阶段提交所有相关评议意见。

委员会可根据附录 SS 决定省略 CD 阶段。

任何图形符号均应提交至负责图形符号注册的相关 ISO 委员会(见附录 SH)。

2.5.2 一旦得到委员会草案,应立即将其分发给技术委员会或分委员会的P成员及O成员供其考虑,并明确注明提交答复的最迟日期。

经技术委员会或分委员会同意,应为国家成员体提供8周、12周或16周的时间对委员会草案进行评论。

分发 CD 文件的默认周期是 8 周。

应根据给定的说明提交评议意见,以便进行汇总。

会前,国家成员体应向其代表充分介绍国家的立场。

2.5.3 答复日期终止后的4周内,秘书处应准备好意见汇总,并将其分发给技术委员会或分委员会的全体P成员及O成员。在准备意见汇总的过程中,秘书处应与技术委员会或分委员会主席协商,必要时与项目负责人协商,提出项目处理意见:

a) 在下次会议上讨论委员会草案及评议意见;

b) 或者，分发修改后的委员会草案供考虑；

c) 或者，登记征询意见阶段用委员会草案（见2.6）。

在b）和c）情况下，秘书处应在评议意见汇总报告中说明针对收到的每条评议意见采取的措施。在将修改的CD提交给委员会考虑［在b）情况下］之前将这些信息提供给所有P成员，或者在向CEO办公室提交最终草案进行征询意见阶段登记时［在c）情况下］同时提供。必要时通过分发修改的评议意见汇总报告方式提供。

各委员须对收到的所有意见作出答复。如果在分发之日起的8周内，2个或更多的P成员不同意秘书处在b）或c）情况下提出的建议，委员会草案应在会议上讨论（见4.2.1.3）。

2.5.4 如果经过会议研究未对委员会草案达成一致意见，应综合会议上的各项决定形成另一草案，在12周内予以分发考虑。经委员会或分委员会同意，应为国家成员体提供8周、12周或16周的时间对该草案及后续的文本进行评论。

2.5.5 应一直研究每一版草案文本，直到技术委员会或分委员会中的P成员达成协商一致，或作出撤销或推迟这个项目的决定。

2.5.6 应在协商一致的原则基础上作出分发征询意见草案（2.6.1）的决定。

技术委员会或分委员会主席有责任与秘书进行协商，必要时与项目负责人进行协商，根据《ISO/IEC指南2：2004》给出的"协商一致"的定义，判断该草案是否得到足够的支持。

"**协商一致**：总体同意，其特点在于利益相关方的任何重要一方对重大问题不坚持反对立场，并具有寻求考虑所有相关方的意见和协调任何冲突的过程。

注：协商一致不意味着一致同意。"

以下内容适用于"协商一致"的定义：

在达成共识的过程中，在文件的制定中不同观点得以表达和处理。然而，"持续反对"是在委员会、工作组或其他小组（例如任务组、咨询组等）的会议上表达的意见，是由相关利益的一个重要部分维持的、不符合委员会的协商一致意见。"相关利益"的概念将视委员会的动态而有所不同，因此应由委员会领导层逐案确定。"持续反对"的概念不适用于成员体就CD、DIS或FDIS进行的投票，因为这些都受适用的投票规则的制约。

表达"持续反对"意见者有权发表意见，并建议在宣布"持续反对"时采取以下办法：

—— 领导层应首先评估反对是否可被视为"持续反对"，即是否得到相关利益的一个重要部分的支持。如果情况并非如此，领导层将登记反对意见（例如以会议纪要、记录等方式），并继续领导有关文件的工作。

—— 如果领导层确定存在持续的反对意见，则需努力真诚地解决这一问题。然而，持续的反对并不等同于否决。处理持续存在的反对意见的义务并不意味着有义务成功地解决这些问题。

评估是否达成共识的责任完全在于领导层。这包括评估是否存在持续的反对意见，或任何持续的反对意见能否在不损害关于文件其余部分现有协商一致意见的情况下得到解决。在这种情况下，领导人将登记反对意见并继续开展工作。

持反对意见的当事方可使用第 5 章详述的上诉机制。

如对协商一致的意见有疑问，经技术委员会或分委员会表决的 P 成员 2/3 多数赞成，可视为足以接受该委员会草案登记为征询意见草案；不过，应尽一切努力解决反对票的问题。

计票时不包括弃权和无技术原因的反对票。

负责委员会草案的技术委员会和分委员会秘书处应确保征询意见草案充分体现在会议上或以通信方式作出的所有决定。

2.5.7 如果某一技术委员会或分委员会已达成协商一致，其秘书处应在最长 16 周内将草案的最终文本用一种适于分发给国家成员体征询意见的电子表格（2.6.1）提交给 CEO 办公室（如果是分委员会，应抄送给技术委员会秘书处一份）。

秘书处应将提议的 DIS 电子版，连同一份完整的解释性报告（ISO 表格 8A），以及针对最终 CD 文件的意见汇总和针对回复意见所采取的处理措施，提交至 ISO 中央秘书处。

2.5.8 关于本阶段的时间限制，见 2.1.6。

2.5.9 当所有的技术问题得到解决，委员会草案作为征询意见草案分发，并由 CEO 办公室登记后，委员会阶段即告结束。登记前，凡是不符合《ISO/IEC 导则 第 2 部分》的文本应退回秘书处进行修改。

2.5.10 如果技术问题不能在适当的时间范围内得到完全解决，技术委员会或分委员会可以考虑在该文件成为国际标准的协议达成之前，以技术规范（见 3.1）形式作为一种中间出版物出版。

2.6 征询意见阶段

2.6.1 在征询意见阶段，征询意见草案（ISO 的 DIS 和 IEC 的 CDV）应由 CEO 办公室分发给所有国家成员体进行为期 12 周的投票。

关于语言使用方面的政策，见附录 E。

CEO 办公室应将其接收投票的截止日期通知国家成员体。

投票终止时，首席执行官应在 4 周内将投票结果及所收到的意见交给技术委员会或分委员会主席和秘书处，以便迅速采取下一步行动。

2.6.2 国家成员体提交的投票应是明确的：赞成、反对或弃权。

赞成票可以附有编辑性或少量的技术性意见，其条件是技术委员会或分委员会的秘书及项目负责人与主席进行磋商，决定如何处理这些意见。

如果某国家成员体发现某一征询意见草案是不可接受的，应投反对票并说明技术理由。该国家成员体可以注明在具体的技术性修改意见获得许可的条件下将反对票改为赞成票，但不得投以接受修改意见为条件的赞成票。

如果成员体投反对票且没有提交理由，该投票将不计算在内。

如果成员体投反对票且提交了不明确属于技术性质的意见，委员会秘书处应在投票结束后 2 周内与 ISO/CS 技术项目经理取得联系。

2.6.3 如果满足下列条件，征询意见草案则可通过：

a) 参加投票的技术委员会或分委员会 P 成员 2/3 多数赞成；

b) 反对票不超过投票总数的 1/4。

计票时，弃权票及未附有技术理由的反对票不计算在内。

将正式投票期过后收到的评议意见提交给技术委员会或分委员会秘书处，待下次国际标准复审时进行研究。

2.6.4 技术委员会或分委员会主席收到投票结果及意见后，应协同其秘书处或项目负责人，并与 CEO 办公室磋商采取下列行动之一：

a) 如符合 2.6.3 的批准准则且无任何技术改动，则直接予以出版（见 2.8）。

b) 如符合 2.6.3 的批准准则但须包括技术改动，则将经修改的征询意见草案登记为国际标准的最终草案。

c) 如果没有满足 2.6.3 的批准准则，则：

   1) 分发修改后的征询意见草案进行投票（见 2.6.1）。

   注：分发修改后的征询意见草案并进行为期 8 周的投票。如有一个或多个相关委员会 P 成员提出要求，可延长至 12 周。

   2) 或者，分发修改后的委员会草案进行评论。

   3) 或者，下次会议讨论和评议征询意见草案。

2.6.5 投票期结束后的 12 周内，技术委员会或分委员会秘书处应准备一份完整报告，并由 CEO 办公室分发到各国家成员体。报告应：

a) 说明投票结果；

b) 陈述技术委员会或分委员会主席的决定；

c) 复制所收到的评议意见内容；

d) 包括技术委员会或分委员会秘书处对所提交的每条评议意见的看法。

应尽最大努力解决反对票问题。

如果自报告分发之日起 8 周内，2 个或更多的 P 成员表示不赞成主席所作出的决定 2.6.4 c) 项中的 1) 或 2.6.4 c) 项中的 2)，该草案则应在会议上进行讨论（见 4.2.1.3）。

各委员会应对收到的所有意见作出答复。

2.6.6 如果主席决定该草案进入批准阶段（见 2.7）或出版阶段（见 2.8），技术委员会或分委员会秘书处应在投票期结束后的最长 16 周内，在编辑委员会的帮助下准备一份最终文本并将其送至 CEO 办公室，以便准备和分发最终国际标准草案。

秘书处应向 CEO 办公室提供可修改的电子格式文本和允许确认修改形式的电子格式文本。

不符合《ISO/IEC 导则　第 2 部分》规定的文本应退回秘书处并在登记前要求进行修改。

修改文本的电子版以及使用 ISO 表 13[3] 记录的关于投票结果的主席决定，应提交至 ISO 中央秘书处，并详细标明针对每条意见所作决定，作为 ISO 表 13 的附表 B。

2.6.7 关于本阶段的时间限制，见 2.1.6。

2.6.8 当 CEO 办公室登记征询意见草案文本，作为最终国际标准草案分发或作为国际标准出版时，即 2.6.4 a) 和 b) 的情况，征询意见阶段即告结束。

2.7 批准阶段

2.7.1 在批准阶段，CEO 办公室应在 12 周内（ISO）将 FDIS 分发给所有国家成员体进行为期 8 周（IEC 6 周）的投票。

CEO 办公室应将其接收投票的截止日期通知国家成员体。

2.7.2 国家成员体提交的投票应明确：赞成、反对或弃权。

国家成员体可就任何关于 FDIS 的投票提出意见。

如果国家成员体认为最终国际标准草案是不可接受的，应投反对票并陈述技术理由，但不能投以接受修改意见为条件的赞成票。

如果成员体投反对票且没有提交理由，则该投票将不计算在内。

如果成员体投反对票且提交了不明确属于技术性质的意见，委员会秘书处应在投票结束后 2 周内与 ISO/CS 技术方案经理取得联系。

2.7.3 如果满足下列要求，分发进行投票的最终国际标准草案则获得通过：

a) 参加投票的技术委员会或分委员会 P 成员的 2/3 多数赞成。

---

[3] ISO 的第 13 号表格，用于汇报 DIS 投票结果和处理意见——译者注。

b) 并且，反对票不超过总数的 1/4。计票时，弃权票和未附有技术理由的反对票不包括在内。

2.7.4 在投票结束前，技术委员会或分委员会秘书处有责任提醒 CEO 办公室注意到草案起草中可能引入的错误。本阶段，不再接受进一步的编辑性或技术性修改意见。

2.7.5 收到的所有意见将被保留，供下次复审用，并在投票表格上记录为"注明供日后考虑"。但 CEO 办公室和秘书处可设法解决明显的编辑错误。不允许对批准的 FDIS 进行技术更改。

CEO 办公室应在投票期结束后的 2 周内，向所有的国家成员体分发报告，在报告中公布投票结果并指明国家成员体正式通过将其发布为国际标准，还是正式否决了该最终国际标准草案。

2.7.6 如果依据 2.7.3 的要求通过了最终国际标准草案，该草案可进入出版阶段（见 2.8）。

2.7.7 如果依据 2.7.3 的要求最终国际标准草案未获通过，则应将此文件退回相关的技术委员会或分委员会，委员会依据支持反对票的技术理由重新考虑。

委员会可以决定：

—— 以委员会草案、征询意见草案或在 ISO 和 JTC 1 中最终国际标准草案的形式再次提交修改后的草案；

—— 出版技术规范（见 3.1）；

—— 发布一项可公开提供的规范（见 3.2）；

—— 取消项目。

2.7.8 分发投票报告，说明 FDIS 已被批准为国际标准（见 2.7.5），或作为技术规范（见 3.1.1.2）出版，或将文件退回委员会，批准阶段即告结束。

2.8 出版阶段

2.8.1 CEO 办公室应在 6 周内更正技术委员会和分委员会秘书处标明的所有错误，并且印刷和分发国际标准。

2.8.2 国际标准出版，出版阶段即告结束。

2.9 可交付使用的成果的维护

在 ISO/IEC 导则独立补充部分给出了可交付使用的文件的维护程序。

2.9.1 引言

ISO 发布的或 ISO 与 IEC 联合发布的每一项国际标准和其他可交付使用文件均应进行系统复审，以确定是否应根据表 S1 进行确认，修订/修正，转换为其他可交付使用文件或撤销。

委员会可在系统复审之间的任何时候通过一项决议,对标准进行修订或修正。现行标准的修订或修正程序见 2.3.1。

对于不影响技术内容的微小修改,如更新和编辑性修改,可采用一种称作"微小修订"的缩短程序。这只包括委员会的微小修订提案(通过决议并完成 ISO 表 8B)、审批和出版阶段(见 2.7 和 2.8)。负责该文件的技术委员会形成决议并与 ISO 技术项目经理协商后,应分发可交付使用文件的最终修订稿,进行为期 8 周的 FDIS 投票,若为根据《维也纳协议》制定的文件,投票周期则为 12 周。可交付使用文件下一版本的前言应表明这是一个微小修订,并列出更新和编辑性修改的内容。

表 S1 系统复审的时间安排

可交付使用文件	系统复审前最长使用时间	可交付使用文件最多可确认次数	最长寿命
国际标准	5 年	无限制	无限制
技术规范(见 3.1.3)	3 年	推荐 1 次	推荐 6 年
可公开提供的规范(见 3.2.4)	3 年	1 次	6 年(如果此期间内未转化,提议撤销此可交付使用文件)
技术报告(见 3.3.3)	未指定	未指定	无限制

下列情况下,通常会进行系统复审:

——(所有可交付使用文件)由负责的委员会秘书处主动提供和负责,通常是在自文件发布或最后一次确认后指定期限届满的情况下;

——或者,(对于国际标准和技术规范)如果国际标准或相关技术规范未经负责的委员会秘书处进行系统复审,则为 ISO 中央办公室默认的行为;

——或者,(所有可交付使用文件)应一个或多个国家成员体的要求;

——或者,(所有可交付使用文件)应 CEO 的要求。

系统复审时间通常根据发布年份,或如果一份文件已经得到确认,则根据最后一次确认年份。当然,无需等到最长时间期限届满才复审文件。

2.9.2 复审

复审期限为 20 周。

注:系统复审由 ISO 中央秘书处以电子方式进行,并邀请所有 ISO 成员体对复审作回应,特定委员会的 P 成员有义务对该委员会负责的所有系统复审投票进行表决。复审目的已扩展到当成员体为使 ISO 标准适于国家采用而需要作修改时的信息获取。委员会需要审议这些修改,以确定是否需要考虑这些修改,以提高标准的全球相关性。修订、确认或撤销一项标准仍由负责委员会的 P 成员决定。

投票结束后，秘书处应使用表21[4]将反映投票结果的建议在TC或SC成员中进行传阅。在投票结束后6个月内，委员会应针对是否修订、确认或撤销标准作出最后决定，随后秘书处应将委员会的决定提交给ISO中央秘书处。

2.9.3 关于投票结果的解释

2.9.3.1 概述

一般情况下，系统复审后采取何种适当的行动，应根据P成员对一项具体行动的简单多数投票来决定。然而，在某些情况下，通过对结果进行更详尽地分析，可能会表明另一种解释更适合。

> 注：在解释投票结果时，由于可能回应、执行程度的多样性和评议意见的相对重要性，不可能就所有情况提供具体的规则。

如果投票结果不确定和/或需根据回应解释作出决定，则秘书处应在规定期限（如两个月）内邀请审批所提出的行动方针。

提出未来行动时，应适当考虑相关可交付使用文件最大可能的确认次数，并指定最长使用寿命（见表S1）。

2.9.3.2 关于国际标准投票结果的解释

方案1：确认（保留，无需技术变更）

如果已核实一份文件得到使用，应继续使用，且无需进行技术变更，则可确认一份可交付使用文件。相关标准如下：

—— 标准被直接采用或修改采用，或至少在5个国家使用（如果不满足该条，应撤销该标准）；

—— 并且，委员会P成员简单多数投票提出确认。

方案2：修正或修订（保留，有变更）

如果已核实一份文件得到使用，应继续提供，但需进行技术变更，则可提出一份可交付使用文件的修正或修订。相关标准如下：

—— 标准被直接采用或修改采用，或至少在5个国家使用（如果不满足该条，应撤销该标准）；

—— 并且，委员会P成员简单多数投票认为有必要进行修正或修订。

在这种情况下，一个项目可登记为获批工作项目（阶段10.99）。

应发起召集专家的活动，但不要求活跃P成员的最低数量。

如果委员会批准后不立即启动修正或修订，建议将该项目首先登记为预研工作项目并将标准登记为已确认。当最终提出登记为阶段10.99时，应参照前一次系统复审结

---

4) ISO的第21号表格，ISO出版物复审结果表——译者注。

果，且委员会应通过决议（见 2.3.1 "关于修正或修订现有标准的程序"）。

经决定需要对国际标准进行修订或修正的，将成为一个新项目，并应列入技术委员会或分委员会的工作方案。修订或修正的步骤与新标准的编制相同（见《ISO/IEC 导则 第 1 部分》，2.3～2.8），并包括设定完成相应阶段的目标日期。

方案 3：撤销

如果一项标准未直接采用或修改采用，或未在至少 5 个国家使用，则需撤销该标准（见上述方案 1 和方案 2）。

如果提出撤销一项国际标准，CEO 应将技术委员会或分委员会的决定通知国家成员体，如果国家成员体反对该决定，邀请其在 8 周内通知 CEO 办公室。

收到的任何反对意见应提交给 ISO 技术管理局，供其审议并作出决定。

2.9.3.3 转化为国际标准（仅限技术规范或可公开提供的规范）

技术规范和可公开提供的规范在系统复审中，除确认、修正或修订、撤销三种基本方案外，第四种方案为将其转化为国际标准。

为了开始转化为国际标准，需按国际标准规定的正常制定程序提交合适的更新文本（见 2.3.1）。

转化程序一般以 DIS 投票开始。如果认为需要作重大变更，以至于需要在 DIS 投票前经委员会全面审查，则应将文件的修订文本提交审查并以 CD 文件投票。

2.9.4 恢复已撤销的标准

如果在撤销一项国际标准后委员会确定仍有需要，则可提出恢复该标准。根据委员会的决定，该标准应作为国际标准草案或最终国际标准草案发布，供成员体投票。应采用常用审批标准，如果审批通过，该标准将作为新版本发布，使用新的发布日期。前言部分应说明该标准是前一版本的恢复。

2.10 更正和修正

2.10.1 概述

已出版的国际标准可通过下列出版物进行后续修改：

—— 技术勘误表（仅在 IEC）；

—— 更正版；

—— 修改单；

—— 或者，修订版（作为 2.9 维护程序的一部分）。

注： 如有修订，将出版国际标准的新版本。

2.10.2 更正

更正只是为了纠正在起草或出版过程中无意间引入并可能导致不正确或不安全地

应用该出版物的错误或歧义。

更正不是发布自出版以来已过时的更新信息。

有疑点的错误应引起相关技术委员会或分委员会秘书处的注意。在秘书处和主席的确认，必要时与项目负责人及技术委员会或分委员会中的 P 成员进行磋商，秘书处应向 CEO 办公室提交有关更正的提案并说明其必要性。

考虑到对组织的财政影响和出版物用户的利益，在与技术委员会和分委员会秘书处磋商后，首席执行官应决定是否出版技术勘误表（仅在 IEC）和（或）现行版本的更正版（见 2.10.4）。委员会秘书处将向委员会成员通报结果。

一般情况下，凡超过 3 年的出版物，将不会出版其更正版。

2.10.3 修改单

修改单是对现行国际标准中原已达成协调一致的技术条款的改动和补充。修改单被视为部分修订，该国际标准的其余部分不开放评论。

修改单通常作为一份单独的文件出版，受影响的国际标准的版本仍在使用中。修改单的起草和出版程序应遵循 2.3（ISO 和 JTC 1）或见 IEC 补充部分中的审查和维护程序以及修正草案（DAM）中 2.4、2.5、2.6、最终修正草案（FDAM）中的 2.7 和 2.8 所述。

在批准阶段（见 2.7），考虑到对组织的财政影响和国际标准用户利益，首席执行官应与技术委员会和分委员会秘书处协商，决定是否出版修改单，或出版包括修正在内的新版国际标准（见 2.10.4）。

注：如果预见到可能对国际标准条款要进行经常性补充，那么在开始起草时，应考虑将这些补充条款作为系列的一部分（见《ISO/IEC 导则　第 2 部分》）。

2.10.4 避免繁多的修改

用以更正现行国际标准的技术勘误表（仅在 IEC）或修改单不得多于 2 个单行本，制定第 3 个这样的文件时应出版国际标准的新版本。

2.11 维护机构

技术委员会或分委员会出版了要求不断修改的标准时，该技术委员会或分委员会可确定成立维护机构。有关指定维护机构的规则，见附录 G。

2.12 注册机构

技术委员会或分委员会制定了含有注册条款的标准时，需要成立一个注册机构。有关指定注册机构的规则，见附录 H。

在 ISO，见附录 SN《注册机构政策》。

2.13 版权

所有草案、国际标准及其他出版物的版权属于各自以 CEO 办公室为代表的 ISO、

IEC 或 ISO 和 IEC 共同所有。

以国际标准为例，其内容可能来自许多渠道，包括国家现行标准、科学或贸易期刊上出版的文章、自主的研发工作或对商业化产品的描述等。这些来源可能拥有一个或多个版权。

在 ISO 和 IEC，有一种理解是提交并成为 ISO、IEC 或 ISO 和 IEC 出版物中一部分的原始材料可以在 ISO 和／或 IEC 系统内（根据情况），作为建立协商一致过程的一部分拷贝和分发。这并不影响原始版权所有者在其他地方对原始文本拥有的权利。如果材料已经受版权保护，ISO 和／或 IEC 应获得复制和分发材料的授权。这通常不需要书面协议，或至多是一个简单的书面同意声明。如果贡献者希望就此向 ISO 和／或 IEC 提交的材料签署正式版权协议，这样的请求必须分别通知 ISO 中央秘书处或 IEC 中央办公室。

需要注意的是，ISO 和 IEC 各自的成员都有权采用 ISO 和 IEC 标准并作为他们的国家标准再出版。确实存在或可能存在类似的采用形式（如区域标准化组织）。

2.14 专利项目的引用（见附录 I）

2.14.1 在例外情况下，如有技术理由证明引用专利项目是合理的，原则上不反对制定含专利权所覆盖的专利项目（基于发明专利、实用新型和其他类似的法定权利，包括前面所提及的出版物中所采用的专利项目）条款的国际标准，即使在这些标准的条款中没有其他符合性方法可供选择。下列规则应适用。

2.14.2 如有技术理由证明起草文件时使用包括专利权所涵盖的项目条款是合理的，则应符合下列程序：

a） 文件的提案方应提请委员会注意其已知的并考虑涵盖其提案的任何项目的专利权，参与起草文件的任何相关方都应提请委员会注意在文件制定的各阶段发现的专利权问题。

b） 如果根据技术理由接受了该提案，起草方应要求已确定的专利权人声明其愿意在合理和非歧视的条款和条件下与全世界的申请人谈判其权利内的全球许可。这种协商任务应由相关团体完成，并在 ISO 和／或 IEC 外进行。专利权人声明的记录应放入 ISO 中央秘书处或视情况放入 IEC 中央办公室登记簿，并在相应文件的引言中提及。如果专利权人不提供这种声明，在没有得到 ISO 理事会或 IEC 理事局授权的情况下，相关委员会不得在文件中包括专利权所涵盖的项目。

c） 除非得到相关理事局授权，否则在收到已确定的专利权持有者发表声明前不得出版该文件。

2.14.3 文件出版后，如果发现不能在合理和非歧视的条款和条件下获取文件所包含的专利权涵盖项目的许可，应将文件退回相关委员会再进一步考虑。

3 其他可交付使用的文件的制定

3.1 技术规范

技术规范可在下列情况和条件下起草和出版。

3.1.1 当所讨论的项目正在研究过程中，或由于其他理由，现在不可能但将来可能达成出版国际标准协议，技术委员会或分委员会可以根据2.3中所规定的程序决定出版技术规范是否适宜。制定这种技术规范的程序应遵循2.4和2.5中的规定。将结果文件以技术规范形式出版的决定要求参加投票的技术委员会或分委员会P成员的2/3多数赞成。

当最终国际标准草案不能得到通过批准阶段所要求的支持率（见2.7），或对协商一致有疑义的情况下，技术委员会或分委员会可通过参加投票的P成员的2/3多数赞成来决定该文件以技术规范形式出版。

3.1.2 如果技术委员会或分委员会P成员已同意出版技术规范，技术委员会或分委员会秘书处应在16周内以电子版形式将技术规范草案提交CEO办公室出版。在不与现有的国际标准相冲突的条件下，提供不同技术解决方案的相互竞争的技术规范是可能的。

3.1.3 技术规范应在出版后3年内由技术委员会或分委员会进行复审，目的是重新检查促使该技术规范出版的原因，以及是否有可能就出版国际标准来取代该技术规范达成必要的协议。在IEC中，复审日期应以技术规范公布前已商定的稳定日期（复审日期）为基础。

3.2 可公开提供的规范（PAS）

2.3中规定的程序适用于PAS的制定。

3.2.1 PAS可以是在正式国际标准制定之前出版的中间性规范，或在IEC中可以是一种与外部组织合作出版的带有"双标识"的出版物。PAS不是完全符合标准要求的文件。

制定一项PAS的决定应在一开始（新工作项目提案获批的同时）即商定下来。

3.2.2 PAS的提案只能由A类联络组织或C类联络组织（见1.17）提交，或由委员会的P成员提交。

3.2.3 提交的PAS文本经过确认、检查，与现行国际标准没有冲突并经相关委员会参加投票的P成员简单多数批准后方可出版。只要不与现有的国际标准相冲突，提供不同技术解决方案的PAS就是可能存在的。

3.2.4 PAS 的初始有效期为 3 年。有效期可以延长，最长可延长 3 年，期满后，不论是否需要修改，都应转换为另一类型的规范性文件，或撤销。

3.3 技术报告

3.3.1 如果技术委员会或分委员会已收集到数据，这些数据不同于正常作为国际标准出版的数据（可以包括诸如国家成员体从所开展的调查中获取的数据，其他国际组织在工作中所获取的数据，或与国家成员体某一专门学科标准有关的"最新技术"数据），技术委员会或分委员会可以通过投票 P 成员的简单多数，决定是否要求首席执行官以技术报告的形式出版这些数据。该文件从本质上说完全是信息性的，而不应有规范性的内容。该文件应明确说明其与主题规范性要素的关系，这些规范性要素正在或将在有关主题的国际标准中处理。如有必要，首席执行官应与技术管理局进行协商，决定是否将本文件作为技术报告出版。

3.3.2 如果技术委员会或分委员会 P 成员同意出版技术报告，技术委员会或分委员会秘书处应在 16 周内以电子版形式将报告草案提交首席执行官出版。

3.3.3 建议由承担工作的委员会对技术报告进行定期复审，以保证其有效性。撤销技术报告的决定由承担工作的技术委员会或分委员会作出。

技术报告不进行系统性复审。

4 会议

4.1 概述

提醒国家成员体不允许向代表／专家收取任何种类的参会费用，也不允许为技术委员会、分委员会、工作组、维护和项目小组的任何会议收取特定旅馆住宿费或房费。基本会议设施应完全由国家成员体和／或自愿赞助者提供资金。关于更多信息，请参阅 IEC 会议指南（http：/www.iec.ch/ members_ Experts/refdocs/iec/IEC_Meeting_Guide 2012.pdf）和 ISO 附录 SF。

4.1.1 技术委员会或分委员会宜使用现代电子手段开展工作，只有需要讨论委员会草案（CD）或其他以通信方式无法解决的重要事宜时，才应召开技术委员会或分委员会会议。

4.1.2 技术委员会秘书处宜与 CEO 办公室协商并考虑其工作计划，预先拟订委员会及其分委员会（如果可能，包括其工作组）至少 2 年的会议计划。

4.1.3 在制定会议计划时，应考虑相关专业技术委员会或分委员会一起召开会议的益处，以增进交流和减少参加若干个技术委员会或分委员会会议代表的参会负担。

4.1.4 在制定会议计划时，应考虑技术委员会或分委员会会议结束后在同一地点立即召开编辑委员会会议，将有利于加快草案的制定工作。

4.2 召开会议的程序

承办会议的相关信息见附录 SF。

4.2.1 技术委员会和分委员会会议

4.2.1.1 召开会议的日期和地点应由相关技术委员会或分委员会主席和秘书处、首席执行官及承办会议的国家成员体共用商定。如果分委员会召开会议，分委员会秘书处首先应与上级技术委员会秘书处协商，确保会议的协调（见 4.1.3）。

4.2.1.2 希望承办某一会议的国家成员体应与首席执行官及相关技术委员会或分委员会秘书处联系。

该国家成员体首先应保证其所在国对以参加会议为目的的技术委员会或分委员会 P 成员代表的入境不受限制。

授权参会代表时，P 成员和 O 成员的国家标准组织应在 ISO 会议申请系统中注册代表信息。

承办会议的国家标准组织可通过 ISO 会议申请系统获取代表名单，以便对会议作出合适的安排。参会代表需要邀请函的，该国家标准组织有责任将参会代表姓名直接发送至承办会议的国家标准组织。

建议承办方核实并提供关于会议设施接入方式的信息。根据 4.2.1.3，应分发一份说明会议后勤情况的文件。除提供地点和交通信息外，还应提供会议设施可接入性的详细情况。

在规划过程中，应要求通知具体的可接入性要求。承办方应尽最大努力满足这些要求。

4.2.1.3 秘书处应做好日程安排，保证会议日程和会务信息在会议召开前 16 周由 CEO 办公室（IEC）分发或由秘书处（ISO）分发并抄送 CEO 办公室。

注： 所有新工作项目提案应经通信方式批准（委员会内部投票——CIB），见 2.3.4。

只有在召开会议前至少 6 周已经得到意见汇总的委员会草案才能纳入会议日程，并可在会上进行讨论。

所有其他工作文件，包括拟在会上进行讨论的草案评议意见汇总，应在会前至少 6 周内分发。

议程应明确说明开始和估计结束时间。

如果会议超过预计结束时间，主席应确保 P 成员愿意作出表决。但是，如果 P 成员离开，他们可以要求主席不采取任何进一步的投票决定。

会议预计结束时间之后或任何 P 成员离开之后作出的任何决定均应在会后通过通信进行确认。

注：与会者在预定旅行时应考虑会议预计时间。

关于委员会和工作组会议前发出的通知、议程和文件，见附录 SK。

4.2.2 工作组会议

4.2.2.1 工作组应尽可能使用现代电子手段开展工作（如电子邮件、远程电话会议）。当需举行会议时，工作组会议的召集人应在会前至少 6 周内将通知交给其成员和上级技术委员会秘书处。

会议安排应由会议召集人和会议承办国工作组的成员共用承担，后者负责所有实际工作的安排。

4.2.2.2 如果工作组会议与上级技术委员会工作会议联合召开，会议召集人应与上级技术委员会秘书处协调安排。特别应确保工作组成员能够收到分发给上级技术委员会会议代表的所有一般性会议资料。

4.2.2.3 工作组（或 IEC 中的 $PT^{5)}$/$MT^{6)}$/$AC^{7)}$）负责人或相关委员会秘书应将在各自国家举行的工作组（或 IEC PT/MT/AC）会议通知国家成员体秘书处。

4.3 会议语言

虽然会议官方语言是英语、法语和俄语，但会议默认语言为英语。

俄罗斯联邦国家成员体提供所有俄语的口译或笔译服务。

适当时，主席和秘书处应依据 ISO 或 IEC 通用规则以参加会议人员能够接受的方式妥善解决会议中使用语言的问题（见附录 E）。

4.4 会议的取消

会议一旦召开，应尽力避免取消或推迟会议。然而，如果不能在 4.2.1.3 所要求的时间内准备好会议议程及基本文件，首席执行官则有权取消会议。

4.5 文件分发

文件分发的要求参见附录 SB。应向 ISO 中央秘书处提供会议议程和召开委员会会议的通知，以供参考。

4.6 远程参加委员会会议

ISO 支持远程参加委员会会议，以实现提高利益相关方参与度和更好地协调委员会工作的目标。所有 ISO 会议和 ISO 委员会均允许远程参与，条件是：

—— 委员会秘书处在会议前与主办方一起核查，并遵守"关于远程参加委员会会议的指南"；

---

5) 项目组——译者注。
6) 维护组——译者注。
7) 顾问委员会——译者注。

—— 主办方同意并能提供必要的技术和支持；
—— 对远程和实际参会者采用相同的注册和认证规则；
—— 在会议前向所有参会者提供"关于远程参加委员会会议的指南"。

5 申诉

5.1 概述

5.1.1 国家成员体有权在 12 周内（ISO）和 8 周内（IEC）对有问题的决定提出申诉：

a) 就分委员会的决定向母技术委员会提出申诉。

b) 就技术委员会的决定向技术管理局提出申诉。

c) 就技术管理局的决定向理事会提出申诉。对任何申诉案例，理事会的决定为最终裁决。

5.1.2 技术委员会或分委员会的 P 成员可以申诉技术委员会或分委员会的任何作为或不作为，如果 P 成员认为这样的作为或不作为：

a) 不符合：

—— 章程和程序规则；

—— ISO/IEC 导则。

b) 或者，不利于国际贸易和商务，或诸如安全、卫生或环境等公共因素。

5.1.3 申诉的问题可以是技术性的或管理性的。

对涉及新工作项目提案、委员会草案、征询意见草案及最终国际标准草案的决定进行的申诉，仅在下述情况下予以考虑：

—— 涉及原则问题；

—— 草案的内容有损于 ISO 或 IEC 的声誉。

5.1.4 应完整记录所有申诉，以支持 P 成员的关注。

5.2 对分委员会决定的申诉

5.2.1 申诉文件应由 P 成员提交给上级技术委员会秘书处，并抄送首席执行官。

5.2.2 上级技术委员会秘书处一收到申诉文件，就应将申诉信息通知其所有 P 成员并立即采取行动，以通信方式或召开会议方式对申诉内容进行研究并作出决定。在此过程中应与首席执行官进行磋商。

5.2.3 如果技术委员会支持其分委员会，那么提出申诉的 P 成员可：

—— 接受技术委员会的决定；

—— 或者，对决定提出申诉。

5.3 对技术委员会决定的申诉

5.3.1 对技术委员会决定的申诉可存在 2 种情况：

—— 起因于上述 5.2.3 的申诉；

—— 或者，对技术委员会的原始决定申诉。

5.3.2 在任何情况下，申诉文件应提交首席执行官，并抄送技术委员会主席和秘书处。

5.3.3 首席执行官进行适宜的协商后，应在收到申诉文件后的 4 周内，将申诉及其意见提交技术管理局。

5.3.4 技术管理局将决定是否对申诉作进一步处理。如果决定予以处理，技术管理局主席应组织一个调解小组。

调解小组应在 12 周内听取申诉意见并努力尽快解决意见分歧。协调小组应在 12 周内提出最终报告。如果调解小组没有成功地解决意见分歧，应向首席执行官报告，并就如何解决这一问题提出建议。

5.3.5 收到调解小组的报告后，首席执行官应通知技术管理局，由技术管理局作出决定。

5.4 对技术管理局决定的申诉

对技术管理局决定的申诉应连同案例各阶段的全部文件一并提交给首席执行官。

首席执行官应在收到申诉文件后 4 周内，将申诉文件及其对申诉的评议意见一并提交理事会成员。

理事会应在 12 周内作出决定。

5.5 申诉期间工作的运行

当申诉针对正在进行的工作时，该项工作应继续，直至进入并包括批准阶段（见 2.7）。

附录 A
（规范性）
指南

A.1 引言

除了技术委员会制定的国际标准、技术规范、可公开提供的规范及技术报告外，ISO 和 IEC 还出版与国际标准化有关的指南。这些指南应根据《ISO/IEC 导则 第 2 部分》起草。

指南不应由技术委员会和分委员会制定，但可以由 ISO 政策制定委员会、IEC 咨询委员会或战略组、向 ISO 技术管理局报告的 ISO 工作组或者 ISO/IEC 联合协调组制定。这些机构以下称为"项目负责委员会或工作组"。

制定和出版指南的程序如下。

A.2 提案阶段

ISO 和／或 IEC 技术管理局批准新指南或指南修订的提案，并决定项目负责委员会或工作组的组成和秘书处。

ISO 和／或 IEC 技术管理局一旦批准某项目，该项目负责委员会或工作组的秘书处应确保 ISO 和 IEC 的利益相关方了解这一信息。

A.3 准备阶段

项目负责委员会或工作组应确保 ISO 和 IEC 的利益相关方有机会参与起草工作草案。

A.4 委员会阶段

一旦工作草案可以作为委员会草案分发，项目负责委员会或工作组的秘书处应将其报送给上级委员会或 ISO 和／或 IEC 技术管理局进行投票、评议、批准其进入征询意见阶段。

A.5 征询意见阶段

A.5.1 CEO 办公室应将修改后指南草案英语和法语文本分发给所有的国家成员体进行为期 16 周的投票。

A.5.2 如果反对票不超过投票的 1/4，该指南草案则被批准作为指南出版。计票时不包括弃权票。

如果是 ISO/IEC 指南，其草案应提交给 ISO 和 IEC 的国家成员体批准。需要两个组织的国家成员体都批准，该文件方可作为 ISO/IEC 指南出版。

如果 ISO 或 IEC 两个组织中只有一个组织满足这一条件，该指南只可以以批准组织的名义出版，除非项目负责委员会或工作组决定采用 A.5.3 规定的程序。

A.5.3 如果指南草案未获批准，或被批准但附有需进一步完善的评议意见，项目负责委员会或工作组的负责人可决定提交修正草案进行为期 8 周的投票。修正草案的批准条件与 A.5.2 相同。

A.6 出版阶段

出版阶段应由项目负责委员会或工作组归属组织的 CEO 办公室负责。

如果是 ISO/IEC 联合工作组，出版的责任应由两个首席执行官之间的协议决定。

A.7 指南的撤销

指南项目的负责委员会或工作组决定该指南是否应撤销。指南的正式撤销应由 TMB 按其正常的程序进行批准。

A.8 将指南转化为 ISO 国际标准的程序

1 提交将指南转化为国际标准的提案

任何成员体、联络组织或向 TMB 或 ISO CEO 汇报的工作组（授权机构），均可提议将经批准的指南转化为 ISO 国际标准。所有提案应附有转化为国际标准的理由。

2 提案的传阅

CEO 办公室将提案分发给所有成员体供其审批。

3 成员体投票

3.1 投票将会提出以下问题：您是否同意将 ×× 指南转化为 ISO 国际标准。

3.2 成员可回答"是"或"否"，反对票需要提供理由。

3.3 因相关指南已按照 ISO 和/或 IEC 协商一致的程序制定并获批准，故技术性和编辑性意见将不予考虑。如果转化指南的提案得到批准，将在未来修订下一版国际标准时考虑这些意见。

3.4 投票时间期限和投票规则与正常 DIS 投票相同。

4 批准条件

4.1 如果在所收到的所有投票中，有超过 1/4 的反对票，则转化为国际标准的提案未获批，批准程序结束。

4.2 如果将指南转化为国际标准的提案获批，则将发布该国际标准，撤销原指南。

5 维护由指南转化的国际标准

ISO 技术管理局将根据具体情况决定如何维护转化后的指南，优先考虑分配给适当的 TC 负责，但在缺少适当 TC 的情况下，将由 TMB 决定何处负责维护，包括建立一个新的 TC 或 PC。

附录 B
（规范性）
ISO/IEC 联络和分工程序

B.1 引言

依据 1976 年 ISO/IEC 协议[1]，ISO 和 IEC 共同形成了一个国际标准化体系。为了使这个体系有效运作，两个组织一致同意下列两个组织的技术委员会和分委员会之间的协调和分工程序。

B.2 总体考虑

ISO 和 IEC 之间的分工是以一致同意的原则为基础，这一原则是指与电气和电子工程领域国际标准化有关的所有问题由 IEC 负责，其他领域由 ISO 负责；对于电气和非电气技术界定目前尚不明确的国际标准化相关事宜的职责分配问题则应由两个组织相互协商解决。

在设立新的 ISO 或 IEC 技术委员会时，或是由于现有技术委员会的活动，有可能产生协调和工作分配的问题。

现有下列几个级别的协调和工作分配协议。对于提出的事宜，只有在较低级别经过努力解决失败后，方可提交到较高级别。

a) 正式联络：为 ISO 和 IEC 委员会之间的正常合作而进行的正式沟通；

b) 组织磋商：包括技术专家、首席执行官代表之间的磋商，针对其技术协调可能对两个组织的未来活动产生比当前更大影响的事宜；

c) 工作分配的决定：
   —— 由技术管理局作出；
   —— 或者，必要时由 ISO/IEC 联合技术咨询局（JTAB）作出。

对 ISO 和 IEC 两个组织都有影响且已证明 ISO 技术管理局和 IEC 标准化管理局不可能作出一致决定的问题，应提交 ISO/IEC JTAB 作出决定（见 1.3.1）。

B.3 设立新的技术委员会

当分别向 ISO 国家成员体或 IEC 国家成员体提出设立新技术委员会的提案时，该提案也应提交给另一个组织征求意见和/或达成协议，协商的结果可能出现以下 2 种情况：

---

[1] ISO 理事会决议 49/1976、50/1976 和 IEC 行政管理通函 13/1977 号。

a) 一致同意该项工作应在其中一个组织中进行；

b) 意见有分歧。

出现 a) 情况时，可根据一致的意见采取正式行动，设立新的技术委员会。

出现 b) 情况时，应安排一次相关领域的专家会议，由两个组织的首席执行官代表参加，旨在达成令人满意的分工（即在组织层面）协议。如果可在这一层面达成协议，可由适宜的组织采取正式行动实施该协议。

如果经过协商后仍不能达成协议，可由任一组织将该问题提交 ISO/IEC JTAB。

B.4 ISO 和 IEC 技术委员会之间工作的协调与分配

B.4.1 TC 层面的正式联络

ISO 和 IEC 各技术委员会之间的大部分协调需求都通过正式技术联络的安排得到成功解决。对于这些安排，当一个组织提出请求时，另一个组织应予以履行。正式联络安排的请求由 CEO 办公室控制，提出请求的组织应明确其所需要的联络类型，诸如：

a) 全部或有选择地交换委员会文件；

b) 联络代表定期或有选择地参加会议；

c) 参加所选的 ISO 和 IEC 技术委员会常设协调（或指导）委员会活动；

d) 组建 JWG。

B.4.2 协议的具体内容

B.4.2.1 通过把某工作领域委托给两个组织中的一个，不断努力将 IEC 和 ISO 之间的重叠性工作降至最低。

在受委托的工作领域中，IEC 和 ISO 应通过 JTAB 就如何将另一组织的观点和利益予以充分考虑的问题达成一致意见。

B.4.2.2 已建立如下 5 种合作工作方式：

方式 1——信息式

一个组织全权负责某一工作领域，并让另一方完全了解工作进展情况。

方式 2——主辅式

一个组织应在工作中起主导作用，另一个组织则在工作进展中以书面形式作出适宜的贡献。这种关系也包括充分交换信息。

方式 3——分包式

一个组织被全权委托负责完成某个确定的项目，但由于另一个组织具有专业特长，一部分工作被分包给第二个组织负责制定。为了保证分包的这部分工作能够正确地融入工作计划的主要部分，双方需作出必要的安排。为此，作为该标准化任务主要承包方的组织负责征询意见和批准阶段。

方式 4——协作式

一个组织在活动中起主导作用，但其工作会议接受另一组织的联络代表。这些联络代表有权参与讨论，但没有表决权。通过这种联络达到信息充分畅通。

方式 5——联合式

联合工作组和联合技术委员会确保召开联席会议，在完全平等参与的原则下共同处理标准，解决问题。

两个组织的技术委员会之间的联合工作组应按照 1.12.6 的规定开展工作。

B.4.2.3 对于可能重叠的领域，ISO 和 IEC 的分工将按相关方商定的工作进程或计划要求列出，并作为双方协议的附录。

该协议的结果是两个组织同意引用对方主管领域的相应标准。

如果被引用的标准已更新，引用该标准的机构有责任关注被引用标准的更新情况。

B.4.2.4 对于一个组织已经承担职责并拟将其分包给其他组织的工作，在确定该工作目的时应充分考虑参加分包工作的利益。

B.4.2.5 征询意见和批准阶段必要的程序应由特定标准化任务的受委托组织来完成，除非两个组织的技术管理局一起作出其他决定。

B.4.2.6 对于按方式 5（联合式）制定的标准，应根据牵头管理的组织的规则，在 ISO 和 IEC 中同步进行委员会、征询意见和批准阶段。承担该项目管理职责的委员会/组织应在草案分发日期前两周将委员会、征询意见和批准阶段用的草案提交给另一个组织。

B.4.2.7 若在一个组织内征询意见草案没有满足批准准则（见 2.6.3），那么：

—— 参加联合工作组的委员会官员可以从 2.6.4 c）给出的选项中选择一项。

—— 或者，在例外情况下，如果参加联合工作组的 ISO 和 IEC 委员会官员与 CEO 办公室达成协议，该项目可以作为批准征询意见草案的组织的单一标志标准继续进行。至此，联合工作组自动解散。

B.4.2.8 如果根据 2.7.3，最终国际标准草案未被批准，那么：

—— 参加联合工作组的委员会可以从 2.7.7 给出的选项中选择一项，注意，在 IEC 不允许传播第二份最终国际标准草案时，应请求 TMB 降低文件等级。

—— 或者，在例外情况下，如果参加联合工作组的 ISO 和 IEC 委员会官员与 CEO 办公室达成协议，该标准就可以作为批准最终国际标准草案的组织的单一标志标准出版。至此，联合工作组自动解散。

B.4.2.9 按方式 5（联合式）由 ISO 和 IEC 联合工作组制定的标准由承担管理职责的委员会所在组织出版。该组织分配标准的参考编号并拥有该标准的版权。该标准配有另一组织的标志并且两个组织都可以销售。这种国际标准的前言应明确所有负责制定的委员会。对于承担管理职责的委员会在 IEC 的那些标准，其前言还要给出 ISO 的投票结果。ISO 主导的文件编号是从 1 到 59999，IEC 主导的文件编号是从 60000 到 79999。如果是多部分标准，一些部分由 ISO 负责和一些部分由 IEC 负责，文件编号分配为 80000 系列（例如，ISO 80000-1，IEC 80000-6）。

B.4.2.10 按方式 5（联合式）制定的标准，其维护程序通常是承担管理职责的委员会所在组织中现行使用的程序。

B.4.2.11 在项目制定期间，如有理由改变运作方式，两个相关的委员会应提出建议并提交给两个技术管理局参考。

B.4.3 秘书处的合作

两个组织相关的技术委员会 / 分委员会秘书处应就本协议的实施进行合作，根据正常程序，应对进行的工作建立完全的信息流，并应要求相互提供工作文件。

附录 C
（规范性）
对制定标准的提案的论证

C.1 概述

C.1.1 由于涉及大量的财力和人力资源，而且应根据需要分配这些资源，因此，在任何标准化活动开始之前，明确需求、确定拟制定标准的目的及可能受影响的利益是非常重要的，而且这将有助于确保制定的标准恰当地涵盖所要求的要素，在其影响的行业具有市场相关性。因此，任何新的活动都应在其开始前进行合理的论证。

C.1.2 很显然，无论根据本附录的规定对提案作出何种结论，开展任何新工作的前提条件都应是有足够数量的利益相关方明确表示愿意安排必要的人力和资金，并积极参加这项工作。

C.1.3 本附录给出了提出和论证新工作的规则，以便提案能够提供尽可能清晰、明确的工作目的和范围，旨在确保相关方正确分配标准化资源，并且最有效地利用这些资源。

C.1.4 本附录不包括实施和监督本附录中包含的指南的程序规则，也不涉及为此应建立的管理机制。

C.1.5 本附录主要针对拟启动各种新工作的提案方，但也可作为一种工具帮助对提案进行分析或提出意见的人员以及负责对提案作出决定的机构。

C.2 术语和定义

C.2.1 新工作提案

新技术活动领域或新工作项目的提案。

C.2.2 新技术活动领域提案

向某组织提出的该组织现有委员会（如技术委员会、分委员会或项目委员会）未包括的工作领域的标准制定提案。

C.2.3 新工作项目提案

向某组织提出的在该组织现有委员会（如技术委员会）领域内的标准或一系列相关标准的制定提案。

C.3 通用原则

C.3.1 任何新工作提案都应在其所提交组织的范围之内。

注：例如，ISO 的提案对象在其章程中予以规定，IEC 的提案对象在其章程第 2 条中予以规定。

C.3.2 论证 ISO 和 IEC 新工作的文件应提供该提案市场相关性的重要案例。

C.3.3 论证 ISO 和 IEC 新工作的文件应提供可靠的资料作为 ISO 或 IEC 国家成员体在知情的前提下投票的基础。

C.3.4 在 ISO 和 IEC 体系内，提案方有义务提供合适的文件以支撑上述 C.3.2 和 C.3.3 阐述的原则。

C.4 提议新技术活动领域或新工作项目时需要说明的要素

C.4.1 新技术活动领域和新工作项目提案应包括下列信息（C.4.2 ~ C.4.13）。

C.4.2 标题

标题应清楚并简洁地表明提案所要涵盖的新技术活动领域或新工作项目。

示例 1：（新技术活动领域提案）"机床"。

示例 2：（新工作项目提案）"电工产品——基本环境测试程序"。

C.4.3 范围

C.4.3.1 对于新技术活动领域

范围应精确地界定活动领域的界限。范围不应重复该组织管理工作的通用目标和原则，而应指明相关的具体领域。

示例："通过切削材料或加压操作加工金属、木材和塑料的各种机床标准化"。

C.4.3.2 对于新工作项目

范围应清楚表明所提议的新工作项目涵盖的内容，必要时为明确起见，应说明不适用的范围。

示例 1：

本标准列举了一系列环境测试程序及其严重性，目的是评估电工产品在预期服务条件下的运行能力。

虽然最初是为这些应用而设计，本标准也可以用于其他需要的领域。

其他个别样品类型专用的环境测试可能被包含在相关的规范里。

示例 2：

渔业和水产养殖业领域标准化，包括但不局限于术语、设备及其操作的技术规范、水产养殖场地特性描述和适宜的物理、化学和生物条件的维护、环境监控、数据报告、溯源性跟踪及废物处理。

不包括：

—— 食品的分析方法（ISO/TC 34 的范围）；

—— 人员防护服（ISO/TC 94 的范围）；

—— 环境监控（ISO/TC 207 的范围）。

C.4.4 提议的初始工作计划（仅针对新技术活动领域提案）

C.4.4.1 提案的工作计划应符合并明确反映标准化活动的目标，并应指明提案的主题之间的关系。

C.4.4.2 工作计划中的每个项目都应采用拟被标准化的主题和内容（例如，就产品而言，项目可以是产品类型、特性、其他要求、拟提供的数据、测试方法等）加以明确。

C.4.4.3 工作计划中的具体项目可附补充论证。

C.4.4.4 提案的工作计划还应建议新工作项目的优先顺序和目标日期（当提议一系列标准时，应建议其优先顺序）。

C.4.5 表明拟制定的可交付使用的文件的优选类型

如果是新技术活动领域的提案，可查看 C.4.4。

C.4.6 国际、区域和国家层面的有关现行文件列表

应列出所有已知的相关文件（如标准和法规），无论其来源如何，并附上其重要性的说明。

C.4.7 与现有工作的关系及其影响

C.4.7.1 应提供一份说明，描述提议的工作与现有工作的关系和对现有工作的影响，尤其是现行 ISO 和 IEC 的可交付使用的文件。提案方需解释其工作与明显类似的工作有何差异或解释如何将重复和冲突降至最低。

C.4.7.2 如果在其组织的其他委员会或在另一组织的范围内已经有看上去类似或相关的工作，提案的范围应将提议的工作与其他工作区分开。

C.4.7.3 提案方应表明是否可以通过扩大现有委员会的范围或设立新的委员会来处理其提案。

C.4.8 相关国家的参与

C.4.8.1 对于新技术活动领域的提案，应提供该提案的主题对其国家商业利益非常重要的相关国家的名单。

C.4.8.2 对于现有委员会内的新工作项目提案，应提供那些已经不是该委员会 P 成员，但该提案的主题对其国家的商业利益非常重要的相关国家名单。

C.4.9 合作与联络

C.4.9.1 应提供在可交付使用的文件制定中约定为联络组织的相关外部国际组织或内部机构（除 ISO 和 / 或 IEC 委员会外）的名单。

C.4.9.2 为避免与其他机构冲突或重复工作，表明所有可能冲突或重复之处很重要。

C.4.9.3 与其他利益相关方的所有沟通结果也应包含在内。

C.4.10 受影响的利益相关方

应提供一个简洁的说明，识别受影响的利益相关方的类别（包括中小企业）并描述他们如何从所提议的可交付使用的文件中获益或受到影响。

C.4.11 基础文件（仅用于新工作项目提案）

C.4.11.1 若提案方认为一个现行完善的文件可以作为标准接受（经修正或不经修正），应附适当理由说明，在提案上附上该文件。

C.4.11.2 所有新工作项目议案都应附上现有文件，作为 ISO 或 IEC 可交付使用的文件、或提议的大纲、目录的初始基础。

C.4.11.3 如果所附的现行文件受版权保护或包含版权保护的内容，提案方应确保已经获得适宜的书面许可，允许 ISO/IEC 使用那些受版权保护的内容。

C.4.12 领导层承诺

C.4.12.1 如果是新技术活动领域提案，提案方应表明其组织是否已做好准备来承担所需要的秘书处工作。

C.4.12.2 如果是新工作项目提案，提案方还应提名一位项目负责人。

C.4.13 目的及论证

C.4.13.1 拟制定的标准的目的和论证应表达清楚，对拟包括在标准中的每个要素（诸如特性）标准化的需求应予以论证。

C.4.13.2 如果提议一系列的新工作项目，其目的和论证是共同的，即可起草一个共用提案，包括拟说明的所有要素并列举每个项目的标题和范围。

C.4.13.3 请注意下列列出的各项是可用于支撑提案的目的和论证的文件的一系列建议或想法。提案方应考虑这些建议，但不限于此，也不需要严格地符合这些建议。最重要的是，提案方编制并提供的立项目的和论证，既要与他们的提案最为相关，又要提供重要的商业案例来说明其提案的市场相关性和需求。编制完善并举证充分的立项的和论证文件有助于提案得到更充分的关注，并有利于其最终在 ISO 和 IEC 系统中成功立项。

— 一份简洁的说明，描述提案需要解决的业务、技术、社会或环境问题，最好与 ISO 或 IEC 相关委员会的战略工作计划相关联。

— 全球相关性文件，说明经济、技术、社会或环境问题的范围或幅度，或新的市场。这可包括对所产生的标准进行未来销售趋势的评估，表明未来使用和全球相关性。

— 技术效益：一份简洁的说明，描写提案在支撑新兴技术与体系的连贯性、合并技术的融合、互操作性、竞争技术的解决、未来的创新等方面的技术影响。

— 经济效益：一份简洁的说明，描述提案在以下方面的趋势，即消除贸易壁垒，改善国际市场准入，支持公共采购，提高广大企业（包括中小企业）的商业效率和/或最终通过灵活、经济可行的方法以符合国际和区域的规则/

惯例等。也可以提供一个简单的成本/效益分析,将产生可交付使用的文件的成本和对全世界企业产生的预期经济效益进行比较分析。

—— 社会效益:一份简洁的说明,描述预期从提议的可交付使用的文件中获得的预期社会效益。

—— 环境效益:一份简洁的说明,描述从提议的可交付使用的文件中获得的预期环境效益或更广泛可持续发展效益。

—— 一份简洁的说明,描述提议的可交付使用的文件的预期用途,例如,该可交付使用的文件是否是支撑合格评定的必要条件,或只是作为指南或者推荐的最佳实践;可交付使用的文件是否是管理体系标准;可交付使用的文件是否拟在技术法规中使用或引用;可交付使用的文件是否拟用来支撑与国际条约和协议有关的法律案件。

—— 一份简洁的指标说明,供委员会跟踪评估已出版标准的效益,能否使 C.4.10 中提及的利益相关方获益。

—— 一份评估最终的可交付使用的文件的前景说明;对 IEC 而言,是否符合 IEC 全球相关性政策:http//www.iec.ch/members_experts/refdocs/ac_cl/AC_200817e_AC.Pdf;对 ISO 而言,是否符合 ISO 全球相关性政策:http://www.iso.org/iso/home/standards_development/governance_of_technical_ work.htm;如相关,是否符合 ISO/TMB 有关可持续发展和可持续性的建议(见下面的注)。

注: 对于 ISO,ISO/TMB 确认下列建议:1)若一个委员会(任一行业)制定一项有关可持续性/可持续发展的标准时,该标准必须保持在该委员会的工作范围内;2)委员会还应尽早地将项目的标题和范围通报给 TMB;3)承担该项工作的委员会应在该标准的引言中阐明其宗旨;4)最广泛使用的可持续发展的定义来自联合国的布伦特兰可持续发展委员会:既满足当代人的需求,而又不损害子孙后代满足其需求的能力的发展。

—— 一份评估该提案是否符合与公共政策制定有关或支持公共政策制定的 ISO 和 IEC 标准制定的原则的说明,(对于 ISO,见附录 SO;对于 IEC 和 ISO,见《采用和引用 ISO/IEC 标准支撑公共政策》:http//www.iso.org/sites/policy/)以及最终可交付使用的文件与公共政策的潜在关系,包括一份有关符合相关法律法规,便利市场准入的说明。

附录 D
（规范性）
秘书处资源和秘书资格

D.1 术语和定义

D.1.1 秘书处

经多方同意并指定，负责向某个技术委员会或分委员会提供技术和管理服务的国家成员体。

D.1.2 秘书

秘书处任命的负责管理秘书处所提供的技术和管理服务的人员。

D.2 秘书处资源

指定为秘书处的国家成员体应认识到，无论其在自己国家为提供所需的服务作出什么安排，最终负责秘书处正常运作的是国家成员体本身。承担秘书处职能的国家成员体应视情况成为 ISO 服务协议的缔约方或 IEC 基础协议的缔约方。

因此，秘书处应具备足够的行政管理和财政手段或支持以确保：

a) 用英语和／或法语进行文字处理的设备，提供电子文本的设备和必要的文件复印设备；

b) 准备足够的技术图表；

c) 识别和使用收到的用官方语言编写的文件，必要时进行翻译；

d) 更新并持续监督委员会及其附属机构（如有）的结构；

e) 接收和快速分发函件和文件；

f) 具备足够的电话、传真及电子邮件等通信设施；

g) 连接互联网；

h) 根据需要，与承办会议的国家成员体协商安排会议期间的笔译、口译和服务设施；

i) 秘书参加任何需要其出席的会议，包括技术委员会和／或分委员会会议、编辑委员会会议、工作组会议以及必要时与主席的协商；

j) 秘书可获取基础国际标准（见《ISO/IEC 导则 第 2 部分》有关参考文件和起草来源），以及相关领域的国际标准、国家标准和／或相关文件；

k) 必要时，秘书可访问能够对委员会领域内技术问题给予指导的专家。

尽管首席执行官争取派其代表参加技术委员会第一次会议、参加有新秘书处的技术委员会会议以及为解决问题而希望他们到会的所有技术委员会或分委员会会议，但是 CEO 办公室不能承担长期或临时为某一秘书处开展工作。

尽管首席执行官尽量派其代表参加技术委员会第一次会议，设立新秘书处的技术委员会会议，以及任何拟定解决问题的技术委员会或分委员会会议，CEO 办公室不能承诺永久或临时性承担秘书处工作。

D.3 对秘书的要求

被任命为秘书的人员应：

a) 通晓英语和 / 或法语；

b) 熟悉《章程和议事规则》，并视情况，熟悉 ISO/IEC 导则（见《ISO/IEC 导则》独立补充部分）。

c) 能够就程序或起草文件向委员会和附属机构提出建议，必要时，先与 CEO 办公室协商。

d) 了解理事局或技术管理局关于技术委员会总体活动的决定，特别是了解关于其归口委员会的活动决定。

e) 是一名优秀的组织者，在技术和管理工作上训练有素，以便组织和开展委员会工作并促进委员会成员和附属机构（如有）积极参与活动。

f) 熟悉 CEO 办公室提供的文件，特别要熟悉使用电子工具和服务。

建议新任命的技术委员会秘书处早日访问位于日内瓦的 CEO 办公室，以便与相关的工作人员讨论程序和工作方法。

附录 E
（规范性）
语言使用通用政策

E.1 国际环境中意见的表达与交流

国际上通常至少使用两种语言来出版可交付使用的文件。许多理由可以说明使用两种语言的优势，例如：

—— 采用语法和句法不同的两种语言表达一个给定的概念，能使意思更清楚更准确。

—— 如果仅以一种语言起草的文本为依据达成一致意见，则当把这个文本翻译成另一种语言时可能会出现一些困难。有些问题可能不得不重新讨论，如果最初一致同意的文本不得不进行改动，则可能会造成时间的延误。如果一个文本已经以第一种语言通过了，之后再转化为第二种语言，常常会出现一些表述上的困难，如果两种文本同时起草、同时修正，这些困难即可避免。

—— 要确保国际会议富有成效，重要的是在达成的协议中杜绝出现意义不明确的表达法，并且必须消除由于语言本身的误解而对协议产生异议的风险。

—— 使用从两个语系中所选的两种语言，扩大了可能被指定参加会议的代表人数。

—— 如果已有两个完美统一的版本，那么就更容易用其他语言来正确地表达某个概念。

E.2 技术工作中使用的语言

官方语言为英语、法语和俄语。

技术委员会在工作和通信中默认使用英语。

为了上述目的，俄罗斯联邦国家成员体负责提供所有俄语的口译和笔译工作。

在 IEC，每个可交付使用的文件的"前言"中应标明本文件制定时使用的一种语言。适用国际电工词汇（IEV）和/或数据库标准的特例除外。

E.3 国际标准

国际标准由 ISO 和 IEC 采用英语和法语两种语言出版（有时是多语种版，也包括俄语和其他语言，尤其是术语标准）。这些版本的国际标准具有同等效力，每个版本都被视为原版。

标准的技术内容从起草程序开始就用英语和法语两种语言表述会很有优势,这样两种版本可同时研究、修正和批准,始终确保语言的等效性(也可见《ISO/IEC 导则 第 2 部分》"语言版本"条款)。

这可由下列机构完成:

—— 由秘书处完成,或在外部帮助下,由秘书处负责;

—— 或者,由归口技术委员会或分委员会的编辑委员会完成;

—— 或者,由母语为英语或法语的国家成员体根据这些国家成员体与相关秘书处达成的协议完成。

当决定出版一个多语种国际标准(例如,术语标准)时,俄罗斯联邦国家成员体负责该文本的俄语部分;同样,当决定用非官方语言出版含有术语或材料的国际标准时,其母语相关的国家成员体负责选择术语或负责起草文本中拟使用这些语言的部分。

E.4 技术委员会制定的其他出版物

其他出版物可以只用一种官方语言出版。

E.5 技术委员会和分委员会会议文件

E.5.1 与会议议程相关的草案和文件

会前起草和分发的文件如下:

a) 议程草案

议程草案应由归口秘书处以会议语言(默认语言为英语)起草并分发。

b) 列入议程的委员会草案

列入会议议程的委员会草案最好以会议语言(默认语言为英语)提供。

征询意见草案应有英语和法语两种版本。ISO 理事会或 IEC 标准化管理局的指南适用于其中一种语言版本不能适时提供的情况。

与会议议程上的事项有关的其他文件(各种建议、评议意见等)可以只用一种语言(英语或法语)起草。

E.5.2 会议期间起草和分发的文件

会议期间起草和分发的文件如下:

a) 会议期间通过的决议

每次会议开始时,由秘书和一名或多名母语为英语和/或法语的代表(只要可能)组成临时起草委员会,编辑所有提出的决议。

b) 每次会后整理的简要记录(如有)

如果准备这种会议记录,则应用英语或法语起草,最好是在特别起草委员会的帮助下（必要时）用两种语言起草。

E.5.3 会后起草和分发的文件

每次技术委员会和分委员会会议后，相关秘书处应起草会议报告。会议报告可以只使用一种语言（英语或法语），并将通过的正式决议，最好用英语和法语，作为附录附上。

E.6 使用非英语或法语起草的文件

其母语既不是英语也不是法语的国家成员体可将秘书处分发的所有文件翻译成本国语言，以便于本国专家对这些文件进行研究，或帮助他们选派的代表参加技术委员会和分委员会会议。

如果一种语言为两个或多个国家成员体所通用，那么其中一个国家可以在任何时间主动将技术文件翻译成该种语言，并向同一语系的其他国家成员体提供副本。

上述两小节的条款可由秘书处依据自身需求实施。

E.7 技术会议

E.7.1 目的

技术会议的目的是对议事日程中的所有事项尽可能达成一致，并尽力确保所有的代表相互理解。

E.7.2 讨论内容口译成英语和法语

虽然基础文本可以提供英语和法语两种版本，但还必须根据情况决定以一种语言表达的发言是否要由下列人员口译成另一种语言：

—— 自愿的代表；

—— 秘书处或东道国成员体的工作人员；

—— 合格的口译员。

还应注意，要使那些母语既不是英语也不是法语的代表能够充分理解会议内容。

对是否有必要口译技术会议上所讨论的内容作出规定是不切实际的，重要的是让所有与会代表能领会讨论的内容，逐字逐句口译每句话并不完全必要。

鉴于上述考虑，除没有必要提供口译的特殊情况外，下列作法是适宜的：

a） 对于要作出程序性决定的会议，可由秘书处的成员或自愿的代表提供简单的口译；

b） 在工作组会议上，只要可能，工作组成员应在工作组召集人的倡议或授权下安排工作组成员之间必要的口译。

为了使负责会议的秘书处能够对口译作出必要的安排，代表们在告知参加会议的同时应将其能够使用的语言和在口译方面上能够提供的帮助告知秘书处。

在主要使用一种语言的会议上,为了帮助使用其他语言的代表,应尽量采用下列作法:

a) 应当用两种语言宣布对某一主题形成的决定之后,然后再进入下一个主题;

b) 凡是批准对现行文本的一种语言版本进行更改时,应给代表留出时间考虑这一更改对另一种语言版本的影响;

c) 如果代表有要求,应提供另一种语言的讲话摘要。

E.7.3 其他语言的发言口译成英语和法语

在技术委员会或分委员会会议上,如果某一与会者在特殊情况下希望用非英语或法语的其他语言发言,在会议具备口译的条件下,会议主席有权批准此项要求。

为了给所有参加技术委员会和分委员会会议的专家提供平等表达其观点的机会,建议灵活运用本条款。

附录 F
（规范性）
项目制定的选择方案

F.1 选择方案的简化流程图

项目阶段	正常程序	与提案同时提交的草案	"快速程序"[a]	"技术规范"[b]	技术报告[c]	可公开提供的规范[d]
提案阶段（见2.3）	受理提案	受理提案	受理提案[a]	受理提案	受理提案	受理提案[g]
准备阶段（见2.4）	起草工作草案	*工作组研究*[e]		起草草案		批准PAS草案
委员会阶段（见2.5）	制定和受理委员会草案	*制定和受理委员会草案*[e]		受理草案	受理草案	
征询意见阶段（见2.6）	起草和受理征询意见草案	起草和受理征询意见草案	受理征询意见草案			
批准阶段（见2.7）	*批准FDIS*[f]	*批准FDIS*[f]	*批准FDIS*[f]			
出版阶段（见2.8）	出版国际标准	出版国际标准	出版国际标准	出版技术规范	出版技术报告	出版PAS

用虚线圈的斜体阶段可以省略。
[a] 见F.2。
[b] 见F.3.1。
[c] 见3.3。
[d] 见3.2。
[e] 根据对新工作项目提案的投票结果，准备阶段和委员会阶段可被省略。
[f] 如征询意见草案得到批准且没有反对票，则可省略此阶段。
[g] 有关PAS提案的细节见ISO和IEC补充部分。

F.2 "快速程序"

F.2.1 采用快速程序的提案可按下述步骤操作。

F.2.1.1 相关技术委员会或分委员会的任何 P 成员及 A 类联络组织可提议将任何来源的现行标准作为征询意见草案提交投票。提案方应在提交提案前得到原组织的同意,是否采用快速程序提交现行标准的准则可由提案方自行决定。

F.2.1.2 ISO 或 IEC 理事局承认的国际标准化机构可建议将该机构制定的标准作为最终国际标准草案提交投票。

F.2.1.3 与 ISO 或 IEC 达成正式技术协议的组织,在征得相应的技术委员会和分委员会的同意后,可提议将该组织制定的标准草案作为征询意见草案在技术委员会或分委员会内提交投票。

F.2.2 提案应由首席执行官受理,并应采取下列行动:

a) 与起草提案文件的组织共用解决版权和 / 或商标问题,以便不受限制地自由复印,并分发到各国家成员体,并建议该组织注意提案文件应遵行 ISO/IEC 知识产权政策,具体见 2.13 和 2.14。

b) 对于 F.2.1.1 和 F.2.1.3 的情况,与相关秘书处磋商后评估具体哪一个技术委员会或分委员会能够胜任提案文件所涉及的主题;如果现有的技术委员会不能解决提案文件所涉及主题的,首席执行官应将提案提交技术管理局,技术管理局可要求首席执行官将文件提交到征询意见阶段,并设立一个特别工作组来解决后续产生的问题。

c) 保证与其他国际标准没有明显的矛盾。

d) 依据 2.6.1 作为征询意见草案（F.2.1.1 和 F.2.1.3）,或依据 2.7.1 作为最终国际标准草案（F.2.1.2 的情况）分发提案文件,同时向技术委员会或分委员会指明（在 F.2.1.1 和 F.2.1.3 的情况下）提案文件所属的领域。

F.2.3 表决期限和批准条件应分别按照 2.6 征询意见草案或 2.7 最终国际标准草案的规定,如果没有技术委员会参与,则最终国际标准草案的批准条件应是反对票少于投票总数的 1/4。

在 ISO 内,作为最终国际标准草案提交的提案投票期限应为 20 周。

F.2.4 对于征询意见草案,如果符合批准条件,则标准草案应进入批准阶段（2.7）;如果不符合批准条件,该提案未通过,依据 F.2.2 b),由归属的技术委员会或分委员会决定采取进一步行动。

委员会领导层可决定是否省略 FDIS 投票,直接出版的情况见 2.6.4。

对于最终国际标准草案,如果符合批准条件,则该文件应进入出版阶段（2.8）;

如果不符合批准条件，该提案未通过，依据 F.2.2 b）由该 FDIS 归属的技术委员会/分委员会应决定采取进一步行动，或者如果没有技术委员会参与，则由原组织与 CEO 办公室讨论决定采取进一步行动。

如果标准已经出版，其维护工作应依据 F.2.2 b），由该文件归属的技术委员会/分委员会负责。如果没有技术委员会参与，当原组织决定需要修改标准时，应重复上面规定的批准程序。

附录 G
（规范性）
维护机构

G.1 技术委员会或分委员会在制定需要有维护机构的国际标准时，应在早期阶段通知首席执行官，以便 ISO/TMB 或 IEC 理事局在出版国际标准之前作出决定。

G.2 ISO 技术管理局或 IEC 理事局根据相关技术委员会的提案，指定与国际标准有关的维护机构及其成员。

G.3 维护机构的秘书处应尽可能地归属制定该国际标准的技术委员会或分委员会的秘书处。

G.4 首席执行官应负责与维护机构工作有关的外部组织联系。

G.5 维护机构的程序规则应经 ISO/TMB 或 IEC 理事局批准，任何与更新国际标准或发布修改单相关的授权都应由 ISO/TMB 或 IEC 理事局特别授予。

G.6 维护机构提供服务的任何收费应由理事局授权。

附录 H
（规范性）
注册机构

H.1 技术委员会或分委员会在制定需要有注册机构的国际标准时，应在早期阶段通知首席执行官，以便进行必要的协商，并允许技术管理局在出版国际标准前作出决定。

H.2 技术管理局根据相关技术委员会的提案，指定与国际标准有关的注册机构。

H.3 注册机构应是具备资格并在国际上得到认可的机构，如果没有这样的组织，可由技术管理局决定将注册任务交给 CEO 办公室完成。

H.4 注册机构应按要求在工作中清楚地表明其是由 ISO 或 IEC 指定（例如，在被指定机构的信笺抬头添加适当的用语）。

H.5 依据相关国际标准的条款，承担注册职能的注册机构不应要求 ISO 或 IEC 或其成员提供财务贡献。但是，如果得到理事局的授权，并不排除注册机构收取服务费。

附录 I
（规范性）
ITU-T/ITU-R/ISO/IEC 共用专利政策实施指南

最新版 ITU-T/ITU-R/ISO/IEC 共用专利政策实施指南（包括 Word 文本或 Excel 表格）参见 ISO 网站：http//www.iso.org/iso/home/standards_developement/governance_of_technical_work/patents.htm。

也可参见 IEC 网站：http：//www.iec.ch/members_experts/tools/patents/patent_policy.htm。

ITU-T/ITU-R/ISO/IEC 共用专利政策实施指南

第 2 版，2015 年 6 月 26 日生效

第 1 版，2012 年 4 月 23 日生效

2007 年 3 月 1 日修订初版

目 录

第 Ⅰ 部分——共用指南

Ⅰ.1 目的

Ⅰ.2 术语解释

Ⅰ.3 专利披露

Ⅰ.4 专利陈述和许可声明表

Ⅰ.4.1 声明表的目的

Ⅰ.4.2 联系人信息

Ⅰ.5 召集会议

Ⅰ.6 专利信息数据库

Ⅰ.7 专利权的转让或转移

第 Ⅱ 部分——各组织的专用条款

Ⅱ.1 ITU 的专用条款

Ⅱ.2 ISO 和 IEC 的专用条款

附件 1：ITU-T/ITU-R/ISO/IEC 共用专利政策

附件 2：ITU-T 或 ITU-R 建议书 /ISO 或 IEC 可交付使用的文件的专利陈述和许可声明表

附件 3：ITU-T 或 ITU-R 建议书的通用专利陈述和许可声明表

ITU-T/ITU-R/ISO/IEC 共用专利政策实施指南

第 2 版，2015 年 6 月 26 日生效

第Ⅰ部分——共用指南

Ⅰ.1 目的

ITU［在其电信标准化局（ITU-T）和无线电通信局（ITU-R）］、ISO 和 IEC 许多年前就制定了专利政策，其目的是为了给他们的技术机构参与者在遇到专利权问题时提供简单明了的实用指导。

考虑到技术专家通常不熟悉专利法的复杂问题，因此 ITU-T/ITU-R/ISO/IEC 共用专利政策（以下简称"专利政策"）的操作部分起草成核查表的形式，涵盖某一建议书/可交付使用的文件要求完全或部分使用或实施专利许可时，可能出现的三种不同的情况。

《ITU-T/ITU-R/ISO/IEC 共用专利政策实施指南》（以下简称"指南"）的目的在于说明该专利政策并为其实施提供便利。专利政策的副本见附件 1，亦可参见各组织网站。

该专利政策鼓励尽早披露和确认那些可能与正在制定的建议书/可交付使用的文件有关的专利。这样做将可能提高标准制定的效率，并且可能避免潜在的专利权问题。

各组织不宜介入有关建议书/可交付使用的文件的专利相关性或必要性评价，不宜干涉专利许可谈判，不宜参与解决关于专利的争端；所有这些都宜如同过去的做法，留给相关方处理。

各组织的专用条款包含在本文件的第Ⅱ部分中。然而，这些专用条款不应与共用专利政策或指南相矛盾。

Ⅰ.2 术语解释

贡献 contribution

技术机构提交审议的任何文件。

免费 free of charge

并不意味着专利权人放弃专利有关的全部权利。更确切地说，是金钱补偿的问题，即专利权人不寻求将任何金钱补偿作为专利许可的一部分（无论这类补偿被称为专利

使用费,还是一次性许可费等)。不过,虽然专利权人承诺不收取任何数量的金钱,但专利权人仍然有权要求相关建议书/可交付使用的文件的实施者签署一项许可协议,包含其他合理的条款和条件,例如有关适用法律、使用领域、担保等。

组织 organizations

ITU、ISO 和 IEC。

专利 patent

基于发明的专利、实用新型和其他类似的法定权利(包括上述任何权利的申请)中包含的和确认的那些权利要求,并且仅指对建议书/可交付使用的文件的实施很必要的那些要求。必要专利是指为实施特定的建议书/可交付使用的文件所必需的专利。

专利权人 patent holder

拥有、控制和/或有能力许可专利的个人或实体。

互惠 reciprocity

只有当未来的专利许可申请人承诺免费实施同一份相关建议书/可交付使用的文件,或在合理的条款或条件下许可其专利,该申请人才能要求专利权人向其授予专利许可。

建议书/可交付使用的文件 recommendations/deliverables

ITU-T 和 ITU-R 建议书统称为"建议书",ISO 可交付使用的文件和 IEC 可交付使用的文件统称为"可交付使用的文件"。在附件 2 的《专利陈述和许可声明表》(以下简称"声明表")中,建议书/可交付使用的文件统称为"文件类型"。

技术机构 technical bodies

ITU-T 和 ITU-R 的研究组、分组和其他小组,以及 ISO 和 IEC 的技术委员会、分委员会和工作组。

I.3 专利披露

专利政策的第 1 段就强调,各组织工作的参与方[2] 任何当事人从一开始就应提请注意任何已知的专利或任何已知的正在处理的专利申请,无论是自己组织的或是其他组织的。

在这种情况下,"一开始"意味着应在建议书/可交付使用的文件制定期间尽可能早地披露上述信息。在制定第一个文本草案时也许不能做到这一点,因为此时的文本也许还不够清晰或其后还需要进行重大修改。此外,应该真诚地尽最大的努力来提供

---

2) 对于 ISO 和 IEC,这包含了在标准制定过程中各阶段的标准草案的接收者。

这方面的信息，但并不要求进行专利检索。

除上述要求外，没有参与技术机构的任何方也可以提请各组织注意任何已知的专利，无论是他们自己的和/或任何第三方的。

在披露他们自己的专利时，专利权人必须使用本指南第4部分规定的《专利陈述和许可声明表》（简称"声明表"）。

任何提请注意第三方专利的信息都宜以书面形式寄给有关组织。然后，如果适用，相关组织的局长/CEO将要求未来的专利权人提交一份声明表。

专利政策和本指南也适用于在建议书/可交付使用的文件批准后披露的或提请各组织注意的任何专利。

无论专利的确认发生在建议书/可交付使用的文件批准之前还是批准之后，如果专利权人不愿意按照专利政策的2.1或2.2授予许可，各组织将迅速通告负责受影响的建议书/可交付使用的文件的技术机构，以便采取适宜的措施。这类措施包括（但不限于）复审该建议书/可交付使用的文件或其草案，以便消除潜在的冲突，或者进一步检查并澄清引起冲突的技术考虑。

I.4 专利陈述和许可声明表

I.4.1 声明表的目的

为了给各组织的专利信息数据库中提供清楚的信息，专利权人必须使用声明表。这个表格可以通过各组织的网站找到（附件2给出了声明表，供参考）。对于ITU，声明表必须提交给电信标准化局（TSB）或无线电通信局（BR）的局长，对于ISO或IEC，必须提交给CEO。声明表的目的是确保专利权人以标准化的形式向各自的组织提交其发表的声明。

声明表为专利权人提供一种为实施特定建议书/可交付使用的文件所需要的专利权有关的许可声明的途径。具体地讲，通过提交这种声明表，提交方声明其愿意按照专利政策（通过选择表格中的选项1或2）/或者不愿意按照专利政策（通过选择表格中的选项3）授予其所有的并且为部分或完全使用或实施某个特定建议书/可交付使用的文件需要的专利许可。

如果专利权人选择了声明表的选项3，那么，对于引用的相关ITU建议书和/或ISO或IEC可交付使用的文件，ITU、ISO和IEC要求专利权人提供允许专利标识的附加信息。

如果专利权人希望对于同一个建议书/可交付使用的文件标识多个专利并将其分类在声明表的不同选项中，或者如果专利权人将某个复杂专利的不同申请归类在同一个声明表中的不同选项中，则使用多重声明表是合适的。

声明表中包含的信息如有明显错误，例如标准或专利引用编号的印刷错误，可予更正。声明表中包含的许可声明持续有效，直到它被另一个包含对申请许可人更有利的许可条款和条件的声明表取代为止，表现在：

a）承诺从选项 3 改成选项 1 或 2；

b）承诺从选项 2 改成选项 1；

c）或者，取消勾选的一个或多个包含在选项 1 或 2 中的子选项。

I.4.2 联系人信息

在填写声明表时，应注意提供长期有效的联系信息。只要可能，通常应提供"姓名和部门"以及电子邮件地址。只要可能，当事人，特别是多国组织，最好在提交的所有声明表上给出同一个联系人。

为了在各组织的专利信息数据库中保持最新信息，要求对过去提交的声明表的任何变更或修改，特别是与联系人有关的信息变更，通知到各组织。

I.5 召集会议

早期披露专利有利于提高建议书 / 可交付使用的文件在制定过程的效率。因此，每个技术机构在制定建议书 / 可交付使用的文件的过程中，都要求披露任何已知的对提议的建议书 / 可交付使用的文件必要的专利。

如果合适，技术机构的主席将在每次会议的适当时间，询问是否有人知道为使用或实施考虑中的建议书 / 可交付使用的文件可能需要使用的专利。应该在会议报告中记录提问情况以及任何肯定的答复。

只要相关组织没有收到专利权人选择专利政策 2.3 的说明，即可运用本组织适宜的相应规则批准该建议书 / 可交付使用的文件。希望在技术机构讨论时考虑建议书 / 可交付使用的文件中包括的专利内容，然而，就主张的任何专利的必要性、范围、有效性或具体的许可期限，技术机构可以不承担负责。

I.6 专利信息数据库

为了促进标准制定过程和建议书 / 可交付使用的文件的应用，各组织都应向公众提供一个专利信息数据库，包含了以声明表的形式通报给各组织的信息。专利信息数据库可能包含特定专利信息，或可能没有包含此种信息而是包含针对某个具体建议书 / 可交付使用的文件符合专利政策的陈述。

专利信息数据库的准确性和完备性未经认证，只反映已经通报给各组织的信息。同样，专利信息数据库可以看成是树起的用于提醒用户的一面旗帜，这些用户可能希望与那些已经向各组织提交声明表的实体联系，以便确定为使用或实施某个具体建议书/可交付使用的文件是否必须获得专利使用许可。

I.7 专利权的转让或转移

指导专利权转让或转移的规则包含在专利陈述和许可声明表中（见附件2和附件3）。通过遵守这些规则，专利权人在转让或转移后完全解除了有关许可承诺的所有义务和责任。本规则不要求专利权人在转让发生后去迫使受让人执行许可承诺。

第Ⅱ部分——各组织的专用条款

Ⅱ.1 ITU 的专用条款

ITU-1 一般性专利陈述和许可声明表

任何人都可以提交《一般性专利陈述和许可声明表》，这种表格可以从 ITU-T 和 ITU-R 的网站上找到（附件 3 中的表格供参考）。这个表格的目的是为专利权人提供自愿选择，可以用它就其任何贡献中包含的受专利保护的内容作出一般性许可声明。具体地讲，当专利权人向组织提交的贡献中包含的任何提案的部分或全部被纳入建议书，并且被纳入的部分包含已经提交申请的专利，使用或实施这些建议书需要获得专利权人许可。专利权人通过提交表格，声明其拥有的专利的许可意愿。

这个《一般性专利陈述和许可声明表》并不是代替每个建议书需提交的那份"专用"声明表（见 I.4）的替代品，而是希望它能提高专利权人对符合专利政策的响应能力和早期披露。因此，除了现有的关于其贡献的一般性专利陈述和许可声明表外，专利权人在适当的时间（例如，当知道其拥有一项用于特定建议书的专利时），还应提交一份"专用"专利陈述和许可声明表：

—— 对于其向组织提交的任何贡献中含有的并被包含在某一建议书中的专利，任何这样的"专用"专利陈述和许可声明表可以包含与《一般性专利陈述和许可声明表》相同的许可条款和条件，也可以含有"专用"声明表（见 I.4.1）中规定的对申请许可人更有利的许可条款和条件；

—— 对于包含在某一建议书中但专利权人并未贡献给组织的那些专利，任何这样的"专用"专利陈述和许可声明表可以包含声明表的三个选项中的任何一种（见 I.4.1），与现行《一般性专利陈述和许可声明表》中的承诺无关。

《一般性专利陈述和许可声明表》持续有效，直到它被另一份包含对申请许可人更有利的许可条款和条件的《一般性专利陈述和许可声明表》取代为止，表现在：

a）承诺从选项 2 改成选项 1；

b）或者，取消勾选一个或多个包含在选项 1 或 2 中的子选项。

ITU 专利信息数据库还包含《一般性专利陈述和许可声明表》的记录。

ITU-2 通告

所有新的和修订的 ITU-T 和 ITU-R 建议书的封面应增加文字说明，在适宜的地方，强烈要求用户查阅 ITU 专利信息数据库。其措辞如下：

"ITU 提请注意，本建议书的使用或实施可能涉及使用已主张的知识产权。ITU 不负责所主张的知识产权的证据、有效性或适用性，无论它们是 ITU 成员宣称的还是本建议书制定过程以外的其他当事人宣称的。

截至本建议书批准之日，ITU（已经/尚未）收到在实施本建议书时可能要求的受专利保护的知识产权的通知。不过，请实施者注意，这可能不代表最新信息，因此强烈要求去查阅 ITU 专利信息数据库"。

Ⅱ.2　ISO 和 IEC 的专用条款

ISO/IEC-1 对可交付使用的文件草案征求意见

所有提交征求意见的草案都应在其封面页上包含如下文字："请本草案的收件人将其评议意见，连同其了解的任何有关专利权的通知一起提交，并提供支撑文件"。

ISO/IEC-2 通告

在起草过程中没有标识专利权的已发布文件应在前言中包含以下通告：

"本文件的某些内容有可能涉及专利权问题，对此应引起注意，ISO 和/或 IEC 不负责识别任何这样的专利权问题"。

在起草过程中已标识专利权的已发布文件应在引言中包含以下通知：

"国际标准化组织（ISO）和/或国际电工委员会（IEC）提请注意以下事实：据称，遵守本文件可能涉及使用（……条……）中规定的关于（……主题内容……）的专利。

ISO 和/或 IEC 对本专利权的证据、有效性和范围不承担责任。

该专利权人已经向 ISO 和/或 IEC 保证，他/她愿意在合理、无歧视的条款和条件下与全世界申请者谈判许可授予事宜。为此，该专利权人向 ISO 和/或 IEC 登记其声明。可以从以下联系方式获得信息：

专利权人姓名……

地址……

提请注意，除了上面标识的专利权外，本文件的某些要素可能是专利权的主题。ISO 和/或 IEC 不应承担标识任一或全部这类专利权的责任"。

ISO/IEC-3 国家采用

ISO、IEC 和 ISO/IEC 可交付使用的文件的专利声明仅适用于声明表中注明的 ISO 和/或 IEC 文件。声明不适用于经过修改的文件（例如通过国家或区域采用）。但是，实施符合相同的国家和区域采用和各自的 ISO 和/或 IEC 可交付使用的文件，可依据向 ISO 和/或 IEC 提交的此类可交付使用的文件声明。

附件 1
ITU-T/ITU-R/ISO/IEC 共用专利政策

下面是有关专利的"行为准则"，这些专利不同程度涵盖 ITU-T 建议书、ITU-R 建议书、ISO 可交付使用的文件和 IEC 可交付使用的文件的主题内容（本文件中，ITU-T 建议书和 ITU-R 建议书统称为"建议书"，把 ISO 可交付使用的文件和 IEC 可交付使用的文件统称为"可交付使用的文件"）。"行为准则"中的规则是简单明了的。建议书/可交付使用的文件是由技术专家而不是专利专家起草的，他们可能不一定很熟悉诸如专利之类知识产权的复杂的国际法状况。

建议书/可交付使用的文件不是约束性的，其目的是确保在世界范围内技术和系统的兼容性。为了达到满足所有参与者共用利益的目的，必须确保每个人都易于获得建议书/可交付使用的文件及其应用和使用等。

因此，随之而来的是，全部或部分被纳入建议书/可交付使用的文件的专利必须是每个人能容易获取的，没有不适当的限制。总之，满足这个要求是行为准则的唯一目的。由于各案协议可能有差异，因此与专利有关的详细安排（专利许可、专利使用费等）留给相关方去处理。

行为准则可以归纳如下：

1 ITU 电信标准化局（ITU-TSB）、ITU 无线电通信局（ITU-BR）以及 ISO 和 IEC 的 CET 办公室不负责就专利或类似权利的证据、有效性或范围给出权威的或全面的信息，但是非常需要披露最充分的可获取信息。因此，任何参与 ITU、ISO 或 IEC 工作的参与方都应该从一开始就分别提请 ITU-TSB 局长、ITU-BR 局长或 ISO 和 IEC 的 CEO 办公室注意其所知晓的自己组织的或是其他组织的任何专利或任何正在处理的专利申请，尽管 ITU、ISO 或 IEC 不能够验证任何这类信息的有效性。

2 如果某建议书/可交付使用的文件已制定，并且上述 1 中言及信息已经披露，则可能出现三种不同状况：

2.1 专利权人愿意与其他方在无歧视基础上以合理的条款和条件谈判免费专利许可事宜。这类谈判留待相关方在 ITU-T/ITU-R/ISO/IEC 外进行。

2.2 专利权人愿意与其他方在无歧视基础上以合理的条款和条件谈判专利许可事宜。这类谈判留待相关方在 ITU-T/ITU-R/ISO/IEC 外进行。

2.3 专利权人不愿意遵循上述 2.1 或 2.2 的条款；在这种情况下，建议书/可交付使用的文件不应包含与该专利有关的条款。

3 无论适用哪种情况（2.1、2.2 或 2.3），专利权人都必须使用相应的"专利陈述和许可声明表"提供书面陈述，分别在 ITU-TSB、ITU-BR 或 ISO 或 IEC 的 CEO 办公室备案。这种陈述除了针对表格中所选方框对应的情况提供信息之外，不得再包含附加的条款、条件，或任何其他免责条款。

附件 2
ITU-T 或 ITU-R 建议书/ISO 或 IEC 可交付使用的
文件的专利陈述和许可声明表

ITU-T 或 ITU-R 建议书/ISO 或 IEC 可交付使用的文件的专利陈述和许可声明

此声明不代表实际授予许可

请把表格按下列每种文件类型的指示返回相关组织：

国际电信联盟（ITU）电信标准化局局长	国际电信联盟（ITU）无线电通信局局长	国际标准化组织（ISO）秘书长	国际电工委员会（IEC）秘书长
地址：Place des Nations CH1211 Geneva 20，Switzerland	地址：Place des Nations CH1211 Geneva 20，Switzerland	地址：8 chemin de Blandonnet CP 401 1214 Vernier，Geneva Switzerland	地址：3 rue de Varembé CH1211 Geneva 20 Switzerland
传真：+41227305853	传真：+41227305785	传真：+41227333430	传真：+41229190300
电邮：tsbdir@itu.int	电邮：brmail@itu.int	电邮：patent.statements@iso.org	电邮：inmail@iec.ch

专利权人：	
依法登记的名称	
申请许可的联系方式：	
名称和部门	
地址	
电话	
传真	
电邮	
网址（可选填）	
文件类型：	
□ ITU-T 建议书（*）□ ITU-R 建议书（*）□ ISO 可交付使用的文件（*）□ IEC 可交付使用的文件（*） （请把所填写的表格返回相关组织） □共用文本或双文本［ITU-T 建议书/ISO/IEC 可交付使用的文件（*）］ （如果是共用文本或双文本，请把所填写的表格返回给三个组织：ITU-T, ISO，IEC） □ ISO/IEC 可交付使用的文件（*） （如果是 ISO/IEC 可交付使用的文件，请把所填写的表格返回 ISO 和 IEC）	
（*）文件号	

续表

(*) 文件标题	
许可声明： 专利权人确认其持有已批准和/或正在申请的专利，要使用这些专利则需要执行上述文件，并在此根据 ITU-T/ ITU-R/ISO/IEC 共用专利政策声明如下（仅能选一个框打勾）： □ 1. 专利权人愿意在全球范围内无歧视地免费向不限制数量的申请人授予专利许可，并在其他合理的条款和条件下许可其生产、使用和销售上述文件的实施。 谈判事宜留待相关方在 ITU-T、ITU-R、ISO 或 IEC 以外进行。 如果专利权人愿意在互惠条件下为上述文件授予许可，亦请标记此处＿＿＿＿＿＿＿。 如果专利权人保留如下权利：即只有申请人愿意在合理的条款和条件下（但不免费）许可其自己的专利（为实施上述文件要求使用的专利），专利权人才在合理的条款和条件下（但不免费）向申请人授予其专利许可，亦请标记此处＿＿＿＿＿＿ □ 2. 专利权人愿意在全球范围内无歧视地向不限制数量的申请人授予专利许可，并在其他合理的条款和条件下许可其生产、使用和销售上述文件的实施。 谈判事宜留待相关方在 ITU-T、ITU-R、ISO 或 IEC 以外进行。 如果专利权人愿意在互惠条件下为上述文件授予许可，亦请标记此处＿＿＿＿＿＿ □ 3. 专利权人不愿意依照上述选项 1 或选项 2 的规定授予许可。 在这种情况下，必须向 ITU、ISO 和/或 IEC 提供下列资料，作为本声明的一部分： ——获批的专利号或专利申请号（如正在申请中）； ——表明上述文件中受影响的部分； ——涵盖上述文件的专利说明 免费： 　　并不意味着专利权人放弃专利有关的全部权利。更确切地说"免费"是指金钱补偿的问题；即专利权人不寻求将任何金钱补偿作为许可协议的一部分（无论这类补偿被称为专利使用费，还是一次性许可费等）。不过，虽然专利权人承诺不收取任何数量的金钱，但专利权人仍然有权要求上述同一份文件的实施者签署一项许可协议，其中包含其他合理的条款和条件，例如，有关适用法律、使用领域、担保等。 互惠： 　　只有当未来的许可申请人承诺免费实施同一份相关建议书/可交付使用的文件，或在合理的条款和条件下许可其专利，该申请人才能要求专利权人向其授予专利许可。 专利： 　　基于发明的专利、实用新型和其他类似的法定权利（包括上述任何权利的申请）中包含和确认的权利要求，并且仅对上述同一种文件的实施很必要的那些要求。必要专利是指为实施特定的建议书/可交付使用的文件所必需的专利。 专利权的转让/转移： 　　根据 ITU-T/ITU-R/ISO/IEC 共用专利政策的 2.1 或 2.2 作出的许可声明应被解释为约束所有与转移的专利有利益关系的继任者的留置权。认识到这种解释可能不适用所有司法管辖区，那些根据共用专利政策已经提交许可声明的专利权人（在专利声明表中选择选项 1 或 2）在通过这样的许可声明转移专利所有权时，应在相关转移文件中包含适当的条款，以确保此类转移的专利的许可声明对受让人具有约束力，并确保受让人在未来转移时也将同样包含适当的条款，目的是约束所有相关继任者	

专利信息 [建议填写选项 1 和 2,但不作要求;ITU、ISO 和 IEC 要求填写选项 3(见"注")]			
序号	状态 [已批准/申请中]	国家	批准的专利号或申请号 (如正在申请中)	文件标题
1				
2				
3				
4				
5				
6				
7				
8				
9				
10				

☐ 如果在附加的页面上提供了附加的专利信息,在此方框中打勾。

注:在选择选项 3 时,还应提供上述选项 3 框内所列出的附加信息。

签名(仅需签在最后一页):	
专利权人	
授权人姓名	
授权人头衔	
签名	
地点、日期	

表格:2015 年 6 月 26 日

附件 3
ITU-T 或 ITU-R 建议书的通用专利陈述和许可声明表

（国际电信联盟 ITU）

ITU-T 或 ITU-R 建议书的通用专利陈述和许可声明表

此声明不代表实际授予许可

请把表格返回相关处局：

国际电信联盟（ITU）
电信标准化局
局长
地址：Place des Nations
CH-1211 Geneva 20 Switzerland
传真：+41 22 730 5853
电邮：tsbdir@itu.int

国际电信联盟（ITU）
无线电通信局
局长
地址：Place des Nations
CH-1211 Geneva 20
Switzerland
传真：+41 22 730 5785
电邮：brmail@itu.int

专利权人：
依法登记的名称
申请许可的联系方式：
名称和部门
地址 _____
电话 _____
传真 _____
电邮：
网址（可选填）
许可声明： 如果 ITU-T / ITU-R 建议书中包含上述专利权人提交的贡献文件的部分或全部提案，且所包含部分包括已提交专利的项目且其专利用途要求实施 ITU-T / ITU-R 建议书，则根据 ITU-T / ITU-R / ISO / IEC 公用专利政策，上述专利权人特此声明如下（仅能选一个框打勾）：

续表

☐ 1. 专利权人愿意在全球范围内无歧视地免费向不限制数量的申请人授予专利许可，并在其他合理的条款和条件下许可其生产、使用和销售相关 ITU-T/ITU-R 建议书的实施。

谈判事宜留待相关方在 ITU-T/ITU-R 以外进行。

如果专利权人愿意在互惠条件下为上述 ITU-T/ITU-R 建议书授予许可，亦请标记此处。

如果专利权人保留如下权利：即只有申请人愿意在合理的条款和条件下（但不免费）许可其自己的专利（为实施上述 ITU-T/ITU-R 建议书要求使用的专利），专利权人才在合理的条款和条件下（但不免费）向申请人授予其专利许可，亦请标记此处

☐ 2. 专利权人愿意在全球范围内无歧视地向不限制数量的申请人授予专利许可，并在其他合理的条款和条件下许可其生产、使用和销售相关 ITU-T/ITU-R 建议书的实施。

谈判事宜留待相关方在 ITU-T/ITU-R 以外进行。

如果专利权人愿意在互惠条件下为上述 ITU-T/ITU-R 建议书授予许可，则也标记此处

免费：
并不意味着专利权人放弃专利有关的全部权利。更确切地说，"免费"是指金钱补偿的问题，即专利权人不寻求将任何金钱补偿作为许可协议的一部分（无论这类补偿被称为专利使用费，还是一次性许可费等）。不过，虽然专利权人承诺不收取任何数量的金钱，但专利权人仍然有权要求上述同一份文件的实施者签署一项许可协议，其中包含其他合理的条款和条件，例如，有关适用法律、使用领域、担保等。

互惠：
只有当未来的许可申请人承诺免费实施同一份相关建议书/可交付使用的文件，或在合理的条款和条件下许可其专利，该申请人才能要求专利权人向其授予专利许可。

专利：
基于发明的专利、实用新型和其他类似的法定权利（包括上述任何权利的申请）中包含和确认的权利要求，并且仅指对上述同一种文件的实施很必要的那些要求。必要专利是指为实施特定的建议书/可交付使用的文件所必需的专利。

专利权的转让/转移：
根据 ITU-T/ITU-R/ISO/IEC 共用专利政策的 2.1 或 2.2 作出的许可声明应被解释为约束所有与转移的专利有利益关系的继任者的留置权。认识到这种解释可能不适用所有司法管辖区，那些根据共用专利政策已经提交许可声明的专利权人（在专利声明表中选择选项 1 或 2）在通过这样的许可声明转移专利所有权时，应在相关转移文件中包含适当的条款，以确保此类转移的专利的许可声明对受让人具有约束力，并确保受让人在未来转移时也将同样包含适当的条款，目的是约束所有相关继任者。

签名：

专利权人	
授权人姓名	
授权人头衔	
签名	
签署地点、日期	

表格：2015 年 6 月 26 日

附录 J
（规范性）
技术委员会和分委员会的制定范围

J.1 引言

技术委员会或分委员会的范围是一份明确界定该委员会工作界限的声明。具有以下（但不限）功能：

—— 帮助对工作领域的疑问和建议人找到合适的委员会；

—— 防止两个或多个 ISO 和/或 IEC 委员会的工作项目重复；

—— 能够帮助防止超越上级委员会授权的活动领域。

J.2 范围的制定

1.5.10 中规定了技术委员会和分委员会制定范围的基本规则。

范围要素的顺序应是：

—— 基本范围；

—— 在 ISO 中，横向功能（适用时）；

—— 在 IEC 中，横向和/或团体安全功能（适用时）；

—— 不适用范围（如果有）；

—— 备注（如果有）。

J.3 基本范围

技术委员会的范围不应涉及国际标准化的总体目标或重复那些管理所有技术委员会的原则。

特殊情况下，如果认为对委员会范围的理解很重要，可以包括说明材料。这类材料应以"备注"的形式给出。

J.4 不适用范围

如果有必要说明某些主题超越该技术委员会的范围，则应列出这些主题，由"不包括……"等词引出。

应清楚地规定不适用范围。

如果不适用范围含在一个或多个其他现有 ISO 或 IEC 技术委员会范围内，还应明确这些委员会。

示例 1 "不包括：那些含在 ISO/TC……范围中的……"

示例 2 "不包括：……（ISO/TC……）……（IEC/TC……）等领域的特定项目的标准化"。

没有必要提及明显的不适用范围。

示例 3 "不包括：其他 ISO 或 IEC 技术委员会涵盖的产品"。

示例 4 "不包括：……电气设备和仪器的规格，这些领域属 IEC 委员会的范围"。

J.5 产品相关委员会的范围

产品相关委员会的范围应清楚表明该委员会拟包括的领域、应用范围或市场部门易于确定一个特定的产品是否在其领域、应用范围或市场部门之内。

示例 1 "用于……的……和……的标准化"。

示例 2 "建筑用材料、组件和设备的标准化和……和……以及……的服务和维护用设备的操作"。

可以通过表明产品的目的或说明产品的特点来明确范围的界限。

范围不宜列举该委员会涵盖的产品类型，因为这样做可能暗示其他类型产品的标准由或正由其他委员会负责。然而，如果有意这样做，最好列出那些不适用范围的项目。

列举诸如术语、技术要求、抽样方法、测试方法、标识、标志、包装、尺寸等说明了范围对那些特定要素的限制，而且其他要素可以由其他委员会负责进行标识。因此，拟进行标准化的产品要素不应包含在范围中，除非有意将该范围限定在特定要素。

如果范围中未提及任何要素，就意味着该主题全部由该委员会涵盖。

注：涵盖范围并不一定意味着有制定标准的需求。只表示在需要的情况下，该委员会（而不是其他委员会）将制定任一要素的标准。

关于不必列举的要素，举例如下：

示例 3 "分类、术语、抽样、物理、化学或其他测试方法、规格等的标准化"。

不管是产品的类型或要素，都不应在范围内提及优先顺序，因为这些优先顺序已在工作计划中标明。

J.6 非产品相关委员会的范围

如果委员会的范围拟限定在某些与产品不相关或仅间接与产品相关的要素，其应只表明拟包括的要素（例如，安全颜色和符号、非破坏性试验、水质）。

术语这个词作为一个可能进行标准化的要素，只有当该要素是委员会负责的唯一要务时才提及。如果情况不是如此，提及术语则是多余的，因为这一要素是任何标准化活动的逻辑组成部分。

附录 K
（规范性）
项目委员会

K.1 提案阶段

不属于现有技术委员会范围的新工作项目提案，应由一个被授权提出新工作提案的机构（见 2.3.2）采用适当的形式提交并进行充分论证（见 2.3.4）。

CEO 办公室可以决定将提案在分发进行投票前退回给提案方作进一步研制。在这种情况下，提案方应按建议修改或提供不作修改的论证。如果提案方不作修改并且按原提交的提案文本分发进行投票，技术管理局将决定采取适当的行动。这可能拦截该提案直到修改完成或接受按收到的文件进行投票。

在所有情况下，CEO 办公室都可以将评论和建议纳入提案表格中。

有关提案论证的详细情况，见附录 C。

如果提议设立制定管理系统标准的项目委员会，见附录 L。

提案应提交给技术管理局秘书处，由其安排把提案提交给所有国家成员体进行投票。

鼓励提案方标明该项目委员会第一次会议的日期（见 K.3）。

如果该提案不是由一个国家成员体提交的，那么在向国家成员体提交时应包括承担项目委员会秘书处的信息。

投票情况应在 12 周内反馈。

提案的通过需符合以下要求：

—— 参加投票的国家成员体中 2/3 多数赞成；

—— 至少有 5 个批准新工作项目的提案并提名技术专家的国家成员体承诺积极参与。

K.2 项目委员会的设立

技术管理局应对新工作项目投票的结果进行复审，如果符合批准准则，就应设立项目委员会（其参考编号应为技术委员会/项目委员会次序中的下一个可用号码）。

项目委员会的秘书处应指派给提交该提案的国家成员体，如果该提案不是源自成员体，技术管理局应在收到的申请中决定秘书处的指派。

批准新工作项目提案并提名技术专家的国家成员体，应被注册为该项目委员会的 P

成员。批准新工作项目提案但未承诺积极参与的国家成员体，应被注册为 O 成员。投反对票但表示如果新工作项目提案通过就愿意积极参与的国家成员体，应被注册为 P 成员。投反对票且没有表示愿意积极参与的国家成员体，应被注册为 O 成员。

CEO 办公室应向国家成员体宣布项目委员会的设立及其成员。

国家成员体将被邀请通过通知 CEO 办公室来确认 / 改变其成员身份。

秘书处将联系在新工作项目提案或国家成员体评论中确定的所有可能的联络组织邀请他们表明是否有兴趣参与这项工作，以及对哪个联络类型有兴趣。联络申请将会按现有程序处理。

K.3 项目委员会的第一次会议

召开项目委员会会议的程序应按照第 4 章进行，例外情况是，如果提交提案通报了第一次会议的日期，可以使用 6 周的通知期限。

项目委员会的主席应是新工作项目提案中提名的项目负责人，如果在新工作的项目提案中没有提名项目负责人，则应由秘书处提名。

第一次会议应确认新工作的项目范围。如果有必要修订（为了说明而不是扩展的目的），修订的范围应提交技术管理局批准。会议还应确认项目计划，在 ISO 的程序，并决定开展工作所需要的任何分支机构。

如果确定该项目需要细分来制定两个或多个出版物，这是可行的，前提是该工作细分完全处于原新工作项目提案的范围内。如果不在其范围内，则需要制定一个新的工作项目供技术管理局考虑。

注： 不要求项目委员会制定战略业务计划。

K.4 准备阶段

应按 2.4 开展准备阶段工作。

K.5 委员会、征询意见、批准和出版阶段

应按 2.5 ~ 2.8 开展委员会、征询意见、批准和出版阶段工作。

K.6 项目委员会的解散

相关标准一旦出版，项目委员会应立即解散。

K.7 项目委员会制定标准的维护

承担秘书处的国家成员体应按照 2.9 的程序承担标准维护的责任，除非项目委员会转换为技术委员会（见 1.10），在这种情况下，该技术委员会应承担该标准的维护责任。

附录 L
（规范性）
管理体系标准提案

L.1 概述

当提议制定新的管理体系标准（MSS）——包括行业专用 MSS，应当按照本附录附件 1 的内容进行合理性研究（JS）。

注：如现有的 MSS 的制定已经得到批准，且范围已经确定（除非第一次制定时未确定范围），对该 MSS 进行修正时不需要进行 JS。

在条件允许的情况下，提案方应尽力确定将用于构成新的或修订的 MSS 系列的全部出版物，并为每个可交付使用文件准备一个 JS。

L.2 术语和定义

下列术语和定义适用于本附录。

L.2.1

管理体系 management system

见本附录的附件 2（3.4）中的定义。

L.2.2

管理体系标准 management system standard

MSS

针对管理体系（L.5.1）的标准。（原文有误，核对是否应为"L.2.1"，目前按照原文"L.5.1"处理）

注：就本文件而言，本定义亦适用于其他 ISO 可交付使用文件（如 TS、PAS）。

L.2.3

通用 MSS generic MSS

能够广泛适用于不同经济部门、不同类型和规模的组织以及不同地理、文化和社会条件的 MSS。

L.2.4

行业专用 MSS sector-specific MSS

将通用 MSS 用于特定的经济或商业部门时，提供附加要求或补充性指导的 MSS。

L.2.5

A 型 MSS　type A MSS

提出要求的 MSS。

示例：

管理体系要求标准（规范）。

管理体系行业专用要求标准。

L.2.6

B 型 MSS　type B MSS

提供指导的 MSS。

示例：

关于使用管理体系要求标准的指南。

关于建立管理体系的指南。

关于改善 / 提高管理体系的指南。

L.2.7

高层次结构　high level structure

HLS

ISO/TMB/JTCG "MSS 联合技术协调小组"的工作成果，其中涉及 HLS、相同的子条款标题、相同的内容以及共同的术语和核心定义。见本附录附件 2。

L.3 提交合理性研究的义务

所有 MSS［包括行业专用 MSS（L.2.4），见附录 M］提案及其合理性研究（JS）应由相关 TC/SC/PC［或 IEC 系统委员会（SyC）］领导层确认，并应在 NP 投票前将 JS 提交 TMB（或其 MSS 特别工作组）进行评估和批准。相关 TC/SC/PC 秘书处负责确认所有 MSS 提案，以确保所有 MSS 提案都能进行 JS（无论是否知晓）或提交 ISO/TMB 进行评估。

若已提交并批准了特定的 A 型 MSS 的 JS，对该 MSS 提供指导的 B 型 MSS 不需要进行 JS。

例如，ISO/IEC 27003：2010《信息技术　安全技术　信息安全管理体系实施指南》不需要提交 JS，因为 ISO/IEC 27001：2013《信息技术　安全技术　信息安全管理体系要求》已经提交了 JS 并获得了批准。

L.4 未提交 JS 的情况

在 NP 投票前未提交 TMB 评估的 MSS 提案将首先提交 TMB 进行评估，在 TMB 公布评估结果之前不得进行新的投票（暂停项目）。建议在将 JS 提交 TMB 之前，先获得 TC/SC/PC（和 / 或 IEC SyC）成员对该 JS 的认可。

注：未提交过 JS 但已发布的 MSS 在修订时作为新的 MSS 处理，即在任何工作开始前提交并批准一个 JS。

L.5 本附录的适用性

上述程序适用于所有 ISO 和 IEC 可交付使用的文件,包括国际研讨会协议(IWA)。

L.6 一般原则

所有新 MSS 项目(或已发布但尚未完成 JS 的 MSS 项目)均应进行 JS(见 L.1 和 L.3 注释)。下列一般原则用于指导评估拟议的 MSS 的市场相关性和 JS 的准备。本附录附件 1 中的评价准则问题基于下列原则。这些问题的答案将作为构成 JS 的一部分。只有在遵守下列所有原则的情况下,才宜发起、发展和维持 MSS。

1) 市场相关性——任何 MSS 都宜满足主要用户和受影响方的需求,并为其增加价值。

2) 兼容性——应该保持不同的 MSS 之间和 MSS 系列之内的兼容性。

3) 主题覆盖范围——通用 MSS(L.5.3)宜具有足够的应用覆盖范围以消除对行业专用差异的需求或使需求最小化(原文有误,请核对是否应为"L.2.3",目前按照原文"L.5.3"处理)。

4) 灵活性——MSS 宜适用于所有相关部门和文化以及任何规模的组织。MSS 不宜限制组织进行竞争性的添加或区别于其他组织,或者超出标准地加强其管理体系。

5) 自由贸易——MSS 宜根据世界贸易组织(WTO)《技术性贸易壁垒协议》的原则,允许货物和服务的自由贸易。

6) 合格评定的适用性——宜评估第一方、第二方、第三方或其任意组合的合格评定的市场需求。所产生的 MSS 宜明确说明其范围内合格评定的适用性。MSS 宜有利于联合审计。

7) 排除范围——MSS 不宜包括直接相关的产品(含服务)规范、测试方法、性能水平(即极限设置)或对执行组织生产产品的其他形式的标准化。

8) 易用性——确保用户可以轻松实现一个或者多个 MSS。MSS 宜易于理解、表达明确、不受文化偏见的影响、易于翻译并适用于一般业务。

L.7 合理性研究过程和标准

L.7.1 概述

本条款描述了合理性研究(JS)过程,用于论证和评估 MSS 提案的市场相关性。本附录的附件 1 提出了需要在合理性研究中处理的一系列问题。

L.7.2 合理性研究过程

JS 过程适用于所有 MSS 项目。JS 过程包括:

a) 由(或代表)MSS 项目的提出者进行 JS;

b）由 TMB（或 ISO 的 ISO/TMB MSS 特别工作组）对 JS 进行批准。

在 JS 过程之后，将按照 ISO 或 IEC 正常投票程序适当批准新工作项。

L.7.3 合理性研究标准

依据附录 C 和上述一般原则，应将一组问题（见本附录的附件1）作为论证和评估一个拟议 MSS 项目的标准，应并由提议人回答。此问题清单并不详尽，宜提供与此项目相关的其他资料。JS 宜阐明已考虑到所有的问题。如果认定它们与某一特定情况无关或者不恰当，宜清楚阐述作出这一决定的理由。为客观评估某一 MSS 独特层面的市场相关性，可能需要考虑其他问题。

L.8 关于 MSS 制定过程和结构的指南

L.8.1 概述

制定 MSS 的影响将与以下方面相关：

—— 这些标准对商业实践的深远影响；

—— 全球范围内支持这些标准的重要性；

—— 许多（如果不是全部）国家成员机构参与的实际可能性；

—— 以及对兼容一致的 MSS 的市场需求。

考虑到以上影响，本条款提供除 ISO/IEC 导则中其他条款规定程序外的指导。

所有的 MSS（不论是 A 型或 B 型，通用或特定部门的 MSS）原则上应结构一致、内容和术语相同，以易于使用和相互兼容。原则上也应遵循本附录附件2中给出的指导和结构。

B 型 MSS 对同一 MSS 系列的另一种 MSS 提供指导，宜遵循相同的结构（即条款编号）。如涉及提供指导的 MSS（即 B 型 MSS），必须清楚界定它们的职能，以及它们与提供需求的 MSS（A 型 MSS）的关系，例如：

—— 关于使用要求标准的指南；

—— 关于建立/实施管理体系的指南；

—— 关于改善/提高管理体系的指南。

当提议行业专用 MSS：

—— 宜与通用 MSS 兼容且一致；

—— 应遵循附录 M 中规定的规则和原则；

—— 负责通用 MSS 的相关委员会可能有需要满足的额外要求或需要遵循的程序（见附录 M）；

—— 可能需要与其他委员会以及 ISO 合格评定委员会（CASCO）和 IEC CAB 就合格评定问题进行磋商。

就特定部门的文件而言,宜明确界定它们的职能及与通用 MSS 的关系(如额外的特定部门的要求和说明,或恰当情况下的两者)。

特定部门的文件宜始终清晰展示(如使用不同的排版样式)所提供的特定部门信息的种类。

注:由于特殊情况导致相同内容或任何要求不能应用于特定 MSS 时,应通过 TMB 秘书(tmb@iso.org)或 IEC/SMB 秘书处向 TMB 报告(见 L.9.3)。

L.8.2 MSS 制定过程

L.8.2.1 概述

除 JS 外,MSS 的制定宜遵循与其他 ISO 和 IEC 可交付使用的文件相同的要求(见第 2 章)。

L.8.2.2 设计规范

为确保维持本标准的意图(如 JS 所示),可在拟订工作草案之前制定一份设计规范书。

归口委员会将决定是否需要设计规格。如认为有必要,归口委员会将决定适合 MSS 的格式和内容,并宜设立必要的组织来执行这项任务。

设计规范通常应解决以下问题:

- 用户需求——确定标准的用户及其相关需求,以及这些用户的成本和收益。
- 范围——标准的范围和用途、标题和适用范围。
- 兼容性——如何实现这个系统内部以及与其他 MSS 系列之间兼容,包括识别具有类似标准的通用元素,以及如何将这些元素纳入推荐的结构(见本附录的附件 2)。
- 一致性——与 MSS 系列中制定(即将制定)的其他文档的一致性。

注:大多数(如果不是全部)有关用户需求和范围的信息都可以从 JS 中获得。

设计规范宜确保:

a) JS 的结果正确转化为 MSS 的要求;

b) 识别并解决与其他 MSS 的兼容性和一致性问题;

c) 在制定过程中的适当阶段存在证实最终 MSS 的基础;

d) 设计规范书的批准为 TC/SC 和 / 或 IEC SyC 成员在整个项目中的所有权提供依据;

e) 考虑通过 NP 投票阶段得到的意见;

f) 考虑到任何限制因素。

制定 MSS 的委员会应根据设计规范监控 MSS 的制定,以确保在项目过程中不会发生偏差。

L.8.2.3 生成可交付使用的文件

L.8.2.3.1 监控输出

在起草过程中，宜监控输出以保证与其他 MSS 兼容及易于使用，要覆盖以下问题：

—— HLS、相同的子条款标题、相同的内容、共同术语和必须明确的核心定义（语言和表达方面）；

—— 且避免重叠和矛盾。

L.8.2.4 MSS 制定过程的透明度

与大多数其他类型的标准相比，MSS 范围更广。它覆盖了人类努力的大片领域，同时影响着广大用户利益。

因此，MSS 编写委员会宜对标准的制定采用高度透明的方法，确保：

—— 明确识别了参与标准制定进程的可能性；

—— 且所有各方都理解所使用的制定过程。

委员会宜提供项目全过程进展情况的相关资料，包括：

—— 迄今为止项目的状况（包括讨论中的条目）；

—— 为获取未来信息的联系点；

—— 全体会议的公报和新闻稿；

—— 常见问题和答案的常规清单。

提供资料时，需考虑到参与国现有的分发设施。

如果预计 A 型 MSS 的用户可能会证明其符合性，则 MSS 宜以书面形式，以确保制造商或供应商（第一方或自我声明）、用户或买方（第二方）或独立机构（第三方，或称认证方或注册方）能够评估该 MSS 的符合性。

应最大限度地利用 ISO 中央秘书处或 IEC 中央办公室的资源，以提高项目透明度，委员会还应考虑设立一个专用的开放访问网站。

委员会应让国家成员机构参与 MSS 制定，以提高国家对 MSS 项目的认识，为不同的相关和受影响的各方（包括审核机构、认证机构、企业和用户团体），适当提供草案，并根据需要提供额外的专用信息。

委员会宜确保各参与成员（特别是来自发展中国家的成员）随时可以获取关于有关正在制定中的 MSS 内容的技术资料。

L.8.2.5 标准的解释过程

委员会可建立来自用户的有关标准解释问题的处理过程，并用合适的方法将解释结果提供给他人。这种机制可有效解决在早期阶段可能产生的误解，识别出在以后的修订过程中可能需要改进的与标准相关的表述。在 ISO 中，此类过程被称为"委员会专用程序"(见前言f)。在IEC中，委员会应称其为"解释单"过程（见IEC补充部分）。

L.9 用于管理体系标准的高层次结构、相同核心内容和通用术语及核心定义

L.9.1 概述

本文件的目的是通过提供一个统一和一致同意的高层次结构、相同的核心内容和通用术语及核心定义来提高MSS的一致性。这一目的是使所有A型MSS（和需要时B型MSS）保持一致，加强这些标准的兼容性。每个MSS都将根据需要增加额外"领域专用"要求。

> 注：L.9.1和L.9.4中使用的"领域专用"用于表明一个管理体系标准涉及的特定主题，如能量、质量、记录、环境等。

本文件的预期读者是TC/SC/PC（和IEC SyC）及参与MSS制定的其他人员。

用以制定新的MSS和未来修订现行标准的通用方法将会提高这些标准对用户的价值。对于那些选择实施能够同时满足两个或者多个MSS要求的单一（有时亦称"聚合"）管理体系的组织，这种方法非常实用。

本附录的附件2列出构成未来和可能修订的A型MSS和B型MSS核心的高层次结构、相同核心内容和通用术语及核心定义。

本附录的附件3陈述了本附录附件2的使用指南。

L.9.2 使用

MSS包括本附录附件2所示的高层次结构和相同核心内容。通用术语及核心定义或直接纳入MSS中，或规范性引用包含这些内容的国际标准。

> 注：高层次结构包括顺序固定的主要条款（第1章～第10章）及其标题。相同核心内容包括编号的分条款（及其标题）以及分条款中的内容。

L.9.3 不适用性

如果由于例外情况导致高层次结构或任何相同核心内容、通用术语及核心定义不能适用于管理体系标准，TC/PC/SC需通过以下方式说明原因，以供评审：

a) 在提交DIS同时向ISO中央秘书处或IEC中央办公室提供初始偏离报告；

b) 在提交最终标准文本进行出版时通过ISO/TMB秘书处（tmb@iso.org）向TMB或IEC/SMB秘书处提供最终偏离报告。

TC/PC/SC（或IEC SyC）应该使用ISO或IEC评论模板提供他们的偏离报告。

> 注1：最终偏离报告可能为初始偏离报告的更新版。
> 注2：TC/PC/SC（或IEC SyC）尽量避免高层次结构或任何核心内容、通用术语及核心定义的不适用性。

L.9.4 使用本附录的附件2

对本附录附件2中增加的领域专用内容的管理如下：

1. 由各 TC/PC/SC（或 IEC SyC）或制定特定管理体系标准的其他团体补充领域专用内容。

2. 领域专用内容不影响协调、不反对和不削弱高层次结构、相同核心内容、通用术语及核心定义的意图。

3. 在相同内容子条款（或次子条款等）前或后插入补充的子条款或次子条款（等），并重新编号。

注1：不允许有悬置段。（见《ISO/IEC 导则 第2部分》）。
注2：注意需要使用交互引用的情况。

4. 在本附录的附件2内增加或插入领域专用内容。增加内容示例如下：

a) 新的箭头；
b) 领域专用说明性内容（如注或实例），以阐明要求；
c) 在相同内容内的子条款（等）中加入领域专用的新段落；
d) 补充内容，以加强本附录附件2中的现有要求。

5. 考虑到上述第2点，通过在相同核心内容中补充内容来避免相同核心内容和领域专用内容之间的重复要求。

6. 起草过程初期就要区分领域专用内容和相同核心内容。这有助于在编制和投票期间识别不同类型的内容。

注1：通过不同颜色、字型、字号、使用斜体或括号等方式加以区分。
注2：对区分内容的识别无需放进已发布的版本中。

7. 对于"风险"概念的理解可能比本附录附件2中3.9的定义更具体。在这种情况下，可能需要一个领域专用定义。领域专用术语和定义有别于核心定义，例如（×××）风险。

注：以上所述也能适用于其他一些定义。

8. 通用术语及核心定义将按标准的概念体系纳入领域专用管理体系标准的术语和定义清单。

L.9.5 实施

制定任何新的管理体系标准和修订任何现行管理体系标准都应遵循如下顺序：高层次结构、相同核心内容、通用术语及核心定义。

L.9.6 指南

见本附录的附件3中的支撑指南。

附件 1
（规范性）
论证准则问题

1 概述

在合理性研究（JS）中拟阐述的问题清单符合 L.6 列出的原则。这一清单并不详尽。需提供与情况相关但未包含在这些问题中的补充性信息。

在编制 JS 时应适当考虑每一条一般原则，提案提出者在回答与原则有关的问题之前，需针对每条原则提供一般理由。

MSS 提案提出者在编制 JS 时需适当注意的原则是：

1. 市场相关性；
2. 兼容性；
3. 主题覆盖范围；
4. 灵活性；
5. 自由贸易；
6. 符合性评定的适用性；
7. 排除范围。

注：没有任何问题直接涉及原则 8（易用性），但是"易用性"宜指导可交付使用文件的制定。

有关 MSS 提案的基本信息

1	提出 MSS 的目的是什么？范围如何？提出的文件是指南文件还是规定要求文件
2	提出建议的目的和范围是否包括产品（含服务）规范、产品试验方法、产品性能水平或其他形式的指南或要求与实施组织生产或提供产品直接相关
3	是否有一个或者多个委员会或非 ISO 和非 IEC 组织能够负责拟议的 MSS？如果有，给出其名称
4	确定了相关参考资料吗？如现行指南或已确定的惯例
5	是否有技术专家支持其标准化工作？这些专家是来自不同地理区域的相关方面的直接代表吗
6	从所需专家和会议的次数/时间方面来看，预计制定本文件需要作出哪些努力
7	预期 MSS 是一份指南性文件、合同规范还是某一组织的管理性规范

原则1：市场相关性

8	已经识别了所有受影响的相关方吗？例如： a）组织（各种类型和规模）：组织内批准实施 MSS 和实现 MSS 符合性的决策者； b）客户/终端用户，即支付或使用来自一个组织的产品（含服务）的个人或团体； c）供方组织，如产品的制造商、销售商、零售商或卖主，或服务、信息的提供商； d）MSS 服务提供者，如 MSS 认证机构、审核机构或咨询者； e）管理机构； f）非政府组织
9	对本 MSS 有何需求？这些需求存在于地方、国家、区域还是全球的层面上吗？这些需求适用于发展中国家吗？这些需求适用于发达国家吗？拥有 ISO 或 IEC 文件会有什么附加值（例如，促进不同国家的组织之间的交流）
10	某些行业有这样的需求吗？是一般性需求吗？如果是，有哪些？小、中或大型组织有这样的需求吗
11	这种需求重要吗？会继续有这种需求吗？如果有，完成拟议的 MSS 的目标日期能满足这种需求吗？确定了可替代的可行方法吗
12	描述如何确定这些需求及其重要性。列出已知的受影响方及其所在的重要地理或经济区域
13	对拟议的 MSS 有已知或预期的支持吗？列出已表明支持该 MSS 的团体。对拟议的 MSS 有已知或预期的反对吗？列出已表明反对该 MSS 的团体。 有反对吗？列出已表明反对的那些团体
14	预测为组织带来什么样的利益和成本？为适用的小、中和大型组织带来的利益和成本有哪些差别？ 描述如何确定利益和成本。提供有关组织的地理或经济重点、工业领域和规模方面的有效信息。提供已知的来源及其基础（如，已证明的范例）、场所、设想和条件（如，推测或理论上的条件）以及其他相关信息
15	预测对其他受影响方（包括发展中国家）带来哪些利益和成本？ 描述如何确定这些利益和成本。提供已表明的受影响方的相关信息
16	对社会带来哪些预期的价值
17	识别了任何其他风险吗（例如，时限或对某一特定企业带来的意外后果）

原则2：兼容性

18	是否存在与其他现行或拟制定的 ISO、IEC、非 ISO 或非 IEC 国际标准、或国家或区域标准重叠或冲突（或者对这些标准有何补充价值）？是否有其他寻求解决确定需求的公共或专用措施、指南、要求和法规（例如技术文件、已证实的惯例、学术或专业研究或已知的任何其他团体）
19	MSS 或相关的合格评定活动（如审核、认证）有可能补充、代替全部或部分、协调和简化、重叠或重复、冲突或损害上述已确认的现有活动吗？考虑采用哪些措施确保兼容性，解决冲突或避免重复
20	拟议的 MSS 有可能促进还是阻碍 MSS 在国家或区域层面或者通过行业部门的迅速扩大

原则 3：主题覆盖范围

21	MSS 是针对某个特定部门吗
22	MSS 是否会参照或纳入现有的非行业特定的 MSS（例如：ISO 中 9000 条质量管理标准）？如果是，MSS 的制定是否符合 ISO/IEC 行业政策（见《ISO/IEC 导则　第 2 部分》），以及任何其他相关的政策及指导程序（如相关 ISO 委员会可提供的政策及指引程序）
23	已采取哪些步骤来消除或减少对通用 MSS 的特定部门偏差的需要

原则 4：灵活性

24	MSS 是否允许组织进行竞争性的增加、区分或鼓励其管理体制标准之外的创新

原则 5：自由贸易

25	MSS 将如何促进或影响全球贸易？MSS 可以建立或阻止技术性贸易壁垒吗
26	MSS 能否建立或阻止针对中小企业或大型组织的技术壁垒
27	MSS 能否为发展中国家或发达中国家建立或阻止技术性贸易壁垒
28	如果拟议的 MSS 拟用于政府法规，是否可能增加、复制、取代、加强或支持现行的政府法规

原则 6：合格的适用性

29	如果预期用途是出于合同或监管目的，有哪些潜在的方法可以证明一致性（如，第一方、第二方或第三方）？MSS 是否使组织能够灵活地选择证明一致性的方法，并适应其操作、管理、地理位置和设备中的更改
30	如果第三方注册/认证是一个潜在的选择，那么组织预期的利益和成本是什么？MSS 是否会促进与其他 MSS 的联合审计或平行评估

原则 7：排除范围

31	拟议的目的或范围是否包括产品（含服务）规格、产品测试方法、产品性能级别，或与实施组织生产或提供的产品直接相关的其他形式指导或要求

附件 2
（规范性）
高层次结构、相同核心文本、通用术语及核心定义

注：在相同的文本提案中，×××=需要插入的 MSS 规程特定限定词（如能源、道路交通安全、IT 安全、食品安全、社会安全、环境、质量）。本附件中斜体文字[3]为标准起草人员的咨询意见注释。

概述

起草说明　领域专用

本文本的编制采用《ISO/IEC 导则　第1部分》中附录 L 的附件2中的"高层次结构"（即条款顺序、相同核心文本和通用术语与核心定义），旨在提高 ISO 和 IEC 管理体系标准的一致性，促进其在需要满足两个及以上此类标准要求的组织中的实施。

文中（第1章～第10章）使用阴影部分字体突出显示 HLS。黑色字体表示 ISO 或 IEC 专用领域文本。删除线用来表示 HLS 文本内同意删除的内容。使用阴影部分文本和删除线仅为了便于分析，不会在国际标准草案编制阶段之后使用。

1　范围

起草说明：领域专用。

2　规范性引用文件

起草说明：应使用条款标题。领域专用。

3　术语与定义

起草说明1：应使用条款标题。术语与定义可位于标准或一个单独的文件中。

通用术语和核心定义以及其他领域专用的术语与定义应予以说明。

术语与定义最好按照概念层次（即系统顺序）排列，最好不要按照字母顺序排列。

本文件适用下列术语与定义。

起草说明2：下列术语与定义作为管理体系标准"通用文本"不可或缺的一部分。可根据需要增加其他术语与定义。为实现每个标准的目的，可增加或修改注。

起草说明3：定义中的斜体字表示对本条款所定义的另一个术语的交叉引用，该引用术语的编号在括号中给出。

起草说明4：在本条款中出现"×××"文本的位置，应根据使用这些术语与定义的

---

[3] 本附件中中文斜体字部分对应英文中的灰色斜体字〔原 ISO/IEC 文件为蓝色斜体字（Blue italicized text）〕——译者注。

上下文插入适当的信息。例如："一个×××目的"可替换为"一个信息安全目标"。

3.1

组织　organization

为实现目标（3.8），由职责、权限和相互关系构成其自身功能的一个人或一组人。

注1：组织的概念包括但不限于：个体经营者、公司、股份有限公司、商号、企业、当局、合伙关系、慈善组织或机构，或上述组织的部分或组合，无论是否有法人、公有或私有。

3.2

相关方（优先术语）　interested party（preferred term）

利益相关方（许用术语）stakeholder（admitted term）

能够影响决定或活动，受决定或活动影响或认为自己受到决定或活动影响的个人或组织（3.1）。

3.3

要求　requirement

表明的需求或期望，包括一般默示的或强制的。

注1："一般默示"即对于组织和相关方，考虑这些隐含的需求和期望是出于惯例或普遍做法。

注2：明确需求是指明确指出的需求，例如文件化信息中指出的需求。

3.4

管理体系　management system

组织的相互关联或相互作用的一套要素（3.1），用于建立政策（3.7）和目标（3.8），以及实现这些目标的过程（3.12）。

注1：一个管理体系可以涉及一个或多个领域。

注2：系统要素包括组织的结构、角色和职责、计划和操作。

注3：一个管理体系的范围可以包括整个组织、组织专用和已确定的职能、组织专用和已确定的部分，或者跨组织群的一项或多项功能。

3.5

高层管理者　top management

在最高层级指导或控制一个组织的一个或一组人员。

注1：高层管理者有权在组织内委托权力或提供资源。

注2：如果管理体系（3.4）的范围仅涵盖组织的一部分，则高层管理者是指指导或控制该部分组织的人员。

3.6

有效性　effectiveness

计划活动的实现程度和计划结果的实现程度。

3.7
政策 policy

组织（3.1）的目的和方向，与其高层管理者（3.5）正式表达的一致。

3.8
目标 objective

达到的结果。

注1：可以是一个战略性、战术性或行动的目标。

注2：目标可以涉及不同的领域（如金融、健康与安全，以及环境目标），且适用于不同的层次[如战略、全组织范围、项目、产品和过程（3.12）]。

注3：目标可以用其他方式表达，例如作为预期成果、目的、操作准则、×××目标，或使用其他具有类似含义（目的、目标或指标）的文字。

注4：为取得具体成效，在×××管理体系下，×××目标由组织设定，符合×××政策。

3.9
风险 risk

不确定的影响。

注1：偏离预期的效果——积极或消极的。

注2：不确定性是缺乏事件、其后果或类似情况的相关信息、理解或知识的状态，即使是部分不确定。

注3：风险的特征通常是指潜在的"事件"（见ISO指南73的定义）和"后果"（见ISO指南73的定义），或其组合。

注4：风险通常指事件（包括环境变化）的结果和发生的"可能性"（见ISO指南73的定义）的组合。

3.10
能力 competence

为达到预期结果，应用知识和技能的能力。

3.11
文件化信息 documented information

需要一个组织（3.1）控制和维护的信息及其所载媒介。

注1：文件化信息可以是任何格式和媒介，并可以来自任何来源。

注2：文件化信息可指：
——管理体系（3.4），包括相关过程（3.12）；
——为组织的运转而建立的信息（文件资料）；
——取得成果的证据（记录）。

3.12
过程 process

将输入转化为输出的一系列相互关联或相互作用的活动。

3.13

绩效　performance

可测量的结果。

注 1：绩效可能涉及定量的或定性的结果。

注 2：绩效可能涉及活动、过程（3.12）、产品（含服务）、体系或组织（3.1）的管理。

3.14

外包（动词）　outsource（verb）

安排外部组织（3.1）执行组织的部分职能或过程（3.12）。

注 1：虽然外包的职能或过程是在管理体系（3.4）覆盖范围内，但是外部组织处在管理体系覆盖范围之外。

3.15

监控　monitoring

确定体系、过程（3.12）或活动的状态。

注 1：确定状态可能需要检查、监督或密切观察。

3.16

测量　measurement

确定数值的过程（3.12）。

3.17

审计　audit

为获得审计证据并对其进行客观的评价，以确定满足审计准则的程度所进行的系统的、独立的并文件化的过程（3.12）。

注 1：审计可以是内部审计（第一方）或外部审计（第二方或第三方），也可以是组合审计（结合两个或多个领域）。

注 2：内部审计由组织本身或由代表其利益的外部方进行。

注 3："审计证据"和"审计准则"在 ISO 19011 中定义。

3.18

合格　conformity

满足要求（3.3）。

3.19

不合格　nonconformity

未满足要求（3.3）。

3.20

纠正措施　corrective action

消除不合格（3.19）原因，并防止再次发生的行动。

3.21

 持续改进　continual improvement

 不断提升绩效（3.13）的活动。

4 组织所处的环境

4.1 理解组织及其所处的环境

 组织应确定与其宗旨相关并影响其实现×××管理体系预期结果的能力的外部和内部问题。

4.2 理解相关方的需求和期望

 组织应确定：

 —— 与×××管理体系有关的相关方；

 —— 这些相关方的相关要求。

4.3 确定×××管理体系的范围

 组织应确定×××管理体系的边界和适用性，以界定其范围。

 确定范围时，组织应考虑：

 —— 4.1所提及的内、外部问题；

 —— 4.2所提及的要求。

 范围应可以文件化信息形式获取。

4.4 ×××管理体系

 组织应按照本文件的要求建立、实施、保持并持续改进×××管理体系，包括所需的过程及其相互作用。

5 领导能力

5.1 领导能力与承诺

 最高管理者应通过以下方式展示其在×××管理体系方面的领导能力和承诺：

 —— 确保建立×××政策和×××目标，并确保其与组织的战略方向相一致；

 —— 确保将×××管理体系的要求融入组织的业务过程；

 —— 确保可获得×××管理体系所需的资源；

 —— 就有效×××管理和符合×××管理体系要求的重要性进行沟通；

 —— 确保×××管理体系实现其预期结果；

 —— 指导并支持员工对×××管理体系的有效性作出贡献；

—— 促进持续改进；

—— 支持其他相关管理人员在其职责范围内展示其领导能力。

注： 在本文件中提到的"业务"可广泛地解释为那些对组织存在的目的具有核心意义的活动。

5.2 政策

最高管理者应制定 ××× 政策，该政策须：

a) 与组织宗旨相适应；

b) 为制定 ××× 目标提供框架；

c) 包括满足适用要求的承诺；

d) 包括持续改进 ××× 管理体系的承诺。

××× 政策应：

—— 可以文件化信息形式获取；

—— 使组织内的员工可沟通；

—— 适当时，可向相关方提供。

5.3 岗位、职责和职权

最高管理者应确保在组织内部分配并沟通相关岗位的责任和权限。

最高管理者应对下列事项分配职责和权限：

a) 确保 ××× 管理体系符合本文件的要求；

b) 向最高管理层报告 ××× 管理体系的绩效。

6 策划

6.1 应对风险和机遇的措施

为 ××× 管理体系策划时，组织应考虑到 4.1 所描述的情况和 4.2 所提及的要求，确定需要应对的风险和机遇，以便：

—— 确保 ××× 管理体系能够实现其预期结果；

—— 预防或减少不良影响；

—— 实现持续改进。

组织应计划：

a) 采取行动应对这些风险和机遇；

b) 如何：

—— 将行动整合并实施到其 ××× 管理体系流程中；

—— 评估这些行动的有效性。

6.2 ×××目标及实现目标的策划

组织应针对其相关职能和层次建立×××目标。

×××目标应：

a) 符合×××政策；

b) 可度量（如可行）；

c) 考虑适用性要求；

d) 可监控；

e) 可沟通；

f) 适当地更新。

组织应保存关于×××目标的文件化信息。

策划如何实现其×××目标时，组织应确定：

—— 要做什么；

—— 需要什么资源；

—— 由谁负责；

—— 何时完成；

—— 如何评价结果。

7 支持

7.1 资源

组织应确定并提供建立、实施、保持和持续改进×××管理体系所需的资源。

7.2 能力

组织应：

—— 确定在其控制下从事影响其×××绩效的工作中人员的必要能力；

—— 基于适当的教育、培训或经历，确保员工能够胜任工作；

—— 适当时，采取措施以培养员工所必需的能力，并评价所采取措施的有效性；

—— 保留适当的文件化信息作为能力的证据。

注：适当的措施可能包括向现有员工提供培训和指导，或重新委派其职务；或聘用、雇佣胜任的人员等。

7.3 意识

在本组织控制下工作的人员应知道：

—— ×××政策；

—— 他们对×××管理体系的有效性作出的贡献，包括提高×××绩效的好处；

—— 不符合×××管理体系要求的可能后果。

7.4 沟通

组织应确定与×××管理体系有关的内外部沟通，包括：

—— 沟通的内容；

—— 何时进行沟通；

—— 与谁进行沟通；

—— 怎样进行沟通。

7.5 文件化信息

7.5.1 概述

组织的×××管理体系应包括：

a) 本文件要求的文件化信息；

b) 组织确定的×××管理体系有效性所必需的文件化信息。

注：不同组织的×××管理体系文件化信息的范围可能不同，取决于：
—— 组织的规模及其活动、过程、产品和服务的类型；
—— 过程的复杂性及其相互作用；
—— 人员的能力。

7.5.2 创建和更新

创建和更新文件化信息时，组织应确保适当的：

—— 识别和描述（如标题、日期、作者或文献编号）；

—— 形式（如语言、软件版本、图表）与载体（如纸质、电子）；

—— 评审和批准，以确保适宜性和充分性。

7.5.3 文件化信息的控制

应控制×××管理体系和本文件要求的文件化信息，以确保：

a) 在需要的时间和场所均可获得并适用；

b) 受到充分的保护（如防止失密、不当使用或完整性受损）。

为了控制文件化信息适用，组织应采取以下措施：

—— 分发、访问、检索和使用；

—— 存储和保护，包括保持易读性；

—— 变更的控制（如版本控制）；

—— 保留和处置。

组织确定的为×××管理体系的策划和运行所必需的外部来源文件化信息应被识别并在适当情况下加以控制。

注：访问可能表明关于仅查看文档化信息的权限，或查看和更改文档化信息的许可和权限的决策。

8 运行

8.1 运行策划和控制

起草说明：若无需添加分条款至第 8 章，删除本分条款题目。

组织应通过以下方式策划、实施和控制满足要求所需的过程，并实施 6.1 中确定的行动：

—— 建立过程准则；

—— 按照准则实施过程控制；

—— 保持必要的文件化信息，以确信已按策划实施过程。

组织应控制计划中的意外变更，并审查意外变更的后果，必要时采取行动减轻任何不利影响。

组织应确保外包过程可控。

9 绩效评价

9.1 监控、测量、分析和评价

组织应确定：

—— 需要监控和测量的内容；

—— 适用时，监控、测量、分析与评价的方法，以确保结果有效；

—— 何时应实施监控和测量；

—— 何时应分析、评价监视和测量结果。

组织应保留适当的文件化信息，作为监控、测量、分析和评价结果的证据。

组织应评价其 ××× 绩效并确定 ××× 管理体系的有效性。

9.2 内部审计

9.2.1 内部审计目标

组织应按计划的时间间隔实施内部审计，以提供 ××× 管理体系相关的以下信息：

a) 是否符合：

—— 组织自身对 ××× 管理体系的要求；

—— 本文件的要求。

b) 是否得到有效实施和维护。

9.2.2 组织应：

a) 策划、建立、实施并保持一个或多个内部审核方案，包括实施审核的频次、方法、职责、策划要求和报告。策划、建立、实施和保持内部审计方案时，应考虑相关过程的重要性和以往审计的结果。

b) 规定每次审计的准则和范围。

c) 选择审计员并实施审计，以确保审计过程的客观性与公正性。

d) 确保向相关管理者报告审计结果。

e) 保留文件化信息，作为审计方案实施和审计结果的证据。

9.3 管理评审

最高管理者应按计划的时间间隔对组织的×××管理体系进行评审，以确保其持续的适宜性、充分性和有效性。

管理评审应包括对下列事项的考虑：

a) 以往管理评审所采取措施的状况。

b) 与×××管理体系相关的内外部问题的变更。

c) ×××绩效方面的信息，包括以下方面的趋势：
—— 不符合和纠正措施；
—— 监控和测量的结果；
—— 审计结果。

d) 持续改进的机遇。

管理评审的输出应包括与持续改进机会和×××管理体系变更的任何需要相关的决策。

组织应保存文件化信息，作为管理评审结果的证据。

10 改进

10.1 不符合和纠正措施

当不符合发生时，组织应：

a) 对不符合作出反应，并且适用时：
—— 采取措施控制并纠正不符合；
—— 处理后果。

b) 通过以下评估采取行动消除不符合原因的必要性，以防止不符合再次发生或在其他地方发生：
—— 评审不符合项；
—— 确定不符合的原因；
—— 确定是否存在或可能发生类似的不符合。

c) 实施任何必要的行动；

d) 评审所采取纠正措施的有效性；

e) 如有必要，对 ××× 管理体系进行变更。

纠正措施应与所遇到的不符合项的影响相适应。

组织应保留文件化信息作为下列事项的证据：

—— 不符合的性质和所采取的任何后续措施；

—— 任何纠正措施的结果。

10.2 持续改进

组织应持续改进 ××× 管理体系的适宜性、充分性与有效性。

附件 3
（资料性）
高层次结构、相同核心文本、通用术语及核心定义指南

高层次结构、相同核心文本、通用术语及核心定义指南在以下 URL 中提供：

附录 L 指南文档（http://isotc.iso.org/livelink/livelink?func=ll&objId=16347818&objAction=browse&viewType=1）。

附录 M
（规范性）
制定行业专用管理标准和管理体系标准（MSS）的政策

M.1 概述

任何技术委员会或分委员会、项目委员会（IEC 体系委员会）或国际研讨会提议制定行业专用管理标准（M.2.2）或行业专用管理体系标准（MSS）（M.2.4）的，均应遵守本附录中规定的导则。其中包括委员会专用政策（M.5），这可不限于行业专用管理标准或管理体系标准。

M.2 术语和定义

M.2.1

通用管理标准 generic management standard

广泛适用于不同经济部门、不同类型和规模的组织以及不同地理、文化和社会条件的管理标准。

M.2.2

行业专用管理标准 sector-specific management standard

就特定经济或业务部门如何适用通用管理标准（M.2.1）提供附加要求或补充性指导的管理标准。

M.2.3

通用管理体系标准 generic management system standard

通用 MSS generic MSS

广泛适用于不同经济部门、不同类型和规模的组织以及不同地理、文化和社会条件的 MSS。

M.2.4

行业专用管理体系标准 MSS sector-sprcific management system standard（MSS）

行业专用 MSS sector-specific MSS

就特定经济或业务部门如何适用通用 MSS（M.2.3）提供附加要求或补充性指导的 MSS。

M.3 行业专用管理标准和管理体系标准

任何要求制定行业专用管理标准（M.2.2）或管理体系标准（M.2.4）的新提议

均应：

—— 透过 ISO 表格 4：新工作项目提案、或 IEC 的表格 NP 完成适当的 ISO 或 IEC 项目批准程序，清楚表明该标准对市场有意义，符合通用标准的要求；

—— ［就行业专用的 MSS（M.2.4）的制定而言］，清楚表明附录 L 中的所有规则和原则均得到遵守，包括关于合理性研究的批准（见附录 L）；

—— 清楚表明与负责相关通用管理标准或通用 MSS 的委员会联络的有效性；

—— 必要时，遵守以下适用于特定委员会的政策。

M.4 起草规则

行业专用管理标准（M.2.2）和行业专用的 MSS（M.2.4）应分别遵守以下规则：

a) 将通用管理标准（M.2.1）或通用 MSS（M.2.3）作为规范性引用文件。也可以逐字复制其条款和分条款。

b) 如果在行业专用标准中复制通用管理标准（M.2.1）或通用 MSS（M.2.3）的文本，则应将其与行业专用标准的其他要素区分开来。

c) 对通用管理标准（M.2.1）或通用 MSS（M.2.3）中规定的术语和定义应作规范性引用或逐字复制。

M.5 委员会专用政策

M.5.1 概述

行业专用管理标准（M.2.2）和行业专用 MSS（M.2.4）不应解释、更改或降低通用管理标准或通用 MSS 的要求。

M.5.2 环境

M.5.2.1 术语和定义

下列术语和定义适用于环境政策。

M.5.2.1.1

行业专用环境管理标准 sector-specific environmental management standard

就特定经济或业务部门如何适用通用环境管理标准提供附加要求或补充性指导的标准。

示例：将环境管理体系（ISO 14001）或生命周期评估（ISO 14044）适用于农业食品或能源行业。

M.5.2.1.2

特定方面的环境管理标准 aspect-specific enviromental management standard

就特定环境方面或其范围内的多个方面如何适用通用环境管理标准提供附加要求或补充性指导的标准。

示例：就温室气体（方面）管理适用环境管理体系（ISO 14001）或产品水足迹（方面）适用生命周期评估（ISO 14044）。

M.5.2.1.3

特定要素的环境管理标准 element-specific environmental management standard

就特定要素或其范围内的多个要素如何适用通用环境管理标准提供附加要求或补充性指导的标准。

> 示例：环境管理体系（ISO 14001）中的通信或应急管理（要素）或生命周期评估（ISO 14044）中的数据收集或关键评审（要素）。

M.5.2.2 概述

任何技术委员会、分委员会、项目委员会（IEC 体系委员会）或国际研讨会提议制定特定行业、方面或要素专用的环境管理标准的，均应透过完成适当的项目批准程序，清楚表明该标准对市场有意义，符合通用标准的要求，包括：

—— 通用环境管理体系标准、环境标识、生命周期评估和温室气体管理标准在特定行业、方面或要素专用特别适用的相关 ISO 表格 4：新工作项目提案；

—— 以及在特定行业、方面或要素专用特别适用通用环境 MSS 的附录 L：管理体系标准提案。

批准文件宜具体说明相关通用 ISO 14000 系列标准不足以满足特定行业、方面或要素专用需求的理由，以及提议的新标准将如何有效解决所识别出的问题。提案方宜批判性地评估是否确实需要附加适用特定行业、方面或要素专用的要求，而不是就通用环境管理标准提供补充性指导。

M.5.2.3 任何技术委员会、分委员会、项目委员会（IEC 体系委员会）或国际研讨会提议制定特定行业、方面或要素专用的环境管理标准的，宜考虑和体现发展中国家、转型经济体、各行各业中小企业和组织的需求。

M.5.2.4 ISO/TC 207 将与制订特定行业、方面或要素专用的环境管理标准的技术委员会、分委员会、项目委员会（IEC 体系委员会）或国际研讨会合作开展联合项目，或在必要情况下根据技术管理局的决定领导开展联合项目，以避免重复工作以促进标准的一致和统一。但这并不意味着要在 ISO/TC 207 以外的委员会限制市场相关标准的制定。

M.5.2.5 技术委员会、分委员会、项目委员会（IEC 体系委员会）或国际研讨会制定特定行业、方面或要素专用的环境管理标准的，应：

—— 将适当的通用 ISO 14000 系列环境管理体系、环境审计、环境标识、生命周期评估和温室气体管理标准作为规范性引用文件；

—— 规范性引用通用 ISO 14050 术语和定义；

—— 将复制的 ISO 14000 系列文本区分开来；

—— 以及不应解释、更改或降低通用 ISO 14000 系列环境管理体系、环境审计、环境标识、生命周期评估和温室气体管理标准的要求。

M.5.2.6 对于任何就该特定行业、方面或要素专用的政策提供指导、或解释通用 ISO 14000 系列标准或 ISO 14050 术语和定义、或就特定行业、方面或要素专用文件提供指导的请求，均应提交给 ISO 中央秘书处以及相关的 TC 207 分委员会。

M.5.3 质量

技术委员会、分委员会、项目委员会（IEC 体系委员会）或国际研讨会希望为特别产品或行业 / 经济部门制定质量管理体系要求的，应遵守以下规则：

a) 应完整规范性引用 ISO 9001。也可以逐字复制其条款和分条款。

b) 如果在行业文件中复制 ISO 9001 的文本，则应将其与行业文件的其他要素区分开 [见 d)]。

c) 对 ISO 9000 中规定的术语和定义应作规范性引用或逐字复制。

d) 无论是在确定是否需要编制行业专用要求或指南文件时，还是在文件编制过程中，均应参照经 ISO/TC 176 批准的《质量管理体系 编制特定产品和行业 / 经济部门专用文件的指南和标准》中规定的指南和标准。

对于本行业政策提供指导、或解释 ISO 9000 术语和定义、ISO 9001 或 ISO 9004 的请求，均应提交给 ISO/TC 176 的秘书处。

M.5.4 资产管理

技术委员会、分委员会、项目委员会（IEC 体系委员会）或国际研讨会希望为特别产品或行业 / 经济部门制定资产管理体系要求的，应遵守以下规则：

a) 应完整规范性引用 ISO 55001。也可以逐字复制其条款和分条款。

b) 如果在行业文件中复制 ISO 55001 的文本，则应将其与行业文件的其他要素区分开。

c) 对 ISO 55000 中规定的术语和定义应作规范性引用或逐字复制。

对于行业专用文件提供指导、或解释 ISO 55000 术语和定义或 ISO 55001 的请求，均应提交给 ISO/TC 251 的秘书处。

M.5.5 风险

技术委员会、分委员会、项目委员会（IEC 体系委员会）或国际研讨会希望为特别产品或行业、经济部门制定风险管理要求或指导的，应遵守以下规则：

a) 应完整引用 ISO 31000。也可以逐字复制其条款和分条款。

b) 如果在行业文件中复制 ISO 31000 的文本，则应将其与行业文件的其他要素区分开。

c) 对 ISO 31000 中规定的术语和定义应作规范性引用或逐字复制。

对于行业专用文件提供指导、或解释 ISO 31000 术语和定义的请求，均应提交给 ISO/TC 262 的秘书处。

M.5.6 社会责任

技术委员会、分委员会、项目委员会（IEC 体系委员会）或国际研讨会希望为特别产品或行业 / 经济部门制定社会责任要求或指导的，应遵守以下规则：

a) 应完整引用 ISO 26000。也可以逐字复制其条款和分条款。

b) 如果在行业文件中复制 ISO 26000 的文本，则应将其与行业文件的其他要素区分开。

c) 对 ISO 26000 中规定的术语和定义应作规范性引用或逐字复制。

以下附录仅适用于 ISO 的附录

附录 SA
（规范性）
ISO 行为准则

本《ISO 行为准则》见官网 www.iso.org 技术工作管理页（http：//www.iso.org/iso/home/standardsdevelopment/govemanceoftechnicalwork.htm），详细地址为：https：//www.iso.org/publication/PUB100397.html。

ISO 的工作环境涉及诸多国家、利益相关方与行业部门，而本准则旨在对 ISO 的工作提供便利。本准则适用于选择加入 ISO 委员会、工作组或其他协商小组的人员。凡在《ISO/IEC 导则》框架内工作的团体，其成员均有义务遵守本准则。

在执行该准则时，可参考相关建议：http：//www.iso.org/iso/suggestionsforimplementationoftheisocodeofconduct.pdf。

附录 SB
（规范性）
文件分发

SB.1 文件分发

表 SB.1 展示了有关文件分发的详细信息。

SB.2 文件送达的电子通知

电子通知应包括以下信息：

—— 主题栏：委员会参考信息；

—— 内容（具体格式可能有所不同）：

1) N——文件号（自动生成）；

2) ISO 官方参考编号（例如：委员会草案的编号可设为 CD 1234，新工作项目提案的编号可设为 NP 2345）；

3) 文件标题；

4) 预期行动；

5) 预期行动的截止日期；

6) 文件类型（信息、投票、评议、会议通知等）；

7) 若为会议文件，需说明日期和地点；

8) 备注字段，说明与特定文档相关的特殊信息；

9) 文档超链接。

表 SB.1 有关文件分发的详细信息

相关方 文件	提案方	技术委员会（TC）或分委员会（SC）秘书处和主席	技术委员会或分委员会的参与成员（P成员）	技术委员会或分委员会的观察成员（O成员）	A&B类或内部联络组织	ISO中央秘书处	工作组（WG）召集人	工作组专家	国家成员体
提案阶段									
新工作项目提案	★	●				○			
新项目投票单		★¹⁾	●	○	○	○			
填写好的投票单		●	★			○			
投票结果	○	★¹⁾	○	○	○	■			
准备阶段									
工作草案（WD）		○				○	★	●	
最终工作草案		●	○			■	★	○	
委员会阶段									
委员会草案（CD）		★	●						
评议意见/投票		●	★						
评议意见和处理建议汇总		★				○			
最终委员会草案		★¹⁾	●			■			
征询意见阶段									
国际标准草案		○			○	★			●
填写好的投票单				★	●				★
投票结果+评议意见 投票报告	●¹⁾					●			○
最终国际标准草案 文本		★¹⁾	○	○	○	■			○
批准阶段									
最终国际标准草案+ 投票单		○	○		○	★			●
填写好的投票单				★	●				★
投票结果		○	○		○	★			●
最终校正稿		●				★			
校正稿修改		★				●			
出版阶段									
国际标准		○				★			○
系统复审									
系统复审投票		○				★			●
投票						●			★
填写好的投票单		●	●	○	○		○		
投票+提案报告		★	○		○	■			

¹⁾ 凡分委员会负责的项目，应向技术委员会秘书处提交一份文件供参考
★ 文件发送方
● 接收并采取行动方
○ 接收信息方
■ 接收文件并注册方
☆ 可选行动

附录 SC
（规范性）
战略业务计划

SC.1 战略业务计划（SBP）的目的

1) 客观说明该技术委员会的工作所创造或预期可创造的具体价值。

注 1：具体价值可能会因委员会涉及领域的不同而有很大差异，可以是经济方面的（节约成本、缩短上市时间、便于进入某些地区市场、降低销售价格），可以是社会方面的（更好地保障工人的安全、减少事故发生），也可以是改善环境的影响。

2) 帮助确定委员会内技术工作的优先次序，并完善技术工作的管理。

注 2：包括指定各项任务的完成日期，规划技术委员会（以及分委员会）在制定新文件和维护已有文件时可使用的资源。

3) 提供更多有关市场各方的信息，说明委员会内部的市场份额代表情况。

如需查阅所有现行战略业务计划，请登陆 http：//www.iso.org/bpo。

SC.2 编制工具和附加指南

为便于制定战略业务计划，ISO 中央秘书处已制定统一的文件模板。此模板和附加指南可登录 ISOTC 服务器下载：www.iso.org/forms。

SC.3 战略业务计划的制定程序

SC.3.1 在制定战略业务计划草案时，委员会秘书应与委员会成员和/或委员会主席合作。SBP 草案可提交技术委员会进行内部磋商。战略业务计划由技术委员会以决议方式正式通过后，可提交技术管理局进行复审。

技术委员会应向 ISO 技术管理局提交（tmb@iso.org）可修改的 SBP 草案（如：Word 和 PDF），技术管理局应根据战略业务计划最低要求审查该草案（见表 SC.1）。

表 SC.1 ISO 技术委员会战略业务计划相关要求

项目	要求说明	分数
1	描述 ISO 委员会工作所涉业务环境中的发展动力以及业务环境相关量化指标的走势、ISO 委员会标准的认可与实施情况	1～5
2	描述标准化工作预期将为此业务环境创造的切实好处	1～5
3	描述 ISO 委员会确定的目标以及为实现目标而实施的策略。这包括 ISO 委员会为顺应业务环境的需求与发展趋势而制定的具体行动或提案	1～5
4	描述可能阻碍 ISO 委员会实现目标及实施策略的因素，包括主要市场力量在委员会内部的代表情况（根据地理位置和类型划分；例如制造商和政府等）	1～5

表 SC.1（续）

项目	要求说明	分数
5	ISO 技术委员会及其工作项目相关的必要客观信息： —— 已添加链接到 ISO 网站技术委员会网页的第 7 条："关于"（秘书处、秘书、主席、设立日期、范围等）、"联系方式（Contact details）""组织架构（Structure）"（分委员会及工作小组）、"联络组织（Liaisons）""会议（Meetings）""工具（Tools）""工作项目（Working Programme）"（已发布标准及正在制订的标准）； —— 工作项目所含子项目的优先次序安排（如委员会分配优先次序），并说明确定优先次序的理由/程序； —— 工作项目与 CEN 的关系； —— 工作组召集人、项目负责人或编辑分配给每个子项目的时间以及预留翻译时间； —— 基于标准化所针对的领域，应召集哪些利益相关方	无分数（是/否）必填信息

SC.3.2 SBP 的平均分数应不低于 10 分，否则不得批准。如未获 TMB 的批准，则 TMB 技术管理局应将评分与意见送交委员会，委员会将会按照要求重新提交战略业务计划。如获批准，则下次委员会全体会议将在战略业务计划审查环节考虑技术管理局的意见。

SC.3.3 战略业务计划一旦获得技术管理局的批准，ISO 中央秘书处应为其设立可公开访问的服务器（http：//www.iso.org/bp），以供公众进行审查和评价。我们鼓励各成员体向所在国家的有关公众通知战略业务计划提供公众审查的事宜，并在适当情况下，建立战略业务计划的链接。

SC.3.4 委员会应尽可能每年对战略业务计划执行一次审查，同时考虑到接收到的所有意见，频次至少不得低于三年一次。更新后的战略业务计划无需提交技术管理局审批，委员会协商一致后即可通过（如果不能协商一致，则需达到 P 成员简单多数投票同意）。

附录 SD
（规范性）
项目阶段的矩阵表（见表 SD.1）

SD.1 协调阶段代码简介

标准化过程包括一系列明确的步骤或阶段，这些步骤或阶段既可用于描述过程，也可用于指示过程中任何一项任务的完成情况。一般来说，无论标准化过程的监督机构如何，相关机构在制定和发布标准时，会遵循国际、区域和国家标准机构运作的正式标准化过程，其操作方式是非常相似的。因此，在较高的层次上，可以树立起有关标准化过程的统一观念，由一系列共同的阶段组成。然而，各成员体实际采用的标准化过程存在差异，导致每个成员体制定不同的阶段体系。

本协调阶段代码（HSC）系统在 ISO 数据库中用于跟踪标准制定项目，其目的是为核心数据的传输提供通用的框架。通过该系统，可以相同的方式在国际、区域、国家等各级 ISO 数据库中追踪某一项目的制定情况，且基于统一追踪方式编制的代码矩阵也可以很容易地满足新的要求。

SD.2 阶段代码矩阵的设计

ISO 按照不同机构共同采用的步骤顺序，建立了一套适用于标准制定的"阶段"组合，它们代表标准制定的主要阶段。

在每个阶段中，依据统一的概念逻辑系统，确立了一系列"子阶段"。因此，术语"阶段""子阶段"是指最终形成的矩阵的各个轴。

每个主阶段、子阶段均配有一个从 00 到 90 以 10 为增量的两位数编号。代码的泛型矩阵中，每个单元格均有 4 位数字编号，由其主阶段坐标和子阶段坐标组成。为便于目视查看（而非便于数据库操作），这对坐标由一个点分隔（例如：编号 10.20，其中 10 代表阶段编号，20 代表子阶段编号）。

所有空缺阶段代码都留作将来使用，如确定需增加某一阶段，则可插入到矩阵中，例如：阶段代码 10、30、40、50 和 80。

SD.3 使用本系统的基本准则

—— 文件来源、文件类型等其他信息应记录在单独的数据库字段中，不应反映在阶段代码中。

—— 不存在表明项目在任何特定阶段处于休眠状态的子代码。建议使用另一个数据库字段来解决这个问题。
—— 协调阶段代码系统考虑到了标准过程的周期性,并考虑到了重复当前阶段或早期阶段的情况。在项目生命周期中,若某一事件发生重复,可以通过重复使用相同的阶段代码来记录。
—— 无论何时皆可使用项目已达到的阶段代码来冻结项目。项目如果暂停,应将暂停情况记录在单独的数据库字段中。
—— 协调阶段代码系统不记录某一阶段完成时已实现的目标,或完成的实际日期。

表 SD.1 项目阶段的矩阵表

阶段	子阶段						
	00 注册	**20** 开始主要行动	**60** 完成主要行动	**90** 决定			
				92 重复上一阶段	**93** 重复当前阶段	**98** 放弃	**99** 继续进行
00 预研阶段	00.00 收到新工作项目提案	00.20 审议新工作项目提案	00.60 结束审议			00.98 放弃新工作项目提案	00.99 批准对新工作项目提案投票表决
10 提案阶段	10.00 注册新工作项目提案	10.20 开始对新工作项目投票	10.60 投票结束	10.92 将提案返至提交者供进一步阐明		10.98 否决新项目	10.99 批准新项目
20 准备阶段	20.00 在 TC/SC 工作计划中注册新项目	20.20 开始研究 WD	20.60 评议期结束			20.98 取消项目	20.99 批准将 WD 注册为 CD
30 委员会阶段	30.00 注册 CD	30.20 开始对委员会草案进行研究/投票	30.60 投票/评议期结束	30.92 将委员会草案退回工作组		30.98 取消项目	30.99 批准将 CD 注册为 DIS 草案
40 征询意见阶段	40.00 注册 DIS	40.20 开始 DIS 投票,持续 12 周	40.60 投票结束	40.92 分发完整报告:将 DIS 退回 TC 或 SC	40.93 分发完整报告:决定重新进行 DIS 投票	40.98 取消项目	40.99 分发完整报告:批准将 DIS 注册为 FDIS

表 SD.1（续）

阶段	子阶段						
	00 注册	**20** 开始主要行动	**60** 完成主要行动	**90** 决定			
				92 重复上一阶段	**93** 重复当前阶段	**98** 放弃	**99** 继续进行
50 批准 阶段	50.00 收到最终文本或注册 FDIS，供正式批准	50.20 将样稿提交至秘书处或开始 FDIS 投票：8 周	50.60 投票结束，由秘书处退还样稿	50.92 将 FDIS 交至 TC 或 SC		50.98 取消项目	50.99 批准发布 FDIS
60 出版 阶段	60.00 国际标准出版中		60.60 国际标准已出版				
90 复审 阶段		90.20 国际标准进行系统性复审	90.60 复审结束	90.92 修订国际标准	90.93 确认国际标准		90.99 TC/SC 提议撤销国际标准
95 撤销 阶段	95.20 启动撤销提案投票	95.60 投票结束	95.92 决定不撤销国际标准				95.99 撤销国际标准

附录 SE
（规范性）
文件编号

SE.1 工作文件（包括委员会草案）

凡与 ISO 技术委员会或分委员会有关的工作文件（包括委员会草案，见 SE.2），在分发至全体或部分成员体之时，均应由系统生成一个"N 编号"。N 编号应标于右页的右上角和左页的左上角。文件制定的日期应标于文件第一页编号的正下方。日期应按文件编制标准 ISO 8601《数据元素和交流格式 信息交流 日期和时间的表示》的规定书写。

一个参考编号一旦用于一份工作文件，就不能再用于措辞和/或内容不同的文件上。然而，提案和投票表格（可选）可共用同一编号。如需以新文件替代早期文件，则应在新文件的第一页标注新的 N 编号。

N 编号由以下两部分组成，以字母 N 隔开：

1）技术委员会（TC）编号，和适当情况下工作文档所属的分委员会（SC）编号；

2）总序号。

参考编号示例：**ISO/TC *a*/SC *b* N*n***。

SE.2 工作草案（WD）、委员会草案（CD）、国际标准草案（DIS）、最终国际标准草案（FDIS）和国际标准

在注册新的工作项目时（见 2.3.5），ISO 中央秘书处将为项目分配一个 ISO 编号。该编号用于项目依次编制的 WD、CD、DIS、FDIS 以及已出版的国际标准，期间始终不变。分配的编号仅为注册和参考编号，而不具有分类或时间顺序方面的含义。分配给撤销项目或国际标准的编号将不再使用。

如某一项目是对现行国际标准的修订或修正，则该注册项目的编号与现行国际标准相同（如该项目是对现行标准的修正，则应添加后缀，以表明文件的性质）。

但是，如果新项目下标准范围发生重大改变，则可分配新的编号。

同一项目下如提交多份 WD 和 CD，将分别标注为"工作草案 1 稿""工作草案 2 稿"等，或"委员会草案 1 稿""委员会草案 2 稿"等。

同一项目下如提交多份国际标准草案（DIS），则使用相同编号，但应增加数字后缀以示区别（例如：.2、.3 等）。

附录 SF
（规范性）
承办会议

SF.1 ISO 会议的承办方

ISO 会议可由以下人员承办：

—— 任何 ISO 成员（成员体、通信成员、注册成员）；

—— 相关委员会的任何联络成员，仅适用于工作组会议。

相关国家内的 ISO 成员体应为技术委员会或分委员会会议的承办方，会议需获其事先批准。

对于工作组会议，相关国家内的 ISO 成员体应在会议确认之前获悉会议。

会议的承办方不必是相应委员会工作的直接参与者，不过通常都是直接参与者。

潜在的会议承办方首先应确定所在国家或会议场所对参加会议的技术委员会或分委员会的任何 P 成员代表入境没有任何限制。在某些情况下，可能有必要取得特别入境许可，承办方应尽可能协助确定是否需取得特别入境许可，但秘书处或工作组负责人以及参会人员应共同确定是否有任何限制。

某些承办单位可能没有足够的会议设施和 / 或资源，此时主要承办方（即某 ISO 成员）可以接受其他组织的帮助，例如：提供会议场所或组织欢迎活动。

无论会议的实际承办方是谁，秘书处（涉及技术委员会或分委员会等类似组织的情况下）或工作组负责人（涉及工作组和临时工作组等情况下）均应对会议的召开负责。因此潜在的承办方应直接与这些人员联络。

SF.2 会议赞助

大型 ISO 会议费用高昂，有时流程复杂，所需资源可能超过 ISO 成员本身能够提供的资源。因此，承办方可以联系一个或多个赞助机构，以协助组织会议和提供资金支持。在会议上，承办方可以展示赞助商的标识，并感谢赞助商的支持，但不应出于商业及其他原因，借 ISO 会议宣传赞助商的任何产品或服务（另见"ISO 会议参会代表费用"）。

SF.3 申请成为会议承办方或撤销申请

《ISO/IEC 导则　第 1 部分》要求承办方提前发出下列通知（见表 SF.1）：

表 SF.1　会议承办方发出的通知

××会议	预先通知会议日期	预先通知会议地点
技术委员会或分委员会	2 年	4 个月
工作组	—	6 周

一般来说，为便于参会者合理安排差旅商务，通知越详细越好，必要时还应提供会议简介。

如会议承办单位需撤销承办申请，也应考虑上述期限。

SF.4　会议地点

在进行会议规划时，为促进交流，减少与会代表出席多个技术委员会和分委员会会议的负担，应考虑是否可以将多个相关主题的技术委员会或分委员会会议集中召开。

SF.5　提供会议设施

任何成员体在提出承办 ISO 技术委员会或分委员会会议时，均承诺根据正常的会议出席情况和秘书处对于并行会议（如工作组、临时工作组、起草委员会）设施的预期要求，为 TC/SC 提供适当的会议设施，包括翻译设备（见 SF.6）。会议期间，不同工作组所需的工作设施（房间除外）并不相同，其中可能包括打印和网络连接（电话会议设施），同时需要行政管理支持。参会者可能会要求承办者提供单独的设施服务，如"主席"或"秘书处"房间，但这并非义务。

应当由相关委员会秘书处或工作组负责人确定具体的会议要求。为确保成员体明确了解承办委员会会议需提供的设施，有些委员会秘书处制定了相关文件并长期沿用。工作组的承办方也按预期提供所有基本的工作设施。

SF.6　会议翻译设施

主席和秘书应根据 ISO 的一般规则，按照实际情况，以参会者能接受的方式解决会议上的语言问题。会议的官方语言是英语、法语和俄语，会议可以使用其中一种或多种语言。典型的语言组合是英语和法语，或只用英语。如仅采用英语，则应取得法语成员的同意，才能将法语省去。因此，会议承办方可能需要承担提供英法翻译的义务。应事先决定是否需要履行这一义务。潜在承办方可能需要向 ISO 中央秘书处或向其他 P 成员寻求协助。俄语口译和笔译服务由俄罗斯国家成员体提供。

就译员（如有必要）的能力而言，需要做到：

—— 对于预期产生标准草案最终决定的会议，通常要求有适宜资格的译员提供服务；

—— 对于预期作出中间或程序性决定的会议，可由秘书处成员或志愿代表提供简单的翻译；

—— 在工作组会议上，经工作组召集人授权，工作组成员应当在可能的情况下主动安排必要的翻译。

SF.7 欢迎活动和社会活动

ISO 受到广泛的认可与尊重，因此在诸如技术委员会会议开幕式等场合，举办相关典礼并非罕见。这类活动是可接受的，但承办方可自行决定是否在会议期间举办此类活动，但一般需要有关工作组秘书和主席的批准。

承办方亦可提供一种或多种社会活动，可能会为此寻找赞助方的支持。是否举办欢迎活动和社会活动完全由承办方自行决定（见 SF.8）。

欢迎活动和社会活动公认大幅增加组织会议的时间和费用，可能也会增加参会费用。鉴于所有的参与者都会感到压力，可以采用简单的会议流程。综上所述，从任何方面讲，承办方都没有义务提供等同或超过以往会议级别的欢迎或社会活动，承办方不应将会议视为互相攀比的场合。

SF.8 ISO 会议代表的费用

不应对授权参会的代表强制收取参加 ISO 会议的费用。然而，在极少数情况下（大型和复杂的会议），某些收费机制可能是必要的，但是这类收费机制需要得到 ISO 技术管理局的单独批准。

承办方没有义务提供 ISO 会议期间的社交活动。但是如果委员会要求承办方组织社交活动，则承办方有权要求参会代表支付此项费用。

SF.9 媒体参会政策

SF.9.1 概述

本条款将介绍与媒体合作以提高 ISO 传播活动有效性的政策。ISO 会议应当主动并持续地向公众提供准确的资讯，确保资讯的透明度和公开性，并支持传播机构通过适当的渠道，尽可能有效且高效地向公众提供资讯。

本政策描述了如何提高公众对 ISO 活动的认识以及确保会议的透明度。重要的是，所有参与 ISO 活动的人员都应秉持善意与责任感，提供准确、适当的信息。媒体有助于公众态度和行为的形成。承办方应建立良好的媒体关系，以通过发布准确的信息树立公众意识，防止孤立且缺乏代表性的观点、意见误导公众。

应通过各组织层次的多种途径寻求媒体的参与。

与 ISO 活动有关的媒体往来需要符合以下条件：

—— 欢迎新闻或其他媒体关注 ISO 工作。ISO 中央秘书处及许多 ISO 成员体掌握公共关系资源，也承担着宣传服务，可以向媒体提供有关 ISO、ISO 标准和工作进展的信息。

—— 近年来，一些媒体对出席某些 ISO 委员会会议的兴趣与日俱增。虽然媒体的参与受到欢迎，但在 ISO 会议期间，媒体的存在可能会妨碍对问题的自由公开讨论。因此，媒体不得出席 ISO 会议的工作会议。但是允许媒体参加 ISO 全体会议的开幕式和闭幕式。

SF.9.2 ISO 中央秘书处的角色

在 ISO 内部，ISO 中央秘书处负责编写、发布所有官方新闻稿。ISO 中央秘书处保留对一切 ISO 新闻稿的最终决定权。

SF.9.3 ISO 委员会和工作组的领导角色

会议领导人只应在 ISO 委员会会议期间举行的新闻发布会上传达代表或专家达成的共识。在会议间歇期，秘书处将尽可能迅速地对新闻界的请求作书面答复。

SF.9.4 成员体及镜像委员会的角色

在地方或国家层面上，应由镜像委员会回答媒体问题。鼓励镜像委员会安排与当地媒体会晤，这有助于促进当地人民的参与和提高其认识。在国家层面上，应适用相关成员体的媒体政策。

SF.9.5 代表、专家和观察员的角色

代表、专家和观察员不得就 ISO 委员会达成的共识向媒体发表意见，也不得就其他代表和专家在 ISO 委员会工作会议上所表达的观点和意见发表评议意见，因其可能会妨碍代表与专家之间的对话和信任，而这种对话和信任是有效制定 ISO 标准所必需的。

SF.9.6 会议及媒体参与

出席会议现场的媒体代表应向 ISO 委员会秘书处和会议主持人注册（姓名、组织机构和联系电话）。

出席会议现场的媒体代表将会得到一枚媒体标识，应佩戴或随身携带。

佩戴标识的已注册媒体代表可出席以下活动：

—— 开幕及闭幕典礼；

—— 新闻发布会；

—— 主办机构邀请的社交活动。

禁止媒体人士出席：

—— 技术委员会或分委员会全体会议；

—— 工作组会议；

—— 主席顾问组（CAG）会议。

新闻发布会应安排在适当的时间（全体会议会前和/或会后），并在指定地点举行，

具体时间地点由 ISO 委员会领导与主办会议的 ISO 成员体的公共关系处协商决定。在记者招待会上，ISO 委员会领导人和 ISO 中央秘书处代表将向新闻界发言。ISO 委员会领导可邀请代表或专家发言；否则，代表、专家和观察员可以出席，但不得在记者招待会上发言。

SF.10 ISO 会议录音

委员会秘书可以对 ISO 会议进行录音，但在会议开始时，应告知全体代表会议将被录音，并且确保会议代表对此无反对意见。如在会议期间出现特殊干扰，代表有权要求关闭录音设备。

录音旨在帮助会议秘书编写会议纪要和报告，也可用以解决会议期间发生的争议。在某些情况下，可能需要准备会议录音的转录。

会议录音和转录属于会议秘书的财产，会议秘书应注意 ISO 会议的保密性。因此，录音和转录不应披露给第三方，且最好在会议纪要获得通过后即予销毁。

附录 SG
（规范性）
ISO 标准的第二（及后续）语言文本

SG.1 引言

《ISO/IEC 导则 第 1 部分》附录 E 说明了 ISO 语言使用的一般政策。本附录介绍了有关获取第二语言及后续语言文本的附加要求（英语、法语和俄语是 ISO 官方语言）。

SG.2 英语、法语、俄语及其他语言

SG.2.1 职责

ISO 委员会秘书处通常负责获得英文文本的编制工作（不一定由委员会自己编制）。通常由法国成员体法国标准化协会（AFNOR）的翻译部门或者其指定的代理机构负责将文件译成法语。

俄罗斯成员体负责决定是否需俄语版本，并提供全部的俄语翻译工作。

如果委员会批准将非官方语言纳入最终的可交付使用文件（仅限于词汇表），则应当由提供非官方语言支持的国家体负责编制该语言文本，并及时交付（最迟在最终国际标准草案投票之前）。

SG.2.2 文本的等效

提供语言支持的国家体在编制 ISO 可交付使用文件时，应负责确保相关语言文本与官方语言文本等效。如有必要，可以寻求其他专家的帮助。建议设立多语言编辑委员会，以确保文本等效。

SG.2.3 文本编制时间

根据《ISO/IEC 导则》，在制定标准时，应尽早编制各语言文本。虽非强制要求，但应尽力提供各种预想语言版本的委员会草案。这不仅有助于相关人士理解标准制定的内容，也可以避免在后续阶段因文本编制而导致拖延。

如需编制第二语言或后续语言文本，则 ISO 中央秘书处应在将第一语言文本（通常为英语版）作为国际标准草案提交至 ISO 中央秘书处之前，向负责第二语言或其他语言文本的成员体提供草案副本。

SG.2.4 语言文本的及时提交

如果必须提供第二语言或后续官方语言文本，则提供相关语言支持的国家体应在第一语言版本（通常为英语）提交后 8 周内向 ISO 中央秘书处提供相应语言文本。

SG.2.5 投票后语言文本的更新

编制国际标准草案或最终国际标准草案的第二语言或后续语言文本时,为促进相关编制工作,技术委员会和分委员会秘书处应向 ISO 中央秘书处提供一份带有修改标记的第一语言国际标准草案文本修订版,以供负责第二语言或后续语言文本的成员体使用。

SG.3 其他非官方语言

非 ISO 官方语言的使用范围仅限于词汇表。在委员会批准某一国家成员体提交的相关任务申请之后,该成员体即可以负责制定相关语言版本的词汇表。

当某一 ISO 项目是 ISO 与 CEN 之间的技术合作协议(《维也纳协议》)下的联合项目时,将要求编制德文文本。相关指导要求可另行参见《主席和秘书处关于实施〈维也纳协议〉的指南》。在所有多语言文件中,文件的封皮、版权及前言页仅提到 ISO 官方语言,只有在《维也纳协议》下与 CEN 的联合项目中,可增加德语。

SG.4 单一语言文本

国际文件有时只需编制单一语言文本。其他官方语言版本可通过已出版文本的翻译得到。

SG.5 翻译

根据《ISO 章程》条款(经《ISO POCOSA 2017——ISO 出版物发行、销售、复制及 ISO 版权保护政策》补充),ISO 成员体可以自行决定将 ISO 出版物和文件翻译成其他语言,并对译稿全权负责。

附录 SH
（规范性）
图形符号的标准化程序

SH.1 引言

本附录针对 ISO 文件中出现的所有图形符号，描述了相应的提交程序以及后续的适时批准、注册程序。

在 ISO 组织框架内，分别由以下两个 ISO 技术委员会负责协调两个主要领域的图形符号制定任务：

—— ISO/TC 145 所有图形符号（用于技术产品文件）（见 ISO/TC 145 网站）；

—— ISO/TC 10 技术产品文件（tpd）的图形符号（见 ISO/TC 10 网站）。

此外，还要与 IEC/TC 3（信息结构、文件和图形符号）和 TC 3/SC 3C（设备用图形符号）进行配合。

图形符号标准化的基本宗旨如下：

—— 满足用户要求；

—— 确保考虑到所有相关 ISO 委员会的利益；

—— 确保图形符号不模糊，并且符合统一的设计准则；

—— 确保图形符号不重复，或避免不必要的扩充。

新图形符号标准化的基本步骤如下：

—— 确定需求；

—— 精心制作；

—— 评价；

—— 批准（适时）；

—— 注册；

—— 出版。

所有步骤均应当以电子形式进行。

—— 新的或经修订的图形符号提案可由 ISO 委员会、ISO 委员会的联络成员或任何 ISO 成员组织提交（以下统称"提案方"）。

—— 每个经过批准的图形符号都将被分配唯一的编号，以便通过统一的注册表实现对图形符号的管理和标识。该注册表内容为电子信息的，可进行检索。

—— 当各方在图形符号相关要求与指导方针方面发生冲突时，ISO/TC 145 或 ISO/TC 10 应与相关产品委员会展开沟通和对话，以尽早解决。

SH.2 技术产品文件用图形符号以外的所有图形符号

SH.2.1 概述

在 ISO 组织框架中，ISO/TC 145 负责图形符号（技术产品文件用图形符号除外）领域标准化的全面协调工作，其职责包括：

—— 图形符号领域以及相关颜色、形状（只要这些要素用于表达符号信息的组成部分）的标准化，如安全标志；

—— 建立图形符号的制定、协调及应用原则：全面负责评价并协调现有、在研或拟建立的图形符号。

文字、数字、标点符号、数学标识与符号、数量以及单位符号的标准化除外。但是，这些要素可以作为图形符号的组成部分。

任何委员会如在本委员会特定范围内开展图形符号的创建和标准化工作，均应交由 ISO/TC 145 进行评审与协调。ISO/TC 145 的责任分工如下：

—— ISO/TC 145/SC 1：公共信息符号；

—— ISO/TC 145/SC 2：安全标识、标志、形状、符号和颜色；

—— ISO/TC 145/SC 3：设备用图形符号。

此外，还应与 ISO/TC 10 和 IEC 保持联络，尤其应与 IEC/SC 3C（在设备用图形符号方面）保持联络。

表 SH.1 展示了每个协调委员会负责的图形符号种类。

表 SH.2 展示了不同类型的图形符号及其使用场景。

表 SH.1 图形符号种类

	基本信息	地点	目标受众	设计原则	概述	负责委员会
公共信息符号	服务或设施地点	公共区域	一般公众	ISO 22727	ISO 7001	ISO/TC 145/SC 1
安全标识（符号）	与人的安全和健康相关	工作场所、公共区域和产品上	a) 一般公众 b) 经授权的和培训的人员	ISO 3864-1 ISO 3864-2 ISO 3864-3	ISO 7010	ISO/TC 145/SC 2
产品安全标签	与人的安全和健康有关	产品上	a) 一般公众 b) 经授权的和培训的人员	ISO 3864-2 ISO 3864-3	——	ISO/TC 145/SC 2

表 SH.1（续）

	基本信息	地点	目标受众	设计原则	概述	负责委员会
设备用图形符号	与设备有关（功能、控制和识别）	设备上	a) 一般公众 b) 经授权的和培训的人员	IEC 80416-1 IEC 80416-2 EC 80416-3	ISO 7000 IEC 60417	ISO/TC145/SC 3 IEC/TC3/SC 3C
技术产品文件符号	（产品标识）	技术产品文件（图纸、示意图等）	经培训的人员	ISO 81714-1	ISO 14617 IEC 60617	ISO/TC10/SC 10 IEC/TC 3

表 SH.2 不同类型的图形符号及其使用场景

公共信息符号	电话 ISO7001-PIPF017	飞机 ISO7001-PI TF 001	体育活动 ISO 7001-PISA 001	加油站 ISO 7001-PICF 009	方向 ISO 7001-PIPF 030
安全标志（符号）	逃生及应急设备标志；安全出口（左手） ISO 7010-E001	消防设备标志：灭火器 ISO 7010-F001	强制性行为标志：一般性强制性行为标志 ISO 7010-M001	禁止标志：禁止吸烟 ISO 7010-P002	警告标志：注意：易爆物品 ISO 7010-W002
设备用图形符号	通风扇：循环风扇 ISO 7000-0089	一般性回收/再循环标志 ISO 7000-1135	重物 ISO 7000-0430	灯、照明设备、照明 IEC 60417-5012	亮度/对比度 IEC 60417-5435
技术产品文件用符号	双向阀门 ISO 14617-8-2101	液体 ISO 14617-15-321	液压发动机 ISO 14617-10-2405	扬声器 ISO 14617-5-892	锅炉/蒸汽发生器 ISO 14617-11-2531

SH.2.2 提案的提交

提案人应尽快填写相关申请表格,将提案提交至 ISO/TC 145 相关的分委员会,以便及时评审和评议。强烈建议在委员会草案阶段进行提交。对于国际标准,提案的提交时间不应晚于首个征询意见阶段(即 DAM)。

提交图形符号提案之前,提案人应当:

—— 能证明图形符号提案的必要性;

—— 已审查相关的 ISO 和/或 IEC 图形符号标准,以免模糊不清和/或与现有的标准化图形符号重叠,并检查提案是否与已经标准化的任何相关图形符号或图形符号一致;

—— 根据相关标准与指南(包括设计原则和验收标准)创建提案的图形符号。

SH.2.3 提案图形符号的标准化程序

收到提案后,ISO/TC 145 相关的分委员会应在两个月内评审申请表,检查申请表填写是否正确以及提案人是否正确提供了有关的图形文件。必要时,委员会将要求提案人修改申请,重新提交。

分委员会一旦收到正确填写的申请表格,就应启动正式评审过程,评审提案是否与现有标准化图形符号一致、是否符合有关设计原则与验收标准。

正式评审过程完成时,应将评审结果通知提案人,并附上改进建议。如有必要,可要求提案人修改提案,再次提交,以供进一步评审。

应遵循 ISO/TC 145 有关分委员会网站上列出的程序:

—— ISO/TC 145/SC 1 公共信息符号(www.iso.org/tcl45/sc1);

—— ISO/TC 145/SC 2 安全标识、标志、形状、符号和颜色(www.iso.org/tc/sc2);

—— ISO/TC 145/SC 3 设备用图形符号(www.iso.org/tc145/sc3)。

网站也提供提交提案的申请表格。

凡 ISO/TC 145 批准的图形符号,应分配最终注册编号,并纳入 ISO/TC 145 相关的标准中。

注:特殊情况下,经技术管理局批准,ISO 标准中可以纳入未经注册的符号。

SH.3 用于技术产品文件(tpd)的图形符号(ISO/TC 10)

ISO/TC 10 全面负责产品技术文件(tpd)图形符号领域的标准化工作。其职责包括:

—— 维护 ISO 81714-1《技术产品文件中图形符号的设计 第 1 部分:基本规则(与 IEC 合作)》;

—— 与 IEC 协作,负责技术产品文件的图形符号标准化;

—— 建立并维护图形符号数据库,包括注册编号的管理。

ISO/TC 10 已将上述职责分配至 ISO/TC 10/SC10。ISO/TC 10/SC10 秘书处下设有维护组,专门负责支持相关工作。

任何委员会在确定需要制定或修订图形符号之后,应尽快将提案交给 ISO/TC 10/SC 10 分委员会秘书处进行评审,一旦获得批准,即分配注册号。

附录 SI
（规范性）
国际研讨会协议（IWA）的制定程序

SI.1 制定 IWA 的提案

包括 ISO 成员体、联络组织、企业在内的任何机构，如需就某个特定议题制定一项或多项"国际研讨会协议"均可提议召开 ISO 专题研讨会。若该机构并非 ISO 成员体或联络组织，或业务范围不具备国际性，则应向其所属国的 ISO 成员体通报其提出提案的意图。

在实际可行的情况下，提案人应根据《ISO/IEC 导则 第 1 部分》附录 C 提供以下相关详情：

—— 提案的目的和理由；

—— 相关的文件；

—— 合作与联络。

此外，如有可能，提案应表示 ISO 成员体愿意向 IWA 专题研讨会提供秘书处的支持。如果成员体认为研讨会需制定参会限制（见本附录的 SI.5.2），还应对此予以说明。

有些情况下，可能需要举办多次会议才能达成一致。在这种情况下，ISO 技术管理局（TMB）应鼓励或要求提案人编写业务计划，具体说明会议日程、文件草案的预期获得日期以及是否可以设立专题研讨会项目组，以推进会议间隔期间的工作进展，并负责处理国际专题研讨会协议的提交日期等有关事项。

SI.2 提案的评审

提案将提交给 TMB 批准。提案应分析不同国家及利益相关方针对 IWA 所提主题方面的全球利益。

如果提案得到认可，则技术管理局应与成员体协商，选出一家愿意作为组织者的候选机构，并负责向提案人提供行政与后勤支持。通常情况下，优先选择：

—— 提案人所在国家的成员体（如提案人非成员体）；

—— 或者，涉及提案相关领域且设有秘书处的成员体。

如超过一个成员体要求承担秘书处工作，则技术管理局应正式指定某一成员体，承担专题研讨会的秘书处工作。被指定的 ISO 成员体可以与提案人签订财务协议，以涵盖专题研讨会涉及的管理与后勤支持费用。若无成员体愿意承担专题研讨会秘书处工作，则技术管理局可授权 ISO 中央秘书处完成此项任务，但相关费用可通过收取专题研讨会注册费获得。

估算 IWA 专题研讨会费用的参考清单见 SI.10。

专题研讨会秘书处与提案人应指定专题研讨会主席。

SI.3　通知

一旦专题研讨会秘书处与提案人就专题研讨会首次会议的日期和地点达成一致意见，即应将结果通知 ISO 成员体。专题研讨会秘书处、ISO 中央秘书处及任何其他相关成员应以最适当的方式进一步发出通知，尽可能广泛地通知会议时间地点（如公众可访问的网址）。上述机构可使用多种方法与媒介进行通知，努力使全球范围内的利益相关方均能获取有关专题研讨会会议通知，并有机会参加专题研讨会会议。

提案人与专题研讨会秘书处应确保，具有主题相关项目的 ISO 委员会将被邀出席该专题研讨会会议。

注册费可用于专题研讨会会议的准备与举行。如有任何注册费，则应在专题研讨会通知中指明。

> **注：** 如果专题研讨会的主题与发展中国家相关，则应提议或者采用注册费以外的筹款机制，或是允许一些国家"免费"注册参加会议，以推动这些国家参加专题研讨会活动。

应在商定的会议日期之前 90 天发出通知，以便给潜在参会人员留出足够的时间安排参加专题研讨会会议。通知中应附有注册表，使拟参会人员进行参会注册。注册表应返至专题研讨会秘书处。

SI.4　专题研讨会信息

应提供一份专题研讨会计划，详述专题研讨会的目标、可交付使用文件、日程、文件草案及其他相关细节，并在专题研讨会日期前 6 个星期内，将计划分发至已注册的参会人员。已注册的参会人员可向专题研讨会秘书处提交相关稿件，以便进一步分发给其他参会人员。

SI.5　参会

SI.5.1　专题研讨会主席

任何专题研讨会的主席均由提案人及专题研讨会秘书指定。主席行事应保持国际性和中立，特别应：

—— 完善地总结、归纳专题研讨会上提出的各种观点，确保所有参会人员清楚明白；

—— 在主持专题研讨会时，以令各方达成共识；

—— 确保会议作出的各项决定充分明确，必要时，在专题研讨会会议结束前告知参会人员。

SI.5.2 已注册的参会人员

所有组织皆可注册参加专题研讨会，只有注册人员方可参会。参会者无需由本国的 ISO 成员体指定。专题研讨会秘书处、主席和提案人均应努力确保会议能代表最广泛的利益相关方，而且具有适当均衡的代表性。如有必要，可对参与人员提出限制条件（例如，同一公司或组织的注册参会人员最多两名）。如果一开始便决定限制参加人员，则应在向 ISO 技术管理局提交的提案中予以说明。如果在专题研讨会发出通知后明确需要添加限制条件，则应与 ISO/TMB 主席协商（如有必要，还应与技术管理局其他成员进行协商）后，由技术管理局秘书处予以授权。

专题研讨会秘书处应负责确保专题研讨会可以取得国际化的成果，即不止一个国家参与会议。

SI.5.3 项目组

如果为达成共识而需要召开多次会议，专题研讨会可设立一个或多个项目组，负责在会议间隔期间推进工作。专题研讨会应指定项目组成员，并确保项目组的工作方法能使所有利益相关方均充分参与会议。

SI.6 专题研讨会的程序和管理监督

应允许专题研讨会采用高度灵活的程序，有效自治地工作。

但是，专题研讨会应遵循一定的 ISO 通用政策，特别是涉及知识产权和国际标准单位制方面的政策。专题研讨会秘书处应保证注册参会人员均了解并遵循相应政策。

应维持最基本的管理监督要求，确保研讨会与现有的标准化活动（如果相关）协调一致，并确保 ISO 体系提供适当的资源。专题研讨会主席应负责判断，参会人员是否对某一特定项目或可交付使用文件达成了一致。为作出正确判断，专题研讨会主席应适用 2.5.6 中的共识定义。

需注意的是，研讨会可能达成一致性意见，认为没有必要制定 IWA 可交付使用文件。

专题研讨会文件应说明专题研讨会所达到的共识情况，包括围绕未来行动或文件修改的建议。专题研讨会上形成的文件应在参会人员协商一致的基础上发布，无需任何其他团体审查或批准，除非存在针对这一文件的申诉（见 SI.7）。

SI.7 申诉

受到专题研讨会所形成的可交付使用文件影响的任何一方，有权以下列原因提出申诉：

—— 专题研讨会和形成文件的过程未遵守规定程序；

—— 专题研讨会文件未代表国际贸易与商业的最佳利益，或安全、卫生、环境等方面的最佳利益；

—— 专题研讨会文件的内容与现有的 ISO 标准、ISO 标准草案可能有冲突，或可能有损于 ISO 的声誉。

这些申诉应在召开专题研讨会会议之日起两个月内提交，并由 ISO 技术管理局进行审议，在此情况下，ISO 技术管理局将就是否发布 IWA 作最后决定。

SI.8 专题研讨会的可交付使用文件和发布

专题研讨会应决定其成果内容。文件应寄送至 ISO 中央秘书处，作为专题研讨会协议发布，并赋予国际专题研讨会协议专用系列编号。在每份国际专题研讨会协议中，应包括一份参与编制协议文件的参会者名单。

国际专题研讨会协议仅发布 ISO 的一种官方语言版本，也允许其他国际专题研讨会就同一主题发布协议。国际专题研讨会协议的技术内容可与现有（或正在制定中的）ISO 或 IEC 标准技术内容相互竞争，但通常不允许有冲突，除非得到 ISO/TM 的明确授权。

SI.9 IWA 的复审

对于已出版的 IWA，若已有 ISO 委员会的工作范围涵盖协议主题，则该协议自动分配给相关委员会负责维护。

国际专题研讨会协议发布 3 年后，承担该专题研讨会秘书处工作的成员体或负责维护协议的委员会应组织复审。复审包括向相关市场主体征询意见。依据《ISO/IEC 导则 第 1 部分》，复审结果可以是确认该国际专题研讨会协议再使用 3 年，也可以是撤销该国际专题研讨会协议，或将协议作为另一种 ISO 可交付适用文件进行提交，以供后续处理。

可根据市场要求进一步处理国际专题研讨会协议，使其转化成可公开获取的规范、技术规范或国际标准。

国际专题研讨会协议最长可维持 6 年，之后应撤销或转换成其他 ISO 可交付使用文件。

SI.10 承办 ISO 国际专题研讨会的预算清单

在提出承办国际专题研讨会时，无须强制填写本清单。该清单仅旨在提供一种工具，帮助提案人和 ISO 指定成员体确定承办专题研讨会所需的主要费用。某些 IWA 研讨会的费用可能与清单中列举的形式有所不同，有时也可能采用不同于下文所述的支付方式。

SI.10.1 需要考虑的综合性措施（见表 SI.1）

表 SI.1 需要考虑的综合性措施

预计可交付使用文件数量	
预计项目组数	
预计参加人数	
预计研讨会（WS）全体会议次数	
预计项目组会议次数	

SI.10.2 计划资源——人力资源要求及职能（见表 SI.2）

表 SI.2 计划资源——人力资源要求及职能

职能	人员	预算	由谁支付	承诺（是/否）
专题研讨会秘书处 —— 组织专题研讨会全体会议； —— 编写专题研讨会和项目会议报告及措施清单； —— 担任专题研讨会项目的管理联络处； —— 管理专题研讨会（和项目组）成员名单； —— 管理专题研讨会（和项目组）文件注册簿； —— 后续措施清单； —— 如专题研讨会工作主要通过电子方式进行，则需协助主席监督并跟进电子方式的讨论				
专题研讨会主席 —— 主持专题研讨会全体会议； —— 负责专题研讨会/项目的全面管理； —— 编写项目管理进展报告； —— 负责联络相关的专题研讨会/项目和联络机构； —— 如专题研讨会工作主要通过电子方式进行，则应跟进并指导会议间发生的电子方式讨论，并作出必要决定				
专题研讨会副主席 如专题研讨会需任命"副主席"，则应编制副主席名单，并明确副主席的职能，将其填在本表的右边				
项目领导人（适用时） —— 主持项目组会议； —— 召开项目组会议； —— 编写项目组会议报告和措施清单； —— 编写项目组工作进展报告； —— 如专题研讨会工作主要通过电子方式进行，则应跟进并指导其所领导的特别项目的电子方式讨论				
国际专题研讨会协议编辑 —— 编辑国际专题研讨会协议文本； —— 参加会议； —— 跟进其正在编辑的国际专题研讨会协议的相关讨论				

SI.10.3 计划资源——材料资源要求（见表 SI.3）

表 SI.3 计划资源——材料资源要求

材料资源费用	编号	估算的每次会议成本	由谁支付	承诺（是/否）
专题研讨会会议				
会议室				
后勤（IT 支持，复印等）				
食品及饮料				
会议期间的翻译				

附录 SJ
（规范性）
表格

表格为电子版（一般为 MS Word 格式），可登录 www.iso.org/forms 下载。

附录 SK
（规范性）
委员会及工作组会议文件邮寄截止日期

SK.1 为何应当遵守截止日期

国家镜像委员会需要时间就会议上讨论的各种问题确定其国家立场。

SK.2 确保在规定的期限内准备好分发的文件

秘书处的责任是应仔细考虑所需的相关步骤，并在会议截止日期之前留出足够的时间，以确保文件已准备妥当。例如，在讨论国际标准草案时，秘书处应在召开委员会会议 7 个月前开始准备文件（ISO 中央秘书处 2 周的文件准备时间 + 20 周时间用于投票和翻译 + 会议前 6 周的截止期限）。

SK.3 对于会议将讨论的问题，如无法提前 6 周发布相关文件，如何处理

如秘书处未按照以下截止日期在会议之前发布相关信息，则委员会有权作出决定。此时，委员会领导层应就以下问题向出席会议的 P 成员征询意见：P 成员原则上是否接受会议将讨论的问题，并准备就此作出决定，或者他们是否更愿意留有时间同其他国家利益相关方进一步协商。委员会领导人应确保尽力使所有 P 成员参与到决定中，即使这些 P 成员不能出席会议。会议结束后，可通过 ISO 委员会 CIB 进行投票，以确认各项决定。

议程及会议后勤（例如：地点、住宿和签证信息），截止日期为会议前 16 周。

供会议上作出决定而编制的文件，例如：

工作组报告，截止日期为会议前 6 周；

联络组织报告，截至日期为会议前 6 周；

领导层选举或更换，截至日期为会议前 6 周；

建立或解散分委员会和工作组，截止日期为会议前 6 周；

战略业务计划，截止日期为会议前 6 周；

CD、DIS、技术报告草案（DTR）以及技术规范草案（DTS）的投票结果与意见汇总，截止日期为会议前 6 周；

需要讨论和跟进的复审（SR）和任何临时征询意见 CIB 投票结果，截止日期为会议前 6 周；

决议，截止日期为会议后 48 小时；

会议纪要（包括出席人员名单），截止日期为会议后 4 周。

SK.4 发送工作组文件的截止日期

《ISO/IEC 导则 第 1 部分》规定，议程应在工作组会议开始的 6 周前发出。其他文件的分发时限将由工作组成员决定。工作组召集人应确保其工作组就会议前分发文件的时限作决定，以确保工作组专家能够作好充分准备。

表 SK.1 工作组文件截止日期

文件、通知或活动类型	截止日期 （- 代表会议前；+ 代表会议后）
TC 作出设立工作组决定后的第一次 WG 会议	+12 周
会议通知	（待工作组决定）
议程	-6 周
工作组会议文件	（待工作组决定） 建议不晚于 -4 ～ -6 周
会议纪要（包括出席人员名单）	+4 周

SK.5 网络会议或电话会议前分发文件的时限（见图 SK.1）

同样，对于任何会议，原则上应提前预留足够的时间，以分发会议议程和各类文件，以便与会者能够对讨论议题和决议事项作好充分准备。技术委员会、项目委员会、分委员会和工作组在针对特定工作内容召开网络或电话会议时，应遵循上述详细要求。相关 P 成员应决定何时可向主席顾问组、咨询组、特设组等分发文件。

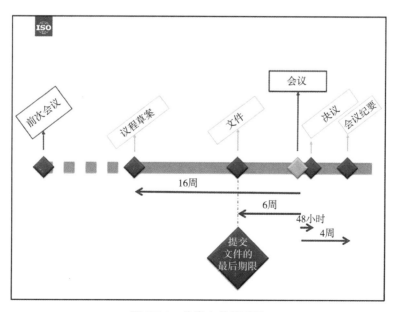

图 SK.1 分发文件的时限

附录 SL
（当前空白——预留给新增附录）

附录 SM
（规范性）
ISO 技术工作及发布文件的全球相关性

SM.1 简介

WTO 成立后，根据 WTO《技术性贸易壁垒协议（WTO/TBT）》的要求，ISO 有义务确保其所制定、采纳和发布的国际标准的全球相关性。根据《协议》第 2 次 3 年期审查附录 4 第 10 段，全球性标准应当遵循如下准则：

—— 有效满足监管及市场需求（国际市场）；
—— 反映各国在科学技术上的新发展；
—— 不扰乱市场；
—— 不破坏公平竞争；
—— 不压制创新和技术发展；
—— 当不同国家或地区间存在需求和利益上的差异时，不应优先考虑个别国家或地区的特点或需求；
—— 以效果为导向而非从设计角度出发。

因此，如果 ISO 制定并采用了未能满足这些要求的国际标准，则将造成自由贸易壁垒，从而面临挑战。

意识到需要向委员会提供更完善的全球相关性建议，同时按照 ISO 理事会的要求，ISO 技术管理局设立了全球相关性工作组。该工作组与 ISO 技术管理局就如下原则达成了一致意见。

SM.2 术语和定义

标准　Standard

经协商一致建立并经权威机构批准，供共同且重复使用，为各种活动或其结果提供规则、指南或特性，其目的是在规定的范围内获得最佳秩序的文件。

注：标准宜依据科学、技术和经验的综合结果，其目的是促进社团的最佳利益。

[来源：《ISO/IEC 指南 2》《ISO/IEC 导则　第 2 部分》]

国际标准　international standard

国际标准化/标准组织采纳的并且可向公众提供的标准。

[来源：《ISO/IEC 指南 2》《ISO/IEC 导则　第 2 部分》]

全球相关性　global relevance

国际标准应具备的特征，即全球市场上的相关产业与其他利益相关方应尽可能广泛地使用或执行该国际标准。

SM.3　原则

SM.3.1　应尊重国际标准的地位和意义

任何国际标准皆应符合上述定义，并应尽可能构成唯一的国际性解决方案。若由于各地区在市场及本质性方面的差异，某一国际标准的特定条款目前还无法构成唯一的国际解决方案，则应在合理的情况下提供多种选择，以包容地区差异。

SM.3.2　在标准制定项目启动时，应说明各国对国际标准制定的参与承诺以及国际标准制定的可行性

人们认识到，就某些问题而言，为满足不同地区与国家的特殊市场情况，存在着各种解决方案。随着全球化与市场一体化，这些市场差异应逐渐消减，演变为统一的全球市场。不能简单地将适用于某一市场（但不包括其他市场）的特定解决方案设立为国际标准，这无助于市场的发展与融合；否则，部分市场及其相关行业将抛弃 ISO，从别处寻找更满足自身需求的标准，而 ISO 将失去与这些市场和行业的关联。为避免出现这种情况，ISO 委员会应在项目之初确认如下问题：

—— 可以制定具有全球相关性的国际标准，其全部条款可构成唯一的国际性解决方案；

—— 在合理情况下，国际标准可以在具体条款中提供不同的解决方法，以包容现有的合理市场差异；

—— 如果无法编制出具有全球相关性的国际标准，则不应该开展相关工作。

ISO 技术管理局编制的《全球相关性实施指南》为委员会领导、代表 / 专家提供了更多实践指南。

SM.3.3　应优先制定性能标准，而非规范性标准

请注意以下内容：

《WTO/TBT 协议》的附录 3 "在适当的情况下，标准化机构应根据产品在性能方面的要求制定标准，而不是设计或描述性的特征要求"。

《ISO/IEC 导则　第 2 部分》5.4 中性能原则（摘录）："只要可能，要求应由性能特性来表示，而不是用设计或描述特性来表示。这种方法为技术发展留有最大余地，并降低了不良市场影响的风险（例如，限制创新解决方案的开发）"。

基于上述引用条款，性能导向原则有利于全球相关性 ISO 标准的制定，因此得到广泛认可。在设计导向标准中，技术创新的自由度受到了最大程度的限制，而性能导向标准则为进一步创新提供了最大的自由度。但是，在实践中，性能导向标准中的某些条款可能包含适当的设计要求。有时也可以完全以设计为导向制定标准，并将其发展成具有全球相关性的 ISO 标准。

因此，选择制定何种标准取决于涉及的技术问题。

ISO 技术管理局的《全球相关性实施指南》文件为委员会领导、代表与专家提供了更多相关实践指南。

SM.3.4 鉴于现有的合理市场差异，一项国际标准在达成最终目标（所有条款共同构成唯一的国际解决方案）之前，或许需要经过不断地发展完善。

根据这一原则，委员会在发布 ISO 可交付使用文件时，或许应考虑如何解决当前存在且可能发生变化的市场差异（立法、经济、社会条件、贸易模式、市场需求、科学理论、设计理念等因素）。ISO 技术管理局的《全球相关性实施指南》文件为委员会领导和代表/专家提供了更多实践指南。

SM.3.5 符合 WTO《技术性贸易壁垒协定》附件 3 的基本差异可纳入国际标准，但若委员会希望在标准中纳入此类差异，则应适用相应的特定规则，当规则不适用时，则需要 ISO 技术管理局作出特别授权。

根据这一原则，委员会可能需要考虑如何解决世界各地市场的基本性差异，即预期不会随时间变化的因素，如固有的技术基础设施、气候、地理或人类学差异。

ISO 技术管理局的《全球相关性实施指南》文件为委员会领导和代表/专家提供了更多实践指南。

SM.3.6 委员会若能了解影响特定标准全球相关性的所有潜在因素，则可确保国际标准的全球相关性。

ISO 技术管理局的《全球相关性实施指南》文件为委员会领导和代表/专家提供了更多实践指南。

附录 SN
（规范性）
注册机构（RA）政策

SN.1 范围

ISO 技术委员会制定的许多国际标准要求分配独特的注册要素，并描述分配这些要素的方法。注册要素本身不是标准的一部分，而由指定的注册机构分配，注册机构还负责保存已分配要素的准确记录。注册机构是有相应能力的实体，其拥有必要的基础设施，可确保有效分配注册要素。ISO 针对特定标准指定唯一的注册机构，这形成了事实上的垄断局面。

本政策为强制性，应结合《ISO/IEC 导则 第 1 部分》中附录 H 一起阅读。

当 ISO 中央秘书处（ISO/CS）发现某一注册机构标准未遵循本政策时，应停止发布流程，待落实本政策之后，再发布该注册机构标准。因此，在项目推进过程中，委员会应该敦促 ISO 技术项目经理（TPM）尽早了解该项目是否需要注册机构，以避免发布延迟。

SN.2 定义

SN.2.1 注册机构标准：注册机构向其提供注册服务的标准。

SN.2.2 注册服务：注册机构在执行注册机构标准过程中提供的服务，应在注册机构标准中作出描述。

SN.2.3 注册机构（RA）：ISO 指定的完成注册机构标准注册服务的实体。

SN.2.4 注册代理：注册机构委托部分注册服务的第三方（例如国家或区域分机构）。委托予注册代理的注册服务仍属于注册机构的责任范围。

SN.2.5 注册机构协议（RAA）：注册机构协议以注册机构及 ISO 秘书长代表 ISO 签署的注册机构协议范本为基础，详述有关各方的职能、角色及法律责任。在发布注册机构标准（包括修订版本）之前，应签署注册机构协议。

SN.2.6 注册要素：具有独特性的要素，标准中描述了分配这些要素的方法，但是注册要素本身并非标准的一部分。

SN.2.7 TPM：ISO 中央秘书处内部人员，被指派支持某一特定委员会工作。

SN.3 程序

SN.3.1 流程顺序

注册机构政策按照典型注册机构的生命周期，依次说明了注册机构的各方面情况，注意某些阶段需要同步开展。各个阶段如下所示：
—— 明确是否需要注册机构（SN.3.2）；
—— 起草注册机构标准（SN.3.3）；
—— 选择注册机构（SN.3.4）；
—— 指定注册机构（SN.3.5）；
—— 签署注册机构协议（SN.3.6）；
—— 终止注册机构工作（SN.3.7）；
—— 执行注册机构标准（SN.3.8）；
—— 注册机构的角色（SN.3.8.1）；
—— 委员会的角色（SN.3.8.2）；
—— ISO 中央秘书处的角色（SN.3.8.2）。

SN.3.2 明确是否需要注册机构

委员会通过决议的方式确认某一标准是否需通过注册机构执行。

委员会秘书处应填写注册机构确认（RAC）表（见附录 SJ），并将其尽快提交技术项目经理，但应在起草（WD）阶段启动之前完成。

SN.3.3 起草注册机构标准

所有注册机构标准中应该包括以下方面：
—— 说明用于生成唯一注册元素的识别方案或机制。
—— 说明注册服务，以及注册机构在提供注册服务方面的责任。
—— iso.org 的网页链接，以供使用者查阅特定标准的注册机构名称及联系方式。还应提供注册机构网站链接，包含更多关于注册服务的信息。

注册机构标准中不应包含以下方面：
—— 根据《ISO/IEC 导则　第 2 部分》中第 4 章："文件不应包括合同要求（例如，有关索赔、担保、费用的支付）以及法律或法定要求"。
—— 有关提供注册服务的程序（例如注册机构提供的手册）不应包括在注册机构标准内。这些资料将由注册机构公开提供（例如，通过网站）。
—— 注册机构名称不应出现在注册机构标准中，而是在 ISO 网站中提供链接。
—— 注册机构标准不应涉及选择或确认注册机构的过程。
—— 若注册机构根据注册机构协议，将注册服务委托给第三方（如被称为"注册机构"的国家或地区分支机构），则注册机构标准可能会提到某些注册服务被委托的事实。但是，有关注册机构的任何细节不应包含在注册机构标准中。

技术项目经理负责与委员会协调,确保注册机构标准所包含的要素正确。围绕注册机构标准中应包含哪些内容,如有任何问题,均应向技术项目经理处反映。

SN.3.4　选择注册机构

在制定新注册机构标准时,或负责现有注册机构标准的注册机构出于任何原因需要终止其身份时,需启动注册机构选择程序。

在修订现有注册机构标准时,委员会应决定是否应继续任用现有注册机构,还是启动相关程序选择其他注册机构候选人。启动选择过程的决定应有依据。委员会应通过决议确认其决定。

选择过程应在项目进入国际标准草案阶段之前完成。国际标准草案应包含有关所需注册服务性质的详细信息。

由委员会确定选择注册机构的标准,并通过决议确认这些标准。选择过程中至少应考虑以下方面:

—— **选拔标准**:应当清楚地解释选拔标准,并附加足够多的细节,以便注册机构候选人评估自身能力是否符合标准,并基于此提出申请。

　　—— 证明文件(如法规),证明该组织是一个法律实体,即根据某一地区的法律成立的组织,因此受有关治理规则的约束。

　　—— 表示愿意承担注册服务的责任。

　　—— 确认注册机构在技术、财务上皆有能力以国际水平开展注册机构标准和注册机构协议中描述的注册机构服务。

　　—— 表示愿意签署注册机构协议,应向注册机构候选人提供 ISO 中央秘书处编制的注册机构协议模板。

　　—— 确认是否计划将部分注册服务委托给注册机构。

　　—— 是否就注册服务收取费用,如果收取费用,则应确认注册费用是否仅旨在收回成本。

—— **公开招募注册机构候选人**:各委员会应采取适当步骤,向尽可能广泛的市场发出公告,招募有能力的注册机构申请人,并可邀请目标注册机构申请。

—— **评估**:注册机构候选人应以书面答复招募公告。委员会(或其分委员会)应以选拔标准(并考虑各选拔标准的相对权重)为基础,确定客观的候选人评价标准。

—— **记录保存**:委员会秘书处应保存选择过程中的所有文件记录,包括对候选人的要求、申请、评估、决定等。

然后,委员会应通过决议确认其推荐的组织可以被任命为注册机构。

SN.3.5 任命注册机构

如果注册机构计划收取费用，则需要根据注册机构确认（RAC）委员会提供的信息（见 SN.3.2），启动任命注册机构的技术管理局投票以及 ISO 委员会投票。如项目为标准修订，而委员会决定继续使用相同的注册机构，则无需技术管理局或 ISO 委员会的批准（见 SN.3.4）。

在《ISO/IEC 导则 第 1 部分》中规定，注册机构可以在 ISO 委员会授权的情况下收取注册服务的费用，只要收费严格按成本回收原则计算即可。

因此，委员会应提供完整信息。如需编制 JTC 1 RA 标准，还应向 IEC 提供注册机构确认表格的副本。因为，凡收费注册机构的任命均应得到 IEC/SMB（与理事会）的确认。

SN.3.6 签署注册机构协议

只有在 ISO 技术管理局和理事会（如收费）（若需编制 JTC 1 RA 标准，还涉及 IEC）指定注册机构后，才能签署注册机构协议。对于所有注册机构，应基于 ISO 中央秘书处模板签署注册机构协议。注册机构协议应在新注册机构标准（或修订版）发布之前签署。如果未签署注册机构协议，则不得发布新注册机构标准（或修订版）。

由于在整个标准制定阶段，可以持续修订注册机构标准，因此注册机构标准的待发布版本完成后，应提交至注册机构处。注册机构协议中引用了注册机构任务列表。

为了确保不同注册机构在标准问题上保持一致且公平，凡 ISO 中央秘书处认为应该提供的、具有重要意义的、但注册机构协议模板未涉及的内容，应提交至技术管理局审批。

SN.3.7 终止注册机构

当注册机构辞任，或依据委员会、ISO 中央秘书处（由于撤销注册机构标准、客户投诉等）的决定而不再继续提供注册服务时，可能会终止注册机构。委员会如发现有任何问题可能导致注册机构终止注册服务，应通知技术项目经理。

在选择替代注册机构时，应遵循 SN.3.4 详述的过程，除非委员会已经确定了符合 SN.3.4 所述选拔标准的替代注册机构候选者，并且如启动选择其他注册机构候选者的程序，会导致注册机构服务的中断。

注册机构协议规定了终止注册机构所需的通知期限。委员会利用通知期来选择新注册机构，以便实现注册机构的无缝过渡。

SN.3.8 实施注册机构标准

SN.3.8.1 注册机构的职能

注册机构通过以下方式提供注册服务：

—— 提供注册机构标准中描述的注册服务；

—— 遵守注册机构协议的规定。

SN.3.8.2 委员会的职能

注册机构协议由注册机构和 ISO 秘书长共同签署，而秘书长签署的注册机构协议对 ISO 系统中的所有组织均有约束力，包括 ISO 成员和 ISO 委员会。委员会扮演着核心角色。除明确是否有必要制定注册机构标准（3.2），起草注册机构标准（3.3）并为注册机构标准（新标准/修订版）选择注册机构（3.4）之外，委员会还承担着一项主要职责，即通过如下方式监督注册机构：

—— **回答问题**：委员会应回答注册机构提出的有关注册机构标准的问题，并解释注册机构在执行注册机构标准上应承担的作用。

—— **评估注册机构的年度报告**：注册机构协议要求注册机构提供包含指定内容的年度报告。委员会应确保注册机构已提交年度报告，并通读报告。

—— **监控**：除年度注册机构报告外，委员会还应向使用注册机构标准的行业与用户收集一切反馈。委员会应当根据如上信息（注册机构报告与其他反馈），向 ISO 中央秘书处进行报告。

—— **向 ISO 中央秘书处报告**：委员会应根据上述监控信息，填写技术项目经理年度委员会报告（ACR）表格，并向负责的技术项目经理提交报告（见附件 SJ），每年至少一次。该报告旨在确认注册机构是否遵循注册机构协议开展工作，或工作中是否存在任何问题（问题可能包括：注册机构标准不符合行业或用户需求，注册服务相关的质量投诉等）。报告频率应不低于每年一次，应提交至负责的技术项目经理；如委员会认为必要，可增加报告提交次数。技术项目经理也可能要求委员会提交临时报告。如果报告指出存在某些问题，则应提供解决问题的计划改进措施。

—— **改进措施**：

—— 通过委员会：委员会负责履行其责任范围内的改进措施，可能包括：修订注册机构标准，为注册机构提供建议和指导，情况严重时审核注册机构，或建议 ISO 中央秘书处终止注册机构协议。

—— 通过 ISO 中央秘书处：技术项目经理协助 ISO 中央秘书处履行其职责范围内的改进措施（例如更新或监督注册机构协议）。技术项目经理还可建议其他改进措施。

—— 通过注册机构：注册机构负责执行其责任范围（包括注册服务和注册机构协议中相关规定）内的任何改进措施。

—— **争议解决**：注册机构如与接受服务的客户发生争议，应尽力解决。注册机构应将其无法解决的争议提交至委员会（如注册机构协议中所述），委员会应向负责的技术项目经理提出建议。委员会（及ISO中央秘书处）的职能仅限于协助注册机构解决争议。委员会不应对争议承担责任，或成为注册机构争议的申诉机构，因为这可能会无意中给人以ISO对注册服务负责的印象。

—— **记录的维护**：来自注册机构的所有关键通信与文件均应归档。委员会秘书处负责将文件保存在电子委员会的单独文件夹中。

委员会可设立专门的小组［通常称为注册管理小组（RMG）］，并帮助小组完成上述工作。委员会（无论是亲自或通过注册管理小组完成上述工作）均不得参与提供注册服务，也不得牵涉其中。

SN.3.8.3 ISO中央秘书处的职能

委员会与ISO中央秘书处通过责任技术项目经理建立联系。技术项目经理的职能包括：

—— 在编制注册机构标准的过程中，如委员会未进行识别，则技术项目经理应进行识别。

—— 为起草注册机构标准提供指导和建议。

—— 为注册机构政策委员会提供培训。

—— 与委员会协作，确保注册机构政策得以履行，保证注册服务质量，保证投诉得到适当处理，保证行业与用户的需求得到满足，包括解决委员会（通过技术项目经理年度委员会报告表格）在年度报告中提出的问题以及提议改进措施，并协助改进措施的执行（见SN.3.8.2）。

—— 维护相关人员参与记录。

附录 SO
（规范性）
与公共政策举措相关或支持性作用的 ISO、IEC 标准的制定原则

SO.1 背景情况

2007 年，ISO 全体大会开幕式上讨论了 ISO 面临的重要情况，即 ISO 标准与公共政策之间的关系，以及一类 ISO 标准主要用户（即政府）的特殊需求和关注点。ISO 的兄弟组织 IEC 负责电工标准化，对该倡议表示赞赏，并希望与 ISO 合作，为两个组织的技术委员会制定共同的 ISO/IEC 原则和指南。ISO 和 IEC 确定了一个合理的目标，即为其各类标准吸引到公共机构的进一步关注，同时确保其标准能够满足公共当局的相关需求，解决其担心的问题。已有事实证明，ISO 和 IEC 标准能为公共政策的实施提供重要支持，因此此次行动将促进二者标准的全球相关性以及标准在世界范围内的适用性。ISO 和 IEC 已经并将继续采取实际行动，提供可支持政府机构工作的自愿性标准，并确保标准既符合 WTO/TBT 准则，也有助于政府出台技术法规和 / 或执行采购行动。

应该指出，作为民间志愿组织，ISO 和 IEC 本身并不直接代表政府利益。ISO 和 IEC 在其标准上达成共识是：标准在起草阶段以一系列利益相关方的一致意见为基础，在审批阶段则反映了不同国家标准机构的共识。ISO 或 IEC 标准的国家立场不一定是政府立场，尽管政府专家可能与私营部门的同行共同确定了标准的立场。

ISO 与 IEC 委员会在制定有关公共政策举措的标准或支持性作用标准时，遵循以下原则。这些原则可确保 ISO 与 IEC 发布的标准能得到公共机构的合理支持与采用。

SO.2 原则

a) ISO 和 IEC 致力于根据全球共识的客观信息与知识，而不是主观判断。创建市场驱动的国际标准，以为政府提供可靠的技术工具，支持监管的实施以及公共政策举措的执行。

b) ISO 和 IEC 致力于制定与市场相关的国际标准，满足所有利益相关方（在适当的情况下，包括公共机构）的需求和关注，而不寻求建立、推动或激励公共政策、法规或社会和政治议程。

在制定 ISO 或 IEC 标准以支持公共政策举措时，所有相关方应清楚地理解标准与公共政策倡议之间的关系。如果公共机构认为标准化与公共政策 / 监管之间可能会相互

影响，则应尽早参与到标准制定过程中。应该指出的是，在许多情况下，监管机构会指派专家代表，作为相关国际委员会的成员和/或 ISO 和 IEC 技术机构的国家级镜像委员会成员，积极参与 ISO 和 IEC 国际标准的制定。

   c) ISO 和 IEC 了解制定法规、公共政策和/或制定解释国际条约是政府或条约组织的工作。

   d) 在制定旨在支持法规、监管合作和公共政策的 ISO 和 IEC 标准时，最好采用 ISO 和 IEC 内部机构证实有效、在 ISO/ IEC 导则中已有详述的操作方法和参与模型。

对于旨在支持监管与公共政策的 ISO/IEC 标准，如使用特殊的委员会架构、程序或参与模式进行制定，可能会损害标准的可信度与适用性。

SO.3 实施

除了向 ISO 和 IEC 委员会、其领导者和参与者普及上述原则之外，ISO 和 IEC 应积极支持和促进 WTO/TBT 协议中确立的国际标准化原则以及 TBT 委员会关于国际标准制定的后续决定。

ISO 成员体和 IEC 国家标准机构应确保各国政府（包括其贸易代表）了解 ISO 和 IEC 的各类标准，掌握标准编制机构 ISO 和 IEC 的相关信息，并在适当时参与 ISO 和 IEC 标准制定，以减少误解和无意的矛盾。

对于公共政策举措相关的 ISO 和 IEC 标准成功案例，ISO 技术管理局和 IEC 标准化管理委员会应补充相关实施指南和案例研究（例如：与医疗设备和温室气体排放相关的 ISO 标准，以及与无线电干扰、家用电器安全性、船舶和海洋技术相关的 IEC 标准）。

附录 SP
（当前空白——预留给新增附录）

附录 SQ
（规范性）
技术工作领导人员的选拔标准

SQ.1 成员体的义务

成员体负责确保领导职位候选人（即主席、召集人和秘书）符合 SQ.3 的要求。成员体还应了解所选领导人与专家在技能或知识方面的任何不足，并举行持续培训，以弥补相关不足。

SQ.2 可用于填补技能或知识空白的资源

ISO 可提供多种资源，以帮助成员体填补任何所发现的技能或知识缺口：

—— ISO 中央秘书处提供各类培训和其他材料，以确保不同机构可以有效地组织或参加会议，包括了解相关的关键概念。

—— 为支持专家培训工作，可在委员会全体会议之前或期间，安排委员会提前会议，并告知现有的资源情况。

—— 技术项目经理（特别是在组建新委员会的情况下）可参与培训技术领导和专家。在组建新委员会时，技术项目经理应参加第一次会议，全面介绍 ISO 及其流程。

—— 成员体应互相组织交流活动，或与 ISO 中央秘书处保持交流。成员体也不妨考虑结对安排。

—— 为协助委员会领导层开发不只适用于 ISO 的技能，如召开有效会议所需的技能、一般领导技能等，可聘请外部培训师提供辅导与培训服务。

SQ.3 技术工作领导人员的选拔标准

任何委员会或工作组的成功皆取决于其领导人员。该选拔标准适用于委员会[4]主席、工作组召集人和委员会秘书。在提名上述职责人员时，成员体应参考此标准，以遵守新的 ISO 技术工作行为准则，并确保正确应用《ISO/IEC 导则》。

SQ.3.1 主席与召集人

SQ.3.1.1 优秀主席与召集人应具备的能力和素质：

—— 目前正在相应领域内供职，且拥有良好声誉；

—— 具有相关专业经验，包括之前有过主席工作经验；

4） 技术委员会、分委员会、项目委员会。

—— 能领导并激励该领域的代表与专家达成共识；
—— 了解 ISO 工作的国际性及其带来的益处；
—— 为其承担的职能投入时间和资源；
—— 通过创新和创造性思维，在共识环境中发现解决方案；
—— 积极行动，巧妙沟通；
—— 促进并重视与其他 ISO、IEC 委员会及合作伙伴的合作，包括来自监管机构的合作伙伴；
—— 保持纯粹的国际立场。

SQ.3.1.2 主席和召集人的岗位说明：
—— 主持会议，获得有效成果，达成协议，并明确立场与决定；
—— 确保所有立场与观点（在会议及通信中）得到平等对待；
—— 根据项目计划，按照约定的目标日期，管理从编制到完成的整个项目过程；
—— 努力确保工作组拥有全方位的技术能力；
—— 充分了解会议主题和市场需求；
—— 根据市场或全球相关性，决定继续推进工作或停止工作；
—— 具备 ISO 及其程序的基本知识。

SQ.3.1.3 此外，委员会主席还应承担以下职能：
—— 负责委员会的整体管理，包括所有分委员会和工作组；
—— 就委员会有关的重要事项，向 ISO 技术管理局提供建议；
—— 确保委员会执行了 ISO 技术管理局的政策和战略决策；
—— 从战略角度思考，以促进 ISO 在该领域的工作。

SQ.3.1.4 此外，工作组召集人还应承担以下职能：

拥有一定的知识和能力，能使用基于 MS Word 的标准起草工具与基于 ISO 网络的应用程序，以支持工作组的工作。

SQ.3.1.5 辅助信息

在提名主席和召集人时，成员体应考虑相关人员的以下信息：
—— 目前在该领域的角色；
—— 学历；
—— 职业生涯；
—— 领导经验；
—— 类似的活动；
—— 语言技能。

SQ.3.2　秘书与秘书处

SQ.3.2.1　选拔秘书与秘书处

ISO 委员会或工作组的成功取决于其秘书处与秘书（如无秘书，则取决于工作组召集人）。如下列表以《ISO/IEC 导则》为基础，列明了秘书处与秘书应承担的工作任务。在任命某一组织、专业人员成为秘书处与秘书时，ISO 成员应参照此列表。

SQ.3.2.2　保证文件质量

为委员会或工作组起草标准草案，安排草案的分发，并处理收到的意见。如果是委员会秘书，则应准备草案、文本和数据，由 ISO 中央秘书处（ISO/CS）分发问询意见，编制最终国际标准草案或发布。在提交此类文档时，应遵循 ISO 中央秘书处的提交要求。

SQ.3.2.3　完成出色的项目管理

协助确定每个项目的优先事项和目标日期；向 ISO 中央秘书处通知所有工作组召集人和项目负责人的姓名；启动投票；当项目显著超时，和／或似乎缺乏足够支持时，提出前瞻性的解决方案。

SQ.3.2.4　做好充分的会议准备

制定会议议程并安排议程的分发，安排议程上提及的所有文件的分发，包括工作组报告，并说明会议期间讨论所需的所有其他文件。记录会议中所作的决定，并以书面形式发布这些决定，以供参会人员确认。会议结束后，及时分发会议纪要。

SQ.3.2.5　对 ISO 流程提供良好建议

向《ISO/IEC 导则》中的主席、项目负责人和召集人提供建议，特别是与项目进展程序相关的建议。联系各分委员会与工作组，了解其活动。

SQ.3.2.6　联络与沟通

与委员会主席或工作组召集人密切联络。与 ISO 中央秘书处以及委员会或工作组成员就其活动保持密切联系。与上级委员会秘书保持密切联络。

SQ.3.2.7　主动跟进行动

确保在会议或通信中商定的所有行动都能够公开、按时完成。

SQ.3.2.8　具备 IT 能力

具有一定的知识和能力水平，可以使用基于 MS Word 的绘图工具和基于 ISO 网络的应用程序（包括 Livelink-e-committee，委员会投票应用程序和提交界面）。

SQ.3.2.9　辅助信息

在任命秘书时，成员体应考虑以下信息：

—— 教育背景；

—— 工作经历；

—— 有标准化工作的经验；

—— 参与过标准化培训计划；

—— 拥有 ISO 的 IT 工具和 IT 基础架构经验；

—— 语言技能。

附录 SR
（规范性）
限制可交付使用文件目的或用途的声明

SR.1 原则

ISO 技术管理局通过了技术管理局第 8/2012 号决议，涉及旨在限制可交付使用文件的目的或用途的声明：

—— 注意，ISO 理事会第 9/2001 号决议确认，ISO 应遵守 WTO《技术性贸易壁垒协议（TBT）》的基本运作与实施原则；

—— 不得发布可能限制上述目的或使用可交付使用文件造成贸易壁垒或障碍的声明；

—— 进一步达成一致意见，不允许发布有关合同义务或政府监管的声明；

—— 在编制可交付使用文件时（即国际标准草案阶段结束之前），应删除一切此类声明，在修订可交付使用文件时，应删除现有可交付使用文件中的一切此类声明；

—— 在特殊情况下，如出现任何例外情况，应寻求 ISO 技术管理局的批准。

ISO 技术管理局也采用了技术管理局第 69/2012 号决议的以下内容，涉及旨在限制可交付使用文件在合格评定方面的目的或用途的声明（《合格评定免责声明》）：

—— 回顾技术管理局第 8/2012 号决议中旨在限制可交付使用文件的目的或用途的声明；

—— 兹决定，ISO 可交付使用文件中不应包括合格评定相关的声明（如认证）；

—— 达成一致意见，在形成可交付使用文件期间（即国际标准草案阶段结束之前），应删除一切合格评定声明，在修订可交付使用文件时，应删除现有可交付使用文件中的一切此类声明，并以前言中的链接替换；

—— 如果出现例外情况，应征求技术管理局的批准。

SR.2 贸易壁垒或障碍相关声明的更多信息

许多 ISO 可交付使用文件中含有不同类型的声明，旨在限制可交付使用文件在贸易壁垒或障碍方面的目的或使用（有时称为"免责声明"）。虽然措辞略有不同，但此类声明均指出，相关可交付使用文件并非旨在为贸易设置壁垒或障碍。

正如 ISO/ IEC 信息中心所解释的那样，WTO《技术性贸易壁垒协议（TBT）》是

WTO 协定所含的法律文本之一。根据《协议》，若不考虑其他标准，世贸组织成员应确保自愿性标准不会对贸易造成不必要的障碍。具体而言，《协议》中附件3《有关标准制定、采用和应用的良好实践守则》第 E 段规定："标准化机构编制、采纳或应用标准时，应确保不以为国际贸易增设不必要障碍为目的，或以确保不会造成此类后果为前提"。

WTO 成员国境内的标准化机构可宣布其已认可《有关标准的制定、采用和应用的良好实践准则》（《协议》中附件3）。宣布接受该准则的国家的标准化机构记录于一份列表中。根据技术性贸易壁垒要求，世贸组织成员有责任确保技术性贸易壁垒协议在其各自的管辖范围内得到遵守。

因此，《协议》中全面包含了避免贸易障碍相关的义务。如果在部分标准中加入《协议》声明，而在其他标准中不添加《协议》声明，便会造成两类标准，而两类标准之间并无明确界线。事实上，如果这种做法继续下去，便可能会在无意中造成这样一种印象：标准对《协议》有两种不同的遵守程度。

此外，ISO 理事会还审查了 ISO 在遵守《协议》基本运行实施原则上的整体情况。通过 ISO 理事会第 9/2001 号决议，ISO 理事会确认，ISO 已遵守《协议》的基本运行实施原则。

SR.3 合同义务或政府监管相关声明的更多信息

技术管理局第 8/2012 号决议是对《ISO/ IEC 导则　第 2 部分》第 4 章的深化。该条款规定："文件不应包括合同要求（例如，有关索赔、担保、费用的支付）以及法律或法定要求"。

此后，ISO 技术管理局还通过了技术管理局第 70/2018 号决议，针对"与合同义务或政府法规有关的声明"作出了解释：

—— 请注意：技术管理局第 8/2012 号决议中有关"进一步达成一致意见，不允许有关合同义务或政府监管的声明"的解释问题。

—— 另注意：

- 许多 ISO 标准中都包含有关遵守合同义务、法律要求和政府法规的文本；
- ISO 可交付使用文件可作为此类要求的补充，并可构成所有相关利益方（可包括政府当局与行业参与者）的有用工具。

—— 注意：导则维护组（DMT）协商会就此事发来的答复。

—— 对于所有 ISO 可交付使用文件，明确要求：

a) 不允许包含明确要求或建议使用者遵守任何特定法律、法规或合同（例如对此类要求的规范性引用）的全文或其中一部分的声明；

b) 允许包含不违反 a) 款的法律和监管要求相关声明；

c) 允许以提供信息为目的，介绍特定法律或法规内容的实际示例；

d) a) 款不存在例外。

此外，文件不应明确要求或建议使用者遵守国家法律，或表明 ISO 标准无意与国家法律相冲突。ISO 标准的应用是自愿的，在发生冲突时，国家法律优先。此外，如果在部分标准中加入《协议》声明，而在其他标准中不添加《协议》声明，便会造成两类标准。这可能在无意中造成一种印象，即 ISO 标准优先于国家法律，除非明确说明国家法律的首要地位。

SR.4 合格评定有关的要求与禁令的更多信息

标准编制者如需进一步了解合格评定的相关信息，请查阅《标准编制者应了解的合格评定——要求与禁令》http：//www.iso.org/iso/PUB100303.pdf 以及《ISO/IEC 导则 第 2 部分》第 33 章。通用文本中列有更多指南，链接见 ISO 可交付使用文件前言（www.iso.org/foreword-info）。

SR.5 例外请求

如有例外情况，需使用旨在限制可交付使用文件的目的或使用的声明，应向 ISO 技术管理局申请，ISO 技术管理局将根据具体情况仔细考虑此类请求。

附录 SS
（规范性）
选择性执行委员会草案（CD）阶段——委员会指南

以下是各委员会关于任择使用委员会草案阶段的指导意见。

SS.1 执行指南

—— 在决定是否省略委员会草案阶段时，应视具体情况而定。

—— 作决定时，应考虑到委员会草案阶段的预期成果（例如能否加深对问题的理解），并对比可能增加的成本和时间（额外的起草工作与会议）。

—— 在与工作组专家磋商达成共识之后，应由工作组召集人/项目负责人提议省略委员会草案阶段。

—— 随后，上级委员会应举行为期 4 周的委员会内部投票，以协商一致的方式作最终决定，或在会议上作最终决定（关于共识的定义，见 2.5.6）。

—— 如担心省略委员会草案阶段可能会严重损害共识，则应避免省略委员会草案阶段。

SS.2 在提议省略 CD 阶段时给召集人与项目负责人可遵守的建议

1) 在预研阶段（Stage 0）形成反映已知利益相关方观点的草案文档。

2) 确保已分发高质量的草案文档，并随附新工作项目提议。

3) 一旦了解到相应 WG 希望省略委员会草案阶段，立即通知上级委员会秘书。

4) 确保对该主题有重大利益的 P 成员已推荐专家至工作组，且工作组就该文档达成共识。

5) 确保适当多次向专家发放工作草案。

6) 定期向委员会秘书告知有关工作状态的最新信息，委员会秘书应将最新信息转告 P 成员，并在必要时分发工作草案以供参考。

7) 要求工作组专家定期向推荐他们的 P 成员通报工作状况，并鼓励工作组专家开展国家层面上的非正式询问。

8) 确保提交给委员会秘书的文件具有较高的技术质量，足以支持国际标准草案投票。

9) 国际标准草案阶段投票结束后，务必确保对国际标准草案阶段的意见进行全面审核。

SS.3 给委员会主席和秘书的建议

—— 一旦了解到工作组希望省略委员会草案阶段，立即告知委员会成员。

—— 在向 ISO 中央秘书处提交国际标准草案以准备投票时，将草案发送给委员会以供参考。

SS.4 其他信息

—— 省略委员会草案阶段不改变目前的标准制定期限，标准制订仍然需要 2、3 或 4 年。

—— 委员会草案阶段可选择性执行，从而有助于在规定时间期限内按时完成标准制定，或提前完成制定。

—— 对于《维也纳协议》项目，也可以选择性执行委员会草案阶段。

附件 ST
（规范性）
结对政策

ST.1 范围

结对旨在增强双方实力，促进参与并提高绩效。结对目标应支持被结对方确定的优先事项，并应纳入其国家发展计划/战略。

此政策是强制性的。

ST.2 定义

ST.2.1 结对：两个 ISO 成员体之间为能力培养而签订的合作协议，其中能力培养是指培养（NSB 级别的国家机构）能力，以顺利完成 ISO 标准化工作。

ST.2.2 结对主导方：提供指导、专业知识与培训的 ISO 成员体。

ST.2.3 被结对方：寻求能力提升的 ISO 成员体。

ST.2.4 领导层结对：领导层之间的结对协议，帮助其在国际标准制定过程中学习与领导职责相关的能力。领导层结对有三种类型：主席、召集人和秘书。通过领导层结对，被结对方有机会学习承担主席/召集人/委员会秘书职责所需的技能和责任。

ST.2.5 主席结对：主席与主席之间的结对协议。

ST.2.6 召集人结对：召集人与召集人之间的结对协议。

ST.2.7 秘书处结对：秘书处与秘书处之间的结对协议。

ST.2.8 P 成员结对：结对主导方和被结对方均需成为相关委员会的 P 成员。被结对方可能尚未具备在领导职位上合作所需的能力水平，但在 P 成员结对之下，被结对方将会积极参与标准制定的技术工作。经过 P 成员结对，P 成员可获得国际和国家级委员会的具体目标经验（包括国家镜像委员会的作用），并深刻理解 P 成员的义务。

ST.3 结对要求

需满足以下要求，才有资格参与结对：

—— ISO 的正式成员（如欲参加 P 成员结对，则应参与 ISO 理事会的新成员权利试点）；

—— 请求成为结对主导方时，应证明自身具备结对主导方的丰富知识与经验（NSB 级别）。

—— 请求成为被结对方时，应证明自身有能力培养的需要，例如，在组织设立秘书处、任用主席/召集人或积极参与技术委员会方面缺乏经验。

选拔标准：

结对主导方应具备如下经验：

—— 秘书处结对：拥有 5 个或以上的委员会秘书处；

—— 主席或召集人结对：拥有 5 名或以上的主席或召集人；

—— P 成员结对：成为 100 个或更多委员会的 P 成员。

被结对方应具备如下经验：

—— 秘书处结对：少于 5 个委员会秘书处；

—— 主席或召集人结对：任命的主席或召集人少于 5 人；

—— P 成员结对：成为不到 100 个委员会中的 P 成员。

ST.4 一般原则

1) **自愿性质**：结对应自愿进行。

2) **目标**：双方都应致力于实现结对的主要目标：增强能力，提高结对伙伴的参与和绩效。应支持双方确定的优先事项，并纳入国家发展计划/战略，以完成计划。

3) **不期望提供财务支持**：不期望结对主导方为被结对方提供财务支持。

4) **"一国一票"**：结对协议不得影响国家立场。不应期望合作伙伴彼此支持对方的立场和观点。结对协议不得危害 ISO 中的"一国一票"原则（不允许代理投票）。

5) **合作**：应该以灵活的方式尽可能广泛地看待合作，包括合作伙伴之间在技术、战略和政策层面上的活动。

6) **承诺**：结对协议里双方的结对承诺在结对期间持续有效。根据结对伙伴的共同协议，除了提供培训和指导之外，结对主导方还可承担其他的特定管理任务。

7) **持续改进**：结对协议应可持续改进。应建立机制，确保能对每项行动跟踪调查，并可在必要时采取纠正措施。应衡量并监控被结对方在 ISO 中的参与、领导水平进展指标，并每年审查进展情况。

8) **可持续性**：被结对方通过结对获得知识、经验之后，可用于培训被结对方的其他工作人员，从而可持续地增强能力。通过参与结对，被结对方对于结对的需求应当逐渐减轻乃至完全消失，一旦结对结束，结对国家即可以独立参与 ISO 进程。

ST.5 程序

ST.5.1 阶段安排

本政策按照典型的结对周期顺序，说明结对的各方面情况，注意某些阶段可以并行完成。各阶段如下：

—— 确定有必要进行结对协议（ST.5.2）；

—— 选择结对伙伴（ST.5.3）；

—— 起草/提交结对申请（ST.5.4）；

—— 批准结对申请（ST.5.5）；

—— 实施结对协议（ST.5.6）；

—— 终止结对协议（ST.5.7）；

—— ISO中央秘书处的角色（ST.5.8）。

ST.5.2 明确结对需要

应自愿参与结对。合作伙伴就结对类型（P成员、秘书处、主席或召集人）达成一致后，应填写申请表（见附录SJ）。对于秘书处或主席结对，申请表应提交给技术管理局秘书处。对于工作组召集人结对，申请表应直接提交至上级委员会与ISO TPM处。对于P成员结对，申请表应提交技术项目经理审查，并提交委员会秘书处以供参考。

ST.5.3 选择结对伙伴

如成员体已经明确需要结对，但尚未确定结对伙伴，则可以联系ISO中央秘书处或相关委员会的技术项目经理（如果已确定委员会）寻求帮助，以确定结对伙伴。

ST.5.4 起草/提交结对申请

以下内容应包含在申请表中：

—— 申请ISO成员体的联系方式；

—— 申请建立结对关系的委员会；

—— 申请结对类型；

—— ISO成员体（被结对方）的声明，表明需要进行能力培养，并解释本次结对协议将以何种能力培养为目标；

—— ISO成员体（结对主导方）的声明，展示可支持能力培养的能力和资源；

—— 结对协议的目标和预期结果；

—— 建议的结对时限；

—— 对于领导层结对协议，应提供秘书/被结对秘书、主席/联合主席以及召集人/被结对召集人的联系方式与简历。

对于技术委员会、项目委员会或分委员会级别的秘书处或主席结对，应取得技术管理局的批准。

对于工作组召集人结对协议，应将结对申请表直接提交上级委员会秘书处批准。

对于P成员结对协议，应直接向ISO中央秘书处技术项目经理提交结对申请表（并将副本提交至委员会秘书处，以供参考）。

ST.5.5　批准结对申请

- 秘书处和主席结对协议应提交至技术管理局秘书处，技术管理局秘书处将向技术管理局成员发送结对申请表和辅助材料，以便通过通信方式进行审批（4周内部投票）。投票用的辅助材料应包括技术项目经理/中央秘书处的评估，以及NSB参与ISO情况和其他委员会领导角色表现的相关统计数据。技术管理局根据资料情况评估结对申请，并参照上述成为"主导"和"被结对"伙伴指导原则（经验建议）开展评估。

  如技术管理局批准申请，则秘书处或主席结对协议将注册至GD中。

- 对于WG召集人结对，申请国家标准机构应将申请表直接提交至上级委员会秘书处批准。随后，委员会应将其决定递交技术项目经理进行处理。

- 对于P成员结对，申请NSB应直接向ISO中央秘书处技术项目经理提交结对申请表（副本提交至委员会秘书处以供参考）。技术项目经理应审查申请表，并在全球目录中注册结对协议。

结对一旦获批，则被结对方应在注册后的3个月内，向其ISO中央秘书处技术项目经理提交完整的结对协议（包括截止日期，最长期限为5年）和协作计划。表格见附录SJ。技术项目经理应审查协议与协作计划，并在必要时与被结对方合作，改进协议与计划。如合作伙伴未在3个月内提交令人满意的结对协议和协作计划，则不得启动结对协议。

ST.5.6　执行结对协议

结对协议应进行年度审查和绩效评估，包括被结对方的自我评估、委员会成员的评估以及技术项目经理的评估。

年度表现评估表：被结对方应每年填写并提交一份表格。该表格的一部分将由技术项目经理完成。

P成员评估将作为会后定期反馈调查过程的一部分。技术项目经理应审查调查结果，并查阅P成员对年度绩效评估表的反馈。

技术管理局秘书处应审查年度绩效评估表，并将填好的表格返回被结对方参考。如无任何问题，表格将被存档，结对协议继续。

ST.5.7　终止结对协议

如果未提交年度绩效评估表，或出现了负面评价或问题，则监督机构应提请技术管理局注意。技术管理局应评估相关信息，决定采取何种行动，可能包括取消结对协议。

一旦结对时长达到5年，则自动终止。不可对结对协议进行延长或续签。

结对协议结束时，不意味着领导职务（秘书处、主席、召集人）自动由结对主导方转移至被结对方。被结对方的领导职务将在全球目录中删除。

对于秘书处结对，如结对主导方欲放弃秘书处，则委员会可以作出决议，支持将秘书处分配给被结对方。对于分委员会秘书处，技术委员会将根据分委员会的决议批准这项分配。对于技术委员会秘书处，技术管理局将根据技术委员会的决议批准这项分配。

领导层角色如有任何其他变更，均应遵循《ISO/IEC 导则 第1部分》中概述的批准流程。

ST.5.8 ISO 中央秘书处的职能

被结对方通过负责的技术项目经理与 ISO 中央秘书处联系。技术项目经理的职能包括：

—— 在尚未确定被结对方/结对主导方的情况下，帮助相关机构确定潜在的被结对方；

—— 就如何执行结对政策提供指导和建议；

—— 向技术管理局汇报潜在问题。

参 考 文 献

以下是 ISO 技术社区关于一些重要主题的参考文献的链接。

— ISO's web page (www.iso.org)

— ISO/IEC Directives, Parts 1 & 2, Consolidated ISO Supplement to ISO/IEC Directives, Part 1, JTC 1 Supplement (www.iso.org/directives)

— Guidance for ISO national standards bodies — engaging stakeholders and building consensus (http://www.iso.org/iso/guidance_nsb.pdf). See also additional guidance from the TMB on stakeholder engagement: (http://www.iso.org/iso/additional_guidance_on_stakeholder_engagement_tmb_peg.pdf)

— Guidance for liaisons organizations — engaging stakeholders and building consensus (http://www.iso.org/iso/guidance_liaison-organizations.pdf)

— Using and referencing ISO and IEC standards to support public policy (https://www.iso.org/publication/PUB100358.html)

— ISO's global relevance policy (http://www.iso.org/iso/home/standards_development/governance_of_technical_work.htm)

— ISO Code of Conduct for the technical work and suggestions for implementation (https://www.iso.org/publication/PUB100397.html)

— Policy for the distribution of ISO publications and the protection of ISO's copyright (ISO POCOSA) (https://connect.iso.org/display/marketing/ISO+POCOSA+2017)

— How to write standards (http://www.iso.org/iso/how-to-write-standards.pdf)

— Model manuscript of a draft International Standard (known as "The Rice Model") (https://www.iso.org/publication/PUB100407.html)

— ISO Guides — adding value to International Standards (http://www.iso.org/iso/iso-iec_guides.pdf)

— Guidelines for TC/SC Chairs and Secretariats for implementation of the Agreement on technical cooperation between ISO and CEN (Vienna Agreement) (http://isotc.iso.org/livelink/livelink/open/4230689)

— Guidance on twinning in ISO standards development activities (https://www.iso.org/publication/PUB100341.html)

— Guidance for writing standards taking into account micro, small and medium-sized enterprises' needs (http://www.iso.org/iso/guidance-for-writing-standards-for-smes.pdf)

— Guide for addressing sustainability in standards ("Guide 82") (http://isotc.iso.org/livelink/livelink?func=ll&objId=8389141&objAction=browse&viewType=1)

— Guidelines on competition law (http://www.iso.org/iso/competition_law_guidelines.pdf)

— Guidelines on remote participation at committee meetings
(http://isotc.iso.org/livelink/livelink/open/17857546)

— Guidelines for the submission of text and drawings to ISO/CS
(http://isotc.iso.org/livelink/livelink/open/18862226)

— Guidance on New Work in ISO
(https://isotc.iso.org/livelink/livelink/fetch/2000/2122/15507012/20090199/Guidance_on_New_Work.pdf?nodeid=20407727&vernum=-2)

ISO/IEC Directives, Part 1

Consolidated ISO Supplement — Procedures specific to ISO

Directives ISO/IEC, Partie 1
Supplément ISO consolidé — Procédures spécifiques à l'ISO

Tenth edition, 2019

[Based on the fifteenth edition (2019) of the ISO/IEC Directives, Part 1]

INTRODUCTION TO THE CONSOLIDATED ISO SUPPLEMENT

0.1 What is the *Consolidated ISO Supplement?*

The ISO/IEC Directives define the basic procedures to be followed in the development of International Standards and other publications. This *Consolidated ISO Supplement* (also referred to as the "ISO Supplement") contains both the ISO/IEC Directives, Part 1 and the procedures specific to ISO which were formerly referred to as the "ISO Supplement".

Part 1 of the ISO/IEC Directives, together with this Consolidated ISO Supplement, provide the complete set of procedural rules to be followed by ISO committees. There are, however, other documents to which reference may need to be made. In particular, for committees cooperating with CEN under the Agreement on technical cooperation between ISO and CEN (Vienna Agreement), reference should be made to the Guidelines for Chairmen and Secretariats on the implementation of the Vienna Agreement.

Attention is also drawn to the fact that ISO/IEC JTC 1 Information technology has adopted ISO/IEC Directives, Part 1 together with the JTC 1 Supplement.

0.2 The structure of this *Consolidated ISO Supplement*

The clause structure of the *Consolidated ISO Supplement* follows that of Part 1 of the *ISO/IEC Directives*. In order to assist in cross-referencing the texts, the ISO specific clauses are highlighted in blue under the clauses of the ISO/IEC Directives. If there is no highlighted blue text under a particular clause, it means that there are no ISO-specific clauses.

Tables and annexes beginning with "S" in this *Consolidated ISO Supplement* are all unique to ISO and do not follow the order in the *ISO/IEC Directives*. Where a reference in an ISO annex (i.e. beginning with "S") is made to an annex in the *ISO/IEC Directives, Part 1*, this is explicitly indicated.

The following clauses that have been modified with respect to the previous edition of the ISO Supplement: Foreword, 1.7, 1.8.1, 1.8.3, 1.9.2, 1.9.3, 1.9.4, 1.12.1, 1.12.6, 2.1.6, 2.9.1, 3.2.2, Annex SD, Annex SH, Annex SI, Annex SK, Annex SQ, Annex SR, Annex ST, and Reference documents. The track change version of this tenth edition should be consulted for the details of the changes made (www.iso.org/directives).

Foreword

The **ISO/IEC Directives** are published in two parts:

— Part 1: Procedures for the technical work

— Part 2: Principles and rules for the structure and drafting of ISO and IEC documents

Furthermore, the International Organization for Standardization (ISO), the International Electrotechnical Commission (IEC) and ISO/IEC Joint Technical Committee (JTC) 1 have published independent supplements to Part 1, which include procedures that are not common.

This part sets out the procedures to be followed within ISO and the IEC in carrying out their technical work: primarily the development and maintenance of International Standards through the activities of technical committees and their subsidiary bodies.

ISO, IEC and ISO/IEC JTC 1 provide additional guidance and tools to all those concerned with the preparation of technical documents on their respective websites (www.iso.org/directives; http://www.iec.ch/members_experts/refdocs/ and http://www.jtc1.org).

This fifteenth edition incorporates changes agreed by the respective technical management boards since publication of the thirteenth edition in 2017. Procedures which are not common to all the ISO/IEC Directives are published separately in the ISO Supplement (also referred to as the Consolidated ISO Supplement), the IEC Supplement or the ISO/IEC JTC 1 Supplement, respectively. The Supplements are to be used in conjunction with this document.

The following clauses have been modified with respect to the previous edition: Foreword, 1.7.3, 1.12.2, 1.13.2, 1.13.6, 1.17.2, 3.2.1, 3.2.2, 3.2.4, Annex B, Annex D, Annex E, Annex F, Annex L, and Annex M. The track changes version of this fifteenth edition should be consulted for the details of the changes made.

These procedures have been established by ISO and IEC in recognition of the need for International Standards to be cost-effective and timely, as well as widely recognized and generally applied. In order to attain these objectives, the procedures are based on the following concepts.

a) **Current technology and project management**

   Within the framework of these procedures, the work may be accelerated and the task of experts and secretariats facilitated both by current technology (e.g. IT tools) and project management methods.

b) **Consensus**

   Consensus, which requires the resolution of substantial objections, is an essential procedural principle and a necessary condition for the preparation of International Standards that will be accepted and widely used. Although it is necessary for the technical work to progress speedily, sufficient time is required before the approval stage for the discussion, negotiation and resolution of significant technical disagreements.

   For further details on the principle of "consensus", see 2.5.6.

c) **Discipline**

   National Bodies need to ensure discipline with respect to deadlines and timetables in order to avoid long and uncertain periods of "dead time". Similarly, to avoid re-discussion, National Bodies have the responsibility of ensuring that their technical standpoint is established taking account of all interests concerned at national level, and that this standpoint is made clear at an early stage of the work rather than, for example, at the final (approval) stage. Moreover, National Bodies need to recognize that substantial comments tabled at meetings are counter-productive, since no opportunity is available for other delegations to carry out the necessary consultations at home, without which rapid achievement of consensus will be difficult.

d) **Cost-effectiveness**

These procedures take account of the total cost of the operation. The concept of "total cost" includes direct expenditure by National Bodies, expenditure by the offices in Geneva (funded mainly by the dues of National Bodies), travel costs and the value of the time spent by experts in working groups and committees, at both national and international level.

Opportunities for remote participation at meetings should be sought to the extent possible.

e) **General principles for voting and decisions**

The committee secretariat shall ensure that all decisions taken by the committee, whether at a plenary meeting or by correspondence, are recorded and in ISO, by committee resolutions.

For votes by correspondence or during a committee meeting, a simple majority of the P-members voting is required for approval unless otherwise specified in the ISO/IEC Directives.

The committee leadership shall ensure that votes submitted in writing, in advance of a committee meeting, are considered at the meeting.

When a document is out for ballot (NP, CD or any later stage) formal discussion during meetings or distribution of National Body positions via formal committee distribution channels are prohibited.

In all votes, abstentions are not counted.

Proxy voting is not permitted in ISO.

f) **Global relevance of ISO International Standards**

It is ISO's aim and expectation that each of its International Standards represents a worldwide consensus and responds to global market needs. In order to achieve this aim, it has been recognized that special measures are needed in particular to ensure that the needs of developing countries are taken into account in ISO's technical work. One such measure is the inclusion of specific provisions for "twinning", i.e. partnerships between two ISO member bodies for the purposes of capacity building, in this ISO Supplement to the ISO/IEC Directives. (See 1.7, 1.8.3, 1.9.2, 1.9.3, 1.9.4 and Annex ST.)

Whilst these provisions are necessarily limited to the technical work, "twinning" may occur at multiple levels, in particular to assist the twinned partner in capacity building in developing countries of their standardization, conformity assessment and IT infrastructures, with the aim of the twinned partner ultimately being self-sufficient in carrying out their activities.

g) **Committee specific procedures**

The ISO/IEC Directives (and Consolidated ISO Supplement) have been developed based on many years of experience with standards development across many committees and fields. Therefore, committees should have little, if any, need to develop committee specific procedures (CSP) (which include committee decisions which are de facto procedures). Where a committee feels that the ISO/IEC Directives (or Consolidated ISO Supplement) do not adequately address a specific issue, the committee should submit a request for a revision/expansion of the ISO/IEC Directives (or Consolidated ISO Supplement) to the ISO Technical Management Board rather than develop a committee specific procedure. In cases where a committee feels the ISO/IEC Directives (or Consolidated ISO Supplement) do not at all address a specific operational issue, the committee may develop a CSP, including a clear rationale for the procedure, and submit it for the approval of the ISO Technical Management Board who will make every effort to respond quickly to the request.

h) **Addressing sustainability in standards**

When developing standards committees are encouraged to refer to the ISO Guide for addressing sustainability in standards ("Guide 82") which is available at: http://isotc.iso.org/livelink/livelink?func=ll&objId=8389141&objAction=browse&viewType=1.

i) Terminology used in this document

NOTE 1　Wherever appropriate in this document, for the sake of brevity the following terminology has been adopted to represent similar or identical concepts within ISO and IEC.

Term	ISO	IEC
National Body	Member Body (MB)	National Committee (NC)
technical management board (TMB)	Technical Management Board (ISO/TMB)	Standardization Management Board (SMB)
Chief Executive Officer (CEO)	Secretary-General	General Secretary
office of the CEO	Central Secretariat (CS)	Central Office (CO)
council board	Council	Council Board (CB)
advisory group	Technical Advisory Group (TAG)	Advisory Committee
secretary (of a committee or subcommittee)	Committee manager	Secretary
For other concepts, ISO/IEC Guide 2 applies.		

NOTE 2　In addition the following abbreviations are used in this document.

JTAB	Joint Technical Advisory Board
JCG	Joint Coordination Group
JPC	Joint Project Committee
JTC	Joint Technical Committee
JWG	joint working group
TC	technical committee
SC	Subcommittee
PC	project committee
WG	working group
PWI	preliminary work item
NP	new work item proposal
WD	working draft
CD	committee draft
DIS	draft International Standard (ISO)
CDV	committee draft for vote (IEC)
FDIS	final draft International Standard
PAS	Publicly Available Specification
TS	Technical Specification
TR	Technical Report

1 Organizational structure and responsibilities for the technical work

1.1 Role of the technical management board

The technical management board of the respective organization is responsible for the overall management of the technical work and in particular for:

a) establishment of technical committees;

b) appointment of chairs of technical committees;

c) allocation or re-allocation of secretariats of technical committees and, in some cases, subcommittees;

d) approval of titles, scopes and programmes of work of technical committees;

e) ratification of the establishment and dissolution of subcommittees by technical committees;

f) allocation of priorities, if necessary, to particular items of technical work;

g) coordination of the technical work, including assignment of responsibility for the development of standards regarding subjects of interest to several technical committees, or needing coordinated development; to assist it in this task, the technical management board may establish advisory groups of experts in the relevant fields to advise it on matters of basic, sectoral and cross-sectoral coordination, coherent planning and the need for new work;

h) monitoring the progress of the technical work with the assistance of the office of the CEO, and taking appropriate action;

i) reviewing the need for, and planning of, work in new fields of technology;

j) maintenance of the ISO/IEC Directives and other rules for the technical work;

k) consideration of matters of principle raised by National Bodies, and of appeals concerning decisions on new work item proposals, on committee drafts, on enquiry drafts or on final draft International Standards.

NOTE 1 Explanations of the terms "new work item proposal", "committee draft", "enquiry draft" and "final draft International Standard" are given in Clause 2.

NOTE 2 For detailed information about the role and responsibilities of the ISO technical management board, see the Terms of reference of the TMB — https://www.iso.org/committee/4882545.html and for the IEC see http://www.iec.ch/dyn/www/f?p=103:47:0::::FSP_ORG_ID,FSP_LANG_ID:3228,25.

1.2 Advisory groups to the technical management board

1.2.1 A group having advisory functions in the sense of 1.1 g) may be established

a) by one of the technical management boards;

b) jointly by the two technical management boards.

NOTE In IEC certain such groups are designated as Advisory Committees.

1.2.2 A proposal to establish such a group shall include recommendations regarding its terms of reference and constitution, bearing in mind the requirement for sufficient representation of affected interests while at the same time limiting its size as far as possible in order to ensure its efficient operation. For example, it may be decided that its members be only the chairs and secretaries of the technical committees concerned. In every case, the TMB(s) shall decide the criteria to be applied and shall appoint the members.

Any changes proposed by the group to its terms of reference, composition or, where appropriate, working methods shall be submitted to the technical management boards for approval.

1.2.3   The tasks allocated to such a group may include the making of proposals relating to the drafting or harmonization of publications (in particular International Standards, Technical Specifications, Publicly Available Specifications and Technical Reports), but shall not include the preparation of such documents unless specifically authorized by the TMB(s).

1.2.4   Any document being prepared with a view to publication shall be developed in accordance with the procedural principles given in Annex A.

1.2.5   The results of such a group shall be presented in the form of recommendations to the TMB(s). The recommendations may include proposals for the establishment of a working group (see 1.12) or a joint working group (see 1.12.6) for the preparation of publications. Such working groups shall operate within the relevant technical committee, if any.

1.2.6   The internal documents of a group having advisory functions shall be distributed to its members only, with a copy to the office(s) of the CEO(s).

1.2.7   Such a group shall be disbanded once its specified tasks have been completed, or if it is subsequently decided that its work can be accomplished by normal liaison mechanisms (see 1.16).

1.3   Joint technical work

1.3.1   Joint Technical Advisory Board (JTAB)

The JTAB has the task of avoiding or eliminating possible or actual overlapping in the technical work of ISO and IEC and acts when one of the two organizations feels a need for joint planning. The JTAB deals only with those cases that it has not been possible to resolve at lower levels by existing procedures. (See Annex B.) Such cases may cover questions of planning and procedures as well as technical work.

Decisions of the JTAB are communicated to both organizations for immediate implementation. They shall not be subject to appeal for at least 3 years.

1.3.2   Joint Technical Committees (JTC) and Joint Project Committees (JPC)

1.3.2.1   JTC and JPC may be established by a common decision of the ISO/TMB and IEC/SMB, or by a decision of the JTAB.

1.3.2.2   For JPC, one organization has the administrative responsibility. This shall be decided by mutual agreement between the two organizations.

Participation is based on the one member/country, one vote principle.

Where two National Bodies in the same country elect to participate in a JPC then one shall be identified as having the administrative responsibility. The National Body with the administrative responsibility has the responsibility of coordinating activities in their country, including the circulation of documents, commenting and voting.

Otherwise the normal procedures for project committees are followed (see 1.10).

1.4   Role of the Chief Executive Officer

The Chief Executive Officer of the respective organization is responsible, *inter alia*, for implementing the ISO/IEC Directives and other rules for the technical work. For this purpose, the office of the

CEO arranges all contacts between the technical committees, the council board and the technical management board.

Deviations from the procedures set out in the present document shall not be made without the authorization of the Chief Executive Officers of ISO or IEC, or the ISO/IEC Joint Technical Advisory Board (JTAB), or the technical management boards for deviations in the respective organizations.

1.5 Establishment of technical committees

1.5.1 Technical committees are established and dissolved by the technical management board.

1.5.2 The technical management board may transform an existing subcommittee into a new technical committee, following consultation with the technical committee concerned.

1.5.3 A proposal for work in a new field of technical activity which appears to require the establishment of a new technical committee may be made in the respective organization by

— a National Body;

— a technical committee or subcommittee;

— a project committee;

— a policy level committee;

— the technical management board;

— the Chief Executive Officer;

— a body responsible for managing a certification system operating under the auspices of the organization;

— another international organization with National Body membership.

1.5.4 The proposal shall be made using the appropriate form (see Annex SJ in the Consolidated ISO Supplement to the ISO/IEC Directives and http://www.iec.ch/standardsdev/resources/docpreparation/forms _templates/), which covers

a) the proposer;

b) the subject proposed;

c) the scope of the work envisaged and the proposed initial programme of work;

d) a justification for the proposal;

e) if applicable, a survey of similar work undertaken in other bodies;

f) any liaisons deemed necessary with other bodies.

For additional informational details to be included in the proposals for new work, see Annex C. The form shall be submitted to the office of the CEO.

1.5.5 The office of the CEO shall ensure that the proposal is properly developed in accordance with ISO and IEC requirements (see Annex C) and provides sufficient information to support informed decision making by National Bodies. The office of the CEO shall also assess the relationship of the proposal to existing work, and may consult interested parties, including the technical management board or committees conducting related existing work. If necessary, an ad hoc group may be established to examine the proposal.

Following its review, the office of the CEO may decide to return the proposal to the proposer for further development before circulation for voting. In this case, the proposer shall make the changes suggested or provide justification for not making the changes. If the proposer does not make the changes and requests that its proposal be circulated for voting as originally presented, the technical management board will decide on appropriate action. This could include blocking the proposal until the changes are made or accepting that it be balloted as received.

In all cases, the office of the CEO may also include comments and recommendations to the proposal form.

For details relating to justification of the proposal, see Annex C.

Proposers are strongly encouraged to conduct informal consultations with other National Bodies in the preparation of proposals.

In some instances, the ISO Technical Management Board may consider it appropriate to carry out an informal exploratory enquiry.

Technical committees are established by the ISO/TMB on a provisional basis. Following the initial meeting of the technical committee, but no later than 18 months, provisionally established technical committees are required to prepare a strategic business plan for review by the ISO/TMB (see Annex SC). The committees are formally established by the ISO/TMB at the time of acceptance of the business plan. This does not preclude the initiation of standardization projects during this 18 month period.

1.5.6 The proposal shall be circulated by the office of the CEO to all National Bodies of the respective organization (ISO or IEC), asking whether or not they

a) support the establishment of a new technical committee providing a statement justifying their decision ("justification statement"), and

b) intend to participate actively (see 1.7.1) in the work of the new technical committee.

The proposal shall also be submitted to the other organization (IEC or ISO) for comment and for agreement (see Annex B).

The replies to the proposal shall be made using the appropriate form within 12 weeks after circulation. Regarding 1.5.6 a) above, if no such statement is provided, the positive or negative vote of a National Body will not be registered and considered.

The form for replies to the proposals has been replaced by an electronic balloting system. Replies not using the electronic balloting system will not be counted.

1.5.7 The technical management board evaluates the replies and either

— decides the establishment of a new technical committee, provided that

   1) a 2/3 majority of the National Bodies voting are in favour of the proposal, and

   2) at least 5 National Bodies who voted in favour expressed their intention to participate actively,

and allocates the secretariat (see 1.9.1), or

— assigns the work to an existing technical committee, subject to the same criteria of acceptance.

1.5.8 Technical committees shall be numbered in sequence in the order in which they are established. If a technical committee is dissolved, its number shall not be allocated to another technical committee.

1.5.9 As soon as possible after the decision to establish a new technical committee, the necessary liaisons shall be arranged (see 1.15 to 1.17).

1.5.10 A new technical committee shall agree on its title and scope as soon as possible after its establishment, preferably by correspondence.

Agreement of the title and scope of the technical committee requires approval by a 2/3 majority of the P-members voting.

The scope is a statement precisely defining the limits of the work of a technical committee.

The definition of the scope of a technical committee shall begin with the words "Standardization of …" or "Standardization in the field of …" and shall be drafted as concisely as possible.

For recommendations on scopes, see Annex J.

The agreed title and scope shall be submitted by the Chief Executive Officer to the technical management board for approval.

1.5.11 The technical management board or a technical committee may propose a modification of the latter's title and/or scope. The modified wording shall be established by the technical committee for approval by the technical management board.

1.5.12 "Stand-by" – a technical committee or subcommittee is said to be in a "stand-by" status when it has no tasks on its work programme but retains its title, scope and secretariat so that it can be reactivated should a new task be assigned to it.

The decision to put a committee on stand-by or to reactivate it is taken by the technical management board on a proposal from the committee in question.

1.6 Establishment of subcommittees

1.6.1 Subcommittees are established and dissolved by a 2/3 majority decision of the P-members of the parent committee voting, subject to ratification by the technical management board. A subcommittee may be established only on condition that a National Body has expressed its readiness to undertake the secretariat.

1.6.2 At the time of its establishment, a subcommittee shall comprise at least 5 members of the parent technical committee having expressed their intention to participate actively (see 1.7.1) in the work of the subcommittee.

1.6.3 Subcommittees of a technical committee shall be designated in sequence in the order in which they are established. If a subcommittee is dissolved, its designation shall not be allocated to another subcommittee, unless the dissolution is part of a complete restructuring of the technical committee.

1.6.4 The title and scope of a subcommittee shall be defined by the parent technical committee and shall be within the defined scope of the parent technical committee.

1.6.5 The secretariat of the parent technical committee shall inform the office of the CEO of the decision to establish a subcommittee, using the appropriate form. The office of the CEO shall submit the form to the technical management board for ratification of the decision.

1.6.6 As soon as possible after ratification of the decision to establish a new subcommittee, any liaisons deemed necessary with other bodies shall be arranged (see 1.15 to 1.17).

1.7 Participation in the work of technical committees and subcommittees

It is recognized that member bodies in developing countries often lack the resources to participate in all committees which may be carrying out work which is important for their national economy. Developing country member bodies are therefore invited to establish P-member twinning arrangements with more experienced P-members. Under such arrangements, the lead P-member will ensure that the views of the twinned P-member are communicated to and taken into consideration by the responsible ISO committee. The twinned P-member shall consequently also have the status of P-member (see note) and be registered as a twinned P-member by the Central Secretariat.

NOTE   It is left to the member bodies concerned to determine the most effective way of implementing twinning. This may include for example the P-member sponsoring an expert from the twinned member body to participate in committee meetings or to act as an expert in a working group, or it may involve the P-member seeking the views of the twinned member body on particular agenda items/documents and ensuring that the twinned member body provides its positions in writing to the committee secretariat.

The details of all twinning arrangements shall be notified to the secretariat and chair of the committee concerned, with the committee members and the office of the CEO being informed accordingly to ensure the greatest possible transparency.

A lead P-member shall twin with only one other P-member in any particular committee.

The twinned P-member shall cast its own vote on all issues referred to the committee for vote by correspondence.

For more information on twinnings, see Annex ST for the Twinning Policy.

Consistent with the ISO Statutes and Rules of Procedure, correspondent and subscriber members are not eligible for P-memberships. Correspondent members of ISO may register as observers of committees but do not have the right to submit comments.

1.7.1   All National Bodies have the right to participate in the work of technical committees and subcommittees.

In order to achieve maximum efficiency and the necessary discipline in the work, each National Body shall clearly indicate to the office of the CEO, with regard to each technical committee or subcommittee, if it intends

— to participate actively in the work, with an obligation to vote on all questions formally submitted for voting within the technical committee or subcommittee, on new work item proposals, enquiry drafts and final draft International Standards, and to contribute to meetings (**P-members**), or

— to follow the work as an observer, and therefore to receive committee documents and to have the right to submit comments and to attend meetings (**O-members**).

In ISO, National Bodies that choose to be P-members of a committee have the additional obligation to vote on all systematic review ballots under the responsibility of that committee.

A National Body may choose to be neither P-member nor O-member of a given committee, in which case it will have neither the rights nor the obligations indicated above with regard to the work of that committee. Nevertheless, all National Bodies irrespective of their status within a technical committee or subcommittee have the right to vote on enquiry drafts (see 2.6) and on final draft International Standards (see 2.7).

National Bodies have the responsibility to organize their national input in an efficient and timely manner, taking account of all relevant interests at their national level.

1.7.2   Membership of a subcommittee is open to any National Body, regardless of their membership status in the parent technical committee.

Members of a technical committee shall be given the opportunity to notify their intention to become a P- or O-member of a subcommittee at the time of its establishment.

Membership of a technical committee does not imply automatic membership of a subcommittee; National Bodies shall notify their intended status in each subcommittee.

1.7.3 A National Body may, at any time, begin or end membership or change its membership status in any technical committee or subcommittee in IEC by informing the office of the CEO and the secretariat of the committee concerned, and in ISO by direct input via the Global Directory, subject to the requirements of clauses 1.7.4 and 1.7.5.

1.7.4 A technical committee or subcommittee secretariat shall notify the Chief Executive Officer if a P-member of that technical committee or subcommittee

— has been persistently inactive and has failed to contribute to 2 successive technical committee/subcommittee meetings, either by direct participation or by correspondence and has failed to appoint any experts to the technical work, or

— In IEC:

has failed to vote on questions formally submitted for voting within the technical committee or subcommittee (see 1.7.1).

— In ISO:

has failed to vote on over 20 % (and at least 2) of the questions formally submitted for voting on the committee internal balloting (CIB) within the technical committee or subcommittee over one calendar year (see 1.7.1).

Upon receipt of such a notification, the Chief Executive Officer shall remind the National Body of its obligation to take an active part in the work of the technical committee or subcommittee. In the absence of a satisfactory response to this reminder, and upon persistent continuation of the above articulated shortcomings in required P-member behaviour, the National Body shall without exception automatically have its status changed to that of O-member. A National Body having its status so changed may, after a period of 12 months, indicate to the Chief Executive Officer that it wishes to regain P-membership of the committee, in which case this shall be granted.

NOTE   This clause does not apply to the development of Guides.

1.7.5 If a P-member of a technical committee or subcommittee fails to vote on an enquiry draft or final draft International Standard prepared by the respective committee, or on a systematic review ballot for a deliverable under the responsibility of the committee, the Chief Executive Officer shall remind the National Body of its obligation to vote. In the absence of a satisfactory response to this reminder, the National Body shall automatically have its status changed to that of O-member. A National Body having its status so changed may, after a period of 12 months, indicate to the Chief Executive Officer that it wishes to regain P-membership of the committee, in which case this shall be granted.

NOTE   This clause does not apply to the development of Guides.

1.8 Chairs of technical committees and subcommittees

1.8.1 Appointment

Chairs of technical committees shall be nominated by the secretariat of the technical committee and approved by the technical management board, for a maximum period of 6 years, or for such shorter period as may be appropriate. Extensions are allowed, up to a cumulative maximum of 9 years.

Chairs of subcommittees shall be nominated by the secretariat of the subcommittee and approved by the technical committee for a maximum period of 6 years, or for such shorter period as may be appropriate. Extensions are allowed, up to a cumulative maximum of 9 years. Approval criterion for both appointment and extension is a 2/3 majority vote of the P-members of the technical committee.

Secretariats of technical committees or subcommittees may submit nominations for new chairs up to one year before the end of the term of existing chairs. Chairs appointed one year before shall be designated as the "chair elect" of the committee in question. This is intended to provide the chair elect an opportunity to learn before taking over as chair of a committee.

Where two NSBs agree to share the leadership of an ISO committee, they may jointly identify a chair and jointly agree the way in which they will support that chair. In such cases, the NSB holding the committee secretariat is encouraged to nominate a chair from the other NSB.

For the purposes of transparency, the NSBs involved shall inform the committee members of any cooperative relationships.

The candidates for chairmanships shall have the competencies and attributes listed in Annex SQ *Selection criteria for people leading the technical work* (see SQ.3.1).

1.8.2 Responsibilities

The chair of a technical committee is responsible for the overall management of that technical committee, including any subcommittees and working groups.

The chair of a technical committee or subcommittee shall

a) act in a purely international capacity, divesting him- or herself of a national position; thus s/he cannot serve concurrently as the delegate of a National Body in his or her own committee;

b) guide the secretary of that technical committee or subcommittee in carrying out his or her duty;

c) conduct meetings with a view to reaching agreement on committee drafts (see 2.5);

d) ensure at meetings that all points of view expressed are adequately summed up so that they are understood by all present;

e) ensure at meetings that all decisions are clearly formulated and made available in written form by the secretary for confirmation during the meeting;

f) take appropriate decisions at the enquiry stage (see 2.6);

g) advise the technical management board on important matters relating to that technical committee via the technical committee secretariat. For this purpose s/he shall receive reports from the chairs of any subcommittees via the subcommittee secretariats;

h) ensure that the policy and strategic decisions of the technical management board are implemented in the committee;

i) ensure the establishment and ongoing maintenance of a strategic business plan covering the activities of the technical committee and all groups reporting to the technical committee, including all subcommittees;

j) ensure the appropriate and consistent implementation and application of the committee's strategic business plan to the activities of the technical committee's or subcommittee's work programme;

k) assist in the case of an appeal against a committee decision.

In case of unforeseen unavailability of the chair at a meeting, a session chair may be elected by the participants.

SC chairs shall attend meetings of the parent committee as required and may participate in the discussion, but do not have the right to vote. In exceptional circumstances, if a chair is prevented from attending, he or she shall delegate the secretary (or in ISO and IEC, another representative) to represent the subcommittee. In the case where no representative from the SC can attend, a written report shall be provided.

1.8.3 Twinned-chairs

Committees are encouraged to establish chair level twinning arrangements where one ISO member body acts as the lead partner and the other ISO member body as the twinned partner (with a limit of one Twinned-chair per committee). Twinned-chairs shall be a P-member (i.e. any P-member whether through a twinning arrangement or not) in the committee concerned. The same rules apply for the appointment and term of chairs and Twinned-chairs. The delineation of responsibilities shall be decided by mutual agreement and recorded in a twinning agreement between the two member bodies nominating the chair and the Twinned-chair, with the committee members and the office of the CEO being informed accordingly.

For more information on twinnings, see Annex ST for the Twinning Policy.

1.9 Secretariats of technical committees and subcommittees

1.9.1 Allocation

The secretariat of a technical committee shall be allocated to a National Body by the technical management board.

The secretariat of a subcommittee shall be allocated to a National Body by the parent technical committee. However, if two or more National Bodies offer to undertake the secretariat of the same subcommittee, the technical management board shall decide on the allocation of the subcommittee secretariat.

Two NSBs may cooperate on the work of an ISO committee secretariat, to support one another on strategic thinking and to share the resource burden of the work. However, only one of these parties will be considered the official committee secretariat and such arrangements shall not be considered "co-secretariats".

For the purposes of transparency, the NSBs involved shall inform the committee members of any cooperative relationships.

For both technical committees and subcommittees, the secretariat shall be allocated to a National Body only if that National Body

a) has indicated its intention to participate actively in the work of that technical committee or subcommittee, and

b) has accepted that it will fulfil its responsibilities as secretariat and is in a position to ensure that adequate resources are available for secretariat work (see D.2).

Once the secretariat of a technical committee or subcommittee has been allocated to a National Body, the latter shall appoint a qualified individual as secretary (see D.1 and D.3).

Secretaries shall have the competencies and attributes listed in Annex SQ *Selection criteria for people leading the technical wor*k (see SQ.3.2).

TC/SC secretariats, at intervals of 5 years, shall normally be subject to reconfirmation by the ISO Technical Management Board.

If, during a year when a particular TC/SC secretariat is due for reconfirmation, TC/SC productivity data show the committee to be experiencing difficulties, this will be drawn to the attention of the ISO Technical Management Board for decision as to whether a reconfirmation enquiry should be conducted. The allocation of those secretariats not notified for detailed examination will be reconfirmed automatically.

A reconfirmation enquiry may also be initiated at any time at the request of the Secretary-General or of a P-member of a committee. Such requests shall be accompanied by a written justification and shall be considered by the ISO Technical Management Board which will decide whether to conduct a reconfirmation enquiry.

The enquiry shall be conducted amongst the P-members of the committee to determine whether the P-members are satisfied that sufficient resources are available to the secretariat and that the performance of the secretariat is satisfactory. Any P-member responding negatively shall be invited to indicate whether it is willing itself to accept the secretariat of the committee.

The enquiry shall be conducted by the ISO Technical Management Board in the case of TC secretariats, and by TC secretariats in the case of SC secretariats. However, in cases where the same member body holds both the TC and SC secretariat, the enquiry shall be conducted by the ISO Technical Management Board.

If there are no negative responses, the secretariat allocation shall be reconfirmed. All negative responses concerning both TC and SC secretariats shall be referred to the ISO Technical Management Board for decision.

1.9.2 Responsibilities

The National Body to which the secretariat has been allocated shall ensure the provision of technical and administrative services to its respective technical committee or subcommittee.

The secretariat is responsible for monitoring, reporting, and ensuring active progress of the work, and shall use its utmost endeavour to bring this work to an early and satisfactory conclusion. These tasks shall be carried out as far as possible by correspondence.

The secretariat is responsible for ensuring that the ISO/IEC Directives and the decisions of the technical management board are followed.

A secretariat shall act in a purely international capacity, divesting itself of a national point of view.

The secretariat is responsible for the following to be executed in a timely manner:

a) Working documents:

   1) Preparation of committee drafts, arranging for their distribution and the treatment of the comments received;

   2) Preparation of enquiry drafts and text for the circulation of the final draft International Standards or publication of International Standards;

   3) Ensuring the equivalence of the English and French texts, if necessary with the assistance of other National Bodies that are able and willing to take responsibility for the language versions concerned. (See also 1.11 and the respective Supplements to the ISO/IEC Directives);

b) Project management

   1) Assisting in the establishment of priorities and target dates for each project;

   2) Notifying the names, etc. of all working group and maintenance team convenors and project leaders to the office of the CEO;

   3) Proposing proactively the publication of alternative deliverables or cancellation of projects that are running significantly overtime, and/or which appear to lack sufficient support;

c) Meetings (see also Clause 4), including:

   1) Establishment of the agenda and arranging for its distribution;

   2) Arranging for the distribution of all documents on the agenda, including reports of working groups, and indicating all other documents which are necessary for discussion during the meeting (see E.5);

3) Regarding the decisions (also referred to as resolutions) taken in a meeting:

   — ensuring that the decisions endorsing working groups recommendations contain the specific elements being endorsed;

   — making the decisions available in writing for confirmation during the meeting (see E.5); and

   — posting the decisions within 48 hours after the meeting in the committee's electronic folder.

4) Preparation of the minutes of meetings to be circulated within 4 weeks after the meeting;

5) Preparation of reports to the technical management board (TC secretariat), in the IEC within 4 weeks after the meeting, or to the parent committee (SC secretariat);

d) Advising

   Providing advice to the chair, project leaders, and convenors on procedures associated with the progression of projects.

In all circumstances, each secretariat shall work in close liaison with the chair of its technical committee or subcommittee.

The secretariat of a technical committee shall maintain close contact with the office of the CEO and with the members of the technical committee regarding its activities, including those of its subcommittees and working groups.

The secretariat of a subcommittee shall maintain close contact with the secretariat of the parent technical committee and as necessary with the office of the CEO. It shall also maintain contact with the members of the subcommittee regarding its activities, including those of its working groups.

The secretariat of a technical committee or subcommittee shall update in conjunction with the office of the CEO the record of the status of the membership of the committee.

The office of the CEO shall also maintain a register of the membership of its working groups.

Member bodies are encouraged to establish twinning arrangements between a secretariat and a twinned secretariat with a limit of one twinned secretariat per committee. The lead partner will hold the secretariat and the twinned partner the twinned secretariat. Twinned secretariats shall be from member bodies holding P-membership in the committee concerned (either directly or via a twinning arrangement). The same rules apply for the allocation of secretariats and twinned secretariats, as well as secretaries and twinned secretaries. The delineation of responsibilities shall be decided by mutual agreement between the member bodies concerned, and recorded in the twinning agreement, with the committee members and the office of the CEO being informed accordingly.

For more information on twinnings, see Annex ST for the Twinning Policy.

1.9.3 Change of secretariat of a technical committee

If a National Body wishes to relinquish the secretariat of a technical committee, the National Body concerned shall immediately inform the Chief Executive Officer, giving a minimum of 12 months' notice. The technical management board decides on the transfer of the secretariat to another National Body.

If the secretariat of a technical committee persistently fails to fulfil its responsibilities as set out in these procedures, the Chief Executive Officer or a National Body may have the matter placed before the technical management board, which may review the allocation of the secretariat with a view to its possible transfer to another National Body.

Concerning the procedure to follow when the lead partner in a twinning arrangement for a technical committee secretariat wishes to relinquish the secretariat, see clause ST.5.7 Terminating a twinning arrangement of Annex ST on the Twinning Policy.

1.9.4 Change of secretariat of a subcommittee

If a National Body wishes to relinquish the secretariat of a subcommittee, the National Body concerned shall immediately inform the secretariat of the parent technical committee, giving a minimum of 12 months' notice.

If the secretariat of a subcommittee persistently fails to fulfil its responsibilities as set out in these procedures, the Chief Executive Officer or a National Body may have the matter placed before the parent technical committee, which may decide, by majority vote of the P-members, that the secretariat of the subcommittee should be re-allocated.

In either of the above cases an enquiry shall be made by the secretariat of the technical committee to obtain offers from other P-members of the subcommittee for undertaking the secretariat.

If two or more National Bodies offer to undertake the secretariat of the same subcommittee or if, because of the structure of the technical committee, the re-allocation of the secretariat is linked with the re-allocation of the technical committee secretariat, the technical management board decides on the re-allocation of the subcommittee secretariat. If only one offer is received, the parent technical committee itself proceeds with the appointment.

Concerning the procedure to follow when the lead partner in a twinning arrangement for a subcommittee secretariat wishes to relinquish the secretariat, see ST.5.7 Terminating a twinning arrangement of Annex ST on the Twinning Policy.

1.10 Project committees

Project committees are established by the technical management board to prepare individual standards not falling within the scope of an existing technical committee.

NOTE   Such standards carry one reference number but may be subdivided into parts.

Procedures for project committees are given in Annex K.

Project committees wishing to be transformed into a technical committee shall follow the process for the establishment of a new technical committee (see 1.5).

1.11 Editing committees

It is recommended that committees establish one or more editing committees for the purpose of updating and editing committee drafts, enquiry drafts and final draft International Standards and for ensuring their conformity to the ISO/IEC Directives, Part 2 (see also 2.6.6).

Such committees should comprise at least

— one technical expert of English mother tongue and having an adequate knowledge of French;

— one technical expert of French mother tongue and having an adequate knowledge of English;

— the project leader (see 2.1.8).

The project leader and/or secretary may take direct responsibility for one of the language versions concerned.

Editing committees shall meet when required by the respective technical committee or subcommittee secretariat for the purpose of updating and editing drafts which have been accepted by correspondence for further processing.

Editing committees shall be equipped with means of processing and providing texts electronically (see also 2.6.6).

1.12 Working groups

1.12.1 Technical committees or subcommittees may establish, by decision of the committee, working groups for specific tasks (see 2.4). A working group shall report to its parent technical committee or subcommittee through a convenor appointed by the parent committee.

Working group convenors shall be appointed by the committee for up to three-year terms ending at the next plenary session of the parent committee following the term. Such appointments shall be confirmed by the National Body (or liaison organization). The convenor may be reappointed for additional terms of up to three-years. There is no limit to the number of terms.

Responsibility for any changes of convenors rests with the committee and not with the National Body (or liaison organization).

The convenor may be supported by a secretariat, as needed.

The nomination of the working group secretary shall be confirmed by his/her National Body.

Committees are encouraged to establish arrangements between a convenor and a twinned convenor, with a limit of one twinned convenor per working group. The lead partner (convenor) and twinned partner (co-convenor) will be decided by mutual agreement. Convenors and twinned convenors shall be from the P-members in the committee concerned. The same rules apply for the appointment of convenors and twinned convenors.

Note that twinned convenors are only possible through twinning arrangements.

Co-convenors are only possible in Joint Working Groups (JWG) — see 1.12.6.

For more information on twinnings, see Annex ST for the Twinning Policy.

A working group comprises a restricted number of experts individually appointed by the P-members, A-liaisons of the parent committee and C-liaison organizations, brought together to deal with the specific task allocated to the working group. The experts act in a personal capacity and not as the official representative of the P-member or A-liaison organization (see 1.17) by which they have been appointed with the exception of those appointed by C-liaison organizations (see 1.17). However, it is recommended that they keep close contact with that P-member or organization in order to inform them about the progress of the work and of the various opinions in the working group at the earliest possible stage.

It is recommended that working groups be reasonably limited in size. The technical committee or subcommittee may therefore decide upon the maximum number of experts appointed by each P-member and liaison organizations.

Once the decision to set up a working group has been taken, P-members and A- and C-liaison organizations shall be officially informed in order to appoint expert(s).Working groups shall be numbered in sequence in the order in which they are established.

When a committee has decided to set up a working group, the convenor or acting convenor shall immediately be appointed and shall arrange for the first meeting of the working group to be held within 12 weeks. This information shall be communicated immediately after the committee's decision to the P-members of the committee and A- and C-liaison organizations, with an invitation to appoint experts within 6 weeks. Additional projects may be assigned, where appropriate, to existing working groups.

1.12.2 The composition of the working group is defined in the ISO Global Directory (GD) or in the IEC Expert Management System (EMS) as appropriate. Experts not registered to a working group in the ISO GD or the IEC EMS respectively, shall not participate in its work. Convenors may invite a specific guest to participate in a single meeting.

In case of missing stakeholder categories, the Convenor shall inform the committee secretariat, who shall launch another call for experts.

1.12.3 Persistently inactive experts, meaning absence of contributions through attendance to working group meetings or by correspondence shall be removed, by the office of the CEO at the request of the technical committee or sub-committee secretary, from working groups after consultation with the P-member.

In case of lack of experts in the working group to meet the minimum number (as defined in clause 2.3.5), the Convenor shall inform the committee secretariat, who shall launch another call for experts.

If the subsequent call for experts fails to recruit the minimum number of experts, the Convenor shall consult with the committee secretariat to decide if the project can or should continue. If the decision is to continue with fewer than the minimum number of experts, the committee secretariat shall request permission from the ISO Technical Management Board to proceed.

1.12.4 On completion of its task(s) — normally at the end of the enquiry stage (see 2.6) of its last project — the working group shall be disbanded by decision of the committee, the project leader remaining with consultant status until completion of the publication stage (see 2.8).

1.12.5 Distribution of the internal documents of a working group and of its reports shall be carried out in accordance with procedures described in the respective Supplements of the ISO/IEC Directives.

1.12.6 In special cases a joint working group (JWG) may be established to undertake a specific task in which more than one ISO and/or IEC technical committee or subcommittee is interested. Committees who receive requests to establish JWG shall reply to such requests in a timely manner.

NOTE   For specific rules concerning JWGs between ISO committees and IEC committees, see Annex B in addition to the following.

The decision to establish a joint working group shall be accompanied by mutual agreement between the committees on:

— the committee/organization having the administrative responsibility for the project;

— the convenor of the joint working group, who shall be nominated by a P-member from one of the committees, with the option to appoint a co-convenor from the other committee;

— the membership of the joint working group (membership may be open to all P-members and category A- and C-liaisons that wish to participate which may be limited to an equal number of representatives from each committee, if agreed).

The committee/organization with the administrative responsibility for the project shall:

— record the project in their programme of work;

— conduct the call for experts in all committees that are part of the JWG;

— be responsible for addressing comments (usually referred back to the JWG) and ensure that the comments and votes at all stages of the project are compiled and handled appropriately (see 2.5, 2.6 and 2.7) — all comments are made available to the leadership of the committees;

— prepare drafts for the committee, enquiry and approval stages according to procedures given in 2.5, 2.6 and 2.7;

— send all relevant documents (minutes, Working drafts, drafts for the committee, enquiry and approval stages) to the secretariat of the other committee(s) for circulation in their respective committee and/or action;

— be responsible for maintenance of the publication.

Approval criteria are based on the Directives used by the committee with the administrative lead. If the lead committee is a JTC 1 committee, the Consolidated JTC 1 Supplement also applies.

For proposal stage (NP)

— For ISO/ISO JWGs, only one NP ballot is needed — if a NP has already been launched or approved in one committee, it cannot be balloted again in another TC, but two (2) NP are launched for ISO/IEC JWG [one (1) in each organization].

— It is possible to establish a JWG at a later stage, in which case its administrative lead will be confirmed by the TCs concerned.

— Once the joint work is agreed, the committee with the administrative lead informs ISO/CS or IEC/CO respectively, of its lead and of the committees participating in the work.

— The other TCs launch a call for experts for participation in the JWG.

For preparatory stage (WD)

— The JWG functions like any other WG: consensus is required to advance to CD.

For committee stage (CD)

— The CD is circulated for review and comment by each committee.

— The final CD requires consensus by all committees, as defined in the ISO/IEC Directives, Part 1.

For DIS and FDIS ballots

— National Bodies are requested to consult all national mirror committees involved to define one position. A statement is included on the cover page to draw attention of NSBs.

— For an ISO/IEC JWG, two DIS/FDIS votes are launched, i.e. one in each organization.

The Foreword identifies all committees involved in the development of the deliverable.

1.13 Groups having advisory functions within a committee

1.13.1 A group having advisory functions may be established by a technical committee or subcommittee to assist the chair and secretariat in tasks concerning coordination, planning and steering of the committee's work or other specific tasks of an advisory nature.

1.13.2 A proposal to establish such a group shall include recommendations regarding its constitution and terms of reference, including criteria for membership, bearing in mind the requirement for sufficient representation of affected interests while at the same time limiting its size as far as possible in order to ensure its efficient operation. Members of advisory groups shall be committee officers, individuals nominated by National Bodies and/or, as relevant, by A-liaison organizations. The parent committee shall approve the final constitution and the terms of reference prior to the establishment of and nominations to the advisory group.

For chair's advisory groups, consideration shall be given to the provision of equitable participation.

In order to achieve greater involvement by National Bodies in developing countries in the governance of ISO committees, it is strongly recommended that special provisions be made to allocate places for representatives of developing countries in any advisory groups established by a committee. Those representatives shall be nominated by member bodies holding P-membership in the committee concerned (either directly or via a twinning arrangement).

1.13.3 The tasks allocated to such a group may include the making of proposals relating to the drafting or harmonization of publications (in particular International Standards, Technical Specifications, Publicly Available Specifications and Technical Reports), but shall not include the preparation of such documents.

1.13.4 The results of such a group shall be presented in the form of recommendations to the body that established the group. The recommendations may include proposals for the establishment of a working group (see 1.12) or a joint working group (see 1.12.6) for the preparation of publications.

1.13.5 The internal documents of a group having advisory functions shall be distributed to its members only, with a copy to the secretariat of the committee concerned and to the office of the CEO.

1.13.6 Such a group shall be disbanded once its specified tasks have been completed and agreed by the parent committee.

1.14 Ad hoc groups

Technical committees or subcommittees may establish ad hoc groups, the purpose of which is to study a precisely defined problem on which the group reports to its parent committee at the same meeting, or at the latest at the next meeting.

The membership of an ad hoc group shall be chosen from the delegates present at the meeting of the parent committee, supplemented, if necessary, by experts appointed by the committee. The parent committee shall also appoint a convenor.

An ad hoc group shall be automatically disbanded at the meeting to which it has presented its report.

1.15 Liaison between technical committees

See Annex SH concerning coordination of the standardization of graphical symbols.

1.15.1 Within each organization, technical committees and/or subcommittees working in related fields shall establish and maintain liaison. Liaisons shall also be established, where appropriate, with technical committees responsible for basic aspects of standardization (e.g. terminology, graphical symbols). Liaison shall include the exchange of basic documents, including new work item proposals and working drafts.

Committees may pass a resolution to decide on the establishment of an internal liaison. Committees receiving requests for internal liaisons cannot refuse such requests and there is no need for the committee receiving the request to pass a resolution confirming its acceptance.

1.15.2 The maintenance of such liaison is the responsibility of the respective technical committee secretariats, which may delegate the task to the secretariats of the subcommittees.

1.15.3 A technical committee or subcommittee may designate a Liaison Representative or Liaison Representatives, to follow the work of another technical committee with which a liaison has been established, or one or several of its subcommittees. The designation of such Liaison Representatives shall be notified to the secretariat of the committee concerned, which shall communicate all relevant documents to the Liaison Representative(s) and to the secretariat of that technical committee or subcommittee. The appointed Liaison Representative shall make progress reports to the secretariat by which s/he has been appointed.

1.15.4 Such Liaison Representatives shall have the right to participate in the meetings of the technical committee or subcommittee whose work they have been designated to follow but shall not have the right to vote. They may contribute to the discussion in meetings, including the submission of written comments, on matters within the competence of their own technical committee and based on feedback that they have collected from their own committee. They may also attend meetings of working groups of the technical committee or subcommittee, but only to contribute the viewpoint of their own technical committee on matters within its competence, and not to otherwise participate in working group activities.

1.16 Liaison between ISO and IEC

1.16.1 Arrangements for adequate liaison between ISO and IEC technical committees and subcommittees are essential. The channel of correspondence for the establishment of liaison between ISO and IEC technical committees and subcommittees is through the offices of the CEOs. As far as the study of new subjects by either organization is concerned, the CEOs seek agreement between the two organizations whenever a new or revised programme of work is contemplated in the one organization which may be of interest to the other, so that the work will go forward without overlap or duplication of effort. (See also Annex B.)

1.16.2 Liaison Representatives designated by ISO or IEC shall have the right to participate in the discussions of the other organization's technical committee or subcommittee whose work they have been designated to follow, and may submit written comments; they shall not have the right to vote.

1.17 Liaison with other organizations

1.17.1 General requirements applicable to all categories of liaisons

In order to be effective, liaison shall operate in both directions, with suitable reciprocal arrangements.

The desirability of liaison shall be taken into account at an early stage of the work.

The liaison organization shall accept the policy based on the ISO/IEC Directives concerning copyright (see 2.13), whether owned by the liaison organization or by other parties. The statement on copyright policy will be provided to the liaison organization with an invitation to make an explicit statement as to its acceptability. The liaison organization is not entitled to charge a fee for documents submitted.

A liaison organization shall be willing to make a contribution to the technical work of ISO or IEC as appropriate. A liaison organization shall have a sufficient degree of representativity within its defined area of competence within a sector or subsector of the relevant technical or industrial field.

A liaison organization shall agree to ISO/IEC procedures, including IPR (see 2.13).

Liaison organizations shall accept the requirements of 2.14 on patent rights.

Technical committees and subcommittees shall review all their liaison arrangements on a regular basis, at least every 2 years, or at every committee meeting.

1.17.2 Different categories of liaisons

1.17.2.1 At the technical committee/subcommittee level (Category A and B liaisons)

The categories of liaisons at the technical committee/subcommittee levels are:

— **Category A**: Organizations that make an effective contribution to the work of the technical committee or subcommittee for questions dealt with by this technical committee or subcommittee. Such organizations are given access to all relevant documentation and are invited to meetings. They may nominate experts to participate in a WG (see 1.12.1).

— **Category B**: Organizations that have indicated a wish to be kept informed of the work of the technical committee or subcommittee. Such organizations are given access to reports on the work of a technical committee or subcommittee.

NOTE   Category B is reserved for inter-governmental organizations.

The procedure for the establishment of Category A and B liaisons is:

— The organization wishing to create a Category A or B liaison shall send an application to the office of the CEO with copies to the technical committee or subcommittee officers and IEC CO Technical Officer or ISO CS Technical Programme Manager giving the following information:

   — Organization is not-for-profit;

   — Organization is open to members worldwide or over a broad region;

   — Its activities and membership demonstrate that it has the competence and expertise to contribute to the development of International Standards or the authority to promote their implementation in the area of the technical committee or subcommittee concerned (Only relevant for category A liaisons);

   — The name of the main contact person.

NOTE   Invariably the organization will have been in contact with the technical committee or subcommittee officers prior to submitting its application and in these cases the technical committee or subcommittee officers should ensure that the organization is aware of their obligations as given in clauses 1.17.1 i.e. copyright, agreeing to ISO/IEC procedures including IPR, and patent rights.

— The Office of the CEO will confirm that the eligibility criteria have been fulfilled and then consult with the National Body where the organization making the application has its headquarters;

— In case of objection from the National Body where the organization making the application has its headquarters, the matter will be referred to the technical management board for decision;

— If there is no objection from the National Body where the organization making the application has its headquarters, the application will be sent to the technical committee or subcommittee secretary with a request to circulate it for vote;

— Approval criterion for category A or B liaisons is a 2/3rds majority of P-members voting to approve.

1.17.2.2 At the working group level (Category C liaisons)

The category of liaisons at the working group level is:

— **Category C**: Organizations that make a technical contribution to and participate actively in the work of a working group. This can include manufacturer associations, commercial associations, industrial consortia, user groups and professional and scientific societies. Liaison organizations shall be multinational (in their objectives and standards development activities) with individual, company or country membership and may be permanent or transient in nature.

1.17.3 Eligibility

1.17.3.1 At the technical committee/subcommittee level (Category A and B liaisons)

When an organization applies for a liaison with a technical committee/subcommittee, the office of the CEO will check with the member body in the country in which the organization is located. If the member body does not agree that the eligibility criteria have been met, the matter will be referred to the TMB to define the eligibility.

The office of the CEO will also ensure that the organization meets the following eligibility criteria:

— it is not-for-profit;

— is a legal entity — the office of the CEO will request a copy of its statutes;

— it is membership-based and open to members worldwide or over a broad region;

- through its activities and membership demonstrates that it has the competence and expertise to contribute to the development of International Standards or the authority to promote their implementation; and

- has a process for stakeholder engagement and consensus decision-making to develop the input it provides (in ISO, see Guidance for ISO liaison organizations — Engaging stakeholders and building consensus http://www.iso.org/iso/guidance_liaison-organizations.pdf).

1.17.3.2 At the working group level (Category C liaisons)

When an organization applies for a liaison with a working group, the office of the CEO will check with the member body in the country in which the organization is located and will ensure that the organization meets the following eligibility criteria:

- it is not-for-profit;

- through its activities and membership demonstrates that it has the competence and expertise to contribute to the development of International Standards or the authority to promote their implementation; and

- has a process for stakeholder engagement and consensus decision-making to develop the input it provides (in ISO, see Guidance for ISO liaison organizations — Engaging stakeholders and building consensus http://www.iso.org/iso/guidance_liaison-organizations.pdf).

1.17.4 Acceptance (Category A, B and C liaisons)

Agreement to establish category A, B and C liaisons requires approval of the application by two-thirds of the P-members voting.

Committees are urged to seek out the participation of all parties at the beginning of the development of a work item. Where a request for category C liaison is submitted late in the development stage of a particular work item, the P-members will consider the value that can be added by the organization in question despite its late involvement in the working group.

1.17.5 Rights and obligations

1.17.5.1 At the technical committee/subcommittee level (Category A and B liaisons)

Technical committees and subcommittees shall seek the full and, if possible, formal backing of the organizations having liaison status for each document in which the latter is interested.

Any comments from liaison organizations should be given the same treatment as comments from member bodies. It should not be assumed that refusal by a liaison organization to provide its full backing is a sustained opposition. Where such objections are considered sustained oppositions, committees are invited to refer to clause 2.5.6 for further guidance.

1.17.5.2 At the working group level (Category C liaisons)

Category C liaison organizations have the right to participate as full members in a working group, maintenance team or project team (see 1.12.1) but not as project leaders or convenors.

Category C liaison experts act as the official representative of the organization by which they are appointed. They may only attend committee plenary meetings if expressly invited by the committee to attend. If they are invited by the committee to attend, they may only attend as observers.

1.17.6 Carrying over liaisons when a project committee is converted into a technical committee or a subcommittee

When a project committee is converted to a technical committee or a subcommittee, the new technical committee or subcommittee shall pass a resolution confirming which category A and B liaisons are carried over. Approval of the resolution requires a 2/3 majority of P-members voting.

Table 1 — Liaison categories

Category	A	B	C
Purpose	To make an effective contribution to the work of the committee.	To be kept informed of the work of the committee.	To make a technical contribution to drafting standards in a Working Group.
Eligibility	— Not for profit — Legal entity — Membership based (worldwide or over a broad region) — Relevant competence and expertise — Process for stakeholder engagement and consensus decision-making (See clause 1.17.3.1 for full details)	Intergovernmental Organizations only — Not for profit — Legal entity — Membership based (worldwide or over a broad region) — Relevant competence and expertise — Process for stakeholder engagement and consensus decision-making (See clause 1.17.3.1 for full details)	— Not for profit — Relevant competence and expertise — Process for stakeholder engagement and consensus decision-making (See clause 1.17.3.2 for full details)
Level	TC/SC	TC/SC	Working Group
Participation	Participate in TC/SC meetings, access to documents, may appoint experts to WGs and these experts may serve as convenors or Project Leaders.	To be kept informed of the work only (access to documents).	Full participation as a member of the WG (but cannot be convenor or Project Leader).
Rights and obligations	No voting rights, but can comment (comments are given the same treatment as comments from member bodies). Can propose new work items (see clause 2.3.2).	No voting rights, but can comment (comments are given the same treatment as comments from member bodies). Cannot propose new work items.	Experts can attend committee meetings if expressly invited by the committee, but only as observers. Cannot propose new work items.

2 Development of International Standards

2.1 The project approach

2.1.1 General

The primary duty of a technical committee or subcommittee is the development and maintenance of International Standards. However, committees are also strongly encouraged to consider publication of intermediate deliverables as described in Clause 3.

International Standards shall be developed on the basis of a project approach as described below.

2.1.2 Strategic business plan

Annex SC describes the objectives of strategic business plans, and the procedure that applies to their development and approval.

Each technical committee shall prepare a strategic business plan for its own specific field of activity,

a) taking into account the business environment in which it is developing its work programme;

b) indicating those areas of the work programme which are expanding, those which have been completed, and those nearing completion or in steady progress, and those which have not progressed and should be cancelled (see also 2.1.9);

c) evaluating revision work needed (see also the respective Supplements to the ISO/IEC Directives);

d) giving a prospective view on emerging needs.

The strategic business plan shall be formally agreed upon by the technical committee and be included in its report for review and approval by the technical management board on a regular basis.

NOTE   In ISO, the TMB approves the first strategic business plan only and committees are responsible for ensuring regular updates and revisions.

2.1.3 Project stages

2.1.3.1 Table 2 shows the sequence of project stages through which the technical work is developed, and gives the name of the document associated with each project stage. The development of Technical Specifications, Technical Reports and Publicly Available Specifications is described in Clause 3.

Table 2 — Project stages and associated documents

Project stage	Associated document	
	Name	Abbreviation
Preliminary stage	Preliminary work item [a]	PWI
Proposal stage	New work item proposal [a]	NP
Preparatory stage	Working draft(s) [a]	WD
Committee stage	Committee draft(s) [a]	CD
Enquiry stage	Enquiry draft [b]	ISO/DIS IEC/CDV
Approval stage	final draft International Standard [c]	FDIS
Publication stage	International Standard	ISO, IEC or ISO/IEC
[a] These stages may be omitted, as described in Annex F.		
[b] Draft International Standard in ISO, committee draft for vote in IEC.		
[c] May be omitted (see 2.6.4).		

2.1.3.2 F.1 illustrates the steps leading to publication of an International Standard.

2.1.3.3 The ISO and IEC Supplements to the ISO/IEC Directives give a matrix presentation of the project stages, with a numerical designation of associated sub-stages.

To facilitate the monitoring of project development, ISO has adopted a systematic approach to project management, based on subdivision of projects into stages and substages.

The project management system is associated with a detailed project tracking system that is a subset of the Harmonized Stage Code system *ISO Guide 69:1999 Harmonized Stage Code system (Edition 2) — Principles and guidelines for use*. Annex SD gives a matrix presentation of this project tracking system, with the numerical designation of associated sub-stages. A project is registered in the ISO Central Secretariat database as having reached each particular step when the action or decision indicated at that point has been taken and ISO Central Secretariat has been duly informed.

2.1.4 Project description and acceptance

A project is any work intended to lead to the issue of a new, amended or revised International Standard. A project may subsequently be subdivided (see also 2.1.5.4).

A project shall be undertaken only if a proposal has been accepted in accordance with the relevant procedures (see 2.3 for proposals for new work items, and the respective Supplements to the ISO/IEC Directives for review and maintenance of existing International Standards).

2.1.5 Programme of work

2.1.5.1 The programme of work of a technical committee or subcommittee comprises all projects allocated to that technical committee or subcommittee, including maintenance of published standards.

2.1.5.2 In establishing its programme of work, each technical committee or subcommittee shall consider sectoral planning requirements as well as requests for International Standards initiated by sources outside the technical committee, i.e. other technical committees, advisory groups of the technical management board, policy level committees and organizations outside ISO and IEC. (See also 2.1.2.)

2.1.5.3 Projects shall be within the agreed scope of the technical committee. Their selection shall be subject to close scrutiny in accordance with the policy objectives and resources of ISO and IEC. (See also Annex C.)

2.1.5.4 Each project in the programme of work shall be given a number (see IEC Supplements to the ISO/IEC Directives for document numbering at the IEC) and shall be retained in the programme of work under that number until the work on that project is completed or its cancellation has been agreed upon. The technical committee or subcommittee may subdivide a number if it is subsequently found necessary to subdivide the project itself. The subdivisions of the work shall lie fully within the scope of the original project; otherwise, a new work item proposal shall be made.

2.1.5.5 The programme of work shall indicate, if appropriate, the subcommittee and/or working group to which each project is allocated.

2.1.5.6 The agreed programme of work of a new technical committee shall be submitted to the technical management board for approval.

2.1.6 Target dates

The technical committee or subcommittee shall establish, for each project on its programme of work, target dates for the completion of each of the following steps:

— completion of the first working draft (in the event that only an outline of a working document has been provided by the proposer of the new work item proposal — see 2.3);

— circulation of the first committee draft;

— circulation of the enquiry draft;

— circulation of the final draft International Standard (in agreement with the office of the CEO);

— publication of the International Standard (in agreement with the office of the CEO).

NOTE   Committees may decide to skip the committee draft (CD) stage in accordance with Annex SS. The final draft International Standard (FDIS) shall be skipped if no technical changes are to be included in accordance with 2.6.4.

These target dates shall correspond to the shortest possible development times to produce International Standards rapidly and shall be reported to the office of the CEO, which distributes the information to all National Bodies. For establishment of target dates, see the respective Supplements to the ISO/IEC Directives.

In establishing target dates, the relationships between projects shall be taken into account. Priority shall be given to those projects intended to lead to International Standards upon which other International Standards will depend for their implementation. The highest priority shall be given to those projects having a significant effect on international trade and recognized as such by the technical management board.

The technical management board may also instruct the secretariat of the technical committee or subcommittee concerned to submit the latest available draft to the office of the CEO for publication as a Technical Specification (see 3.1).

All target dates shall be kept under continuous review and amended as necessary, and shall be clearly indicated in the programme of work. Revised target dates shall be notified to the technical management board. The technical management board will cancel all work items which have been on the work programme for more than 5 years and have not reached the approval stage (see 2.7).

2.1.6.1   General

When a proposed new project is approved (whether for a new deliverable or for the revision of an existing deliverable), when submitting the results to the ISO Central Secretariat the committee secretariat shall also indicate the selected standards development track, as follows (all target dates are calculated from the date of adoption as an approved project, AWI (approved work item), stage 10.99):

NOTE   The deadlines for the various stages within the development tracks shall be established on a case-by-case basis.

SDT 18 standards development track — 18 months to publication

SDT 24 standards development track — 24 months to publication

SDT 36 standards development track — 36 months to publication

SDT 48 standards development track — 48 months to publication

NOTE   Projects using the 18-month development track shall be eligible for the 'Direct publication process' offered by ISO/CS if they successfully complete the DIS ballot within 13 months of the project's registration. This process reduces publication processing time by approximately one third.

Committee secretariats are reminded to perform risk assessments during project planning in order to identify potential problems in advance and set the target dates accordingly. The target dates shall be kept under continuous review by committee secretariats which shall ensure that they are reviewed and either confirmed or revised at each committee meeting. Such reviews shall also seek to confirm that projects are still market relevant and in cases in which they are found to be no longer required, or if the likely completion date is going to be too late, thus causing market players to adopt an alternative solution, the projects shall be cancelled or transformed into another deliverable (see 2.1.6.2).

NOTE   Time spent on round-robin testing during the development of a standard shall not be counted in the overall development time. The standards development track is paused on request from the secretariat to ISO/CS during round-robin testing.

2.1.6.2 Automatic cancellation of projects (and their reinstatement)

If the target date for DIS (stage 40.00) or publication (stage 60.60) is exceeded, the committee shall decide within 6 months on one of the following actions:

a) projects at the preparatory or committee stages: submission of a DIS — if the technical content is acceptable and mature; projects at the enquiry stage: submission of a second DIS or FDIS — if the technical content is acceptable and mature;

b) publication of a TS — if the technical content is acceptable but unlikely sufficiently mature for a future International Standard;

c) publication of a PAS — if the technical content is acceptable but unlikely sufficiently mature for a future International Standard or a TS;

d) publication of a TR — if the technical content is not considered to be acceptable for publication as a TS or for a future International Standard, is informative in nature and is considered to be of interest to the public;

e) submission of a request for extension to the ISO/TMB — if no consensus can be reached but there is strong interest from stakeholders to continue — a committee may be granted one extension of up to 9 months for the total project duration but the publication of intermediary deliverables (such as PAS and TS) is recommended;

f) cancellation of the work item — if the committee is unable to find a solution.

If, at the end of the six month period, none of the above actions has been taken, the project shall be automatically cancelled by the ISO Central Secretariat. Such projects may only be reinstated with the approval of the ISO Technical Management Board.

2.1.7 Project management

The secretariat of the technical committee or subcommittee is responsible for the management of all projects in the programme of work of that technical committee or subcommittee, including monitoring of their progress against the agreed target dates.

If target dates (see 2.1.6) are not met and there is insufficient support for the work (that is, the acceptance requirements for new work given in 2.3.5 are no longer met), the committee responsible shall cancel the work item.

2.1.8 Project leader

For the development of each project, a project leader (the WG convenor, a designated expert or, if appropriate, the secretary) shall be appointed by the technical committee or subcommittee, taking into account the project leader nomination made by the proposer of the new work item proposal (see 2.3.4). It shall be ascertained that the project leader will have access to appropriate resources for carrying out the development work. The project leader shall act in a purely international capacity, divesting him- or herself of a national point of view. The project leader should be prepared to act as consultant, when required, regarding technical matters arising at the proposal stage through to the publication stage (see 2.5 to 2.8).

The secretariat shall communicate the name and address of the project leader, with identification of the project concerned, to the office of the CEO.

2.1.9 Progress control

Periodical progress reports to the technical committee shall be made by its subcommittees and working groups (see also ISO and IEC Supplements to the ISO/IEC Directives). Meetings between their secretariats will assist in controlling the progress.

The office of the CEO shall monitor the progress of all work and shall report periodically to the technical management board. For this purpose, the office of the CEO shall receive copies of documents as indicated in the ISO and IEC Supplements to the ISO/IEC Directives.

To enable ISO Central Secretariat to monitor the progress of all work and to report periodically to the ISO Technical Management Board, the committee secretariat shall ensure that the ISO Central Secretariat is notified each time a new document is distributed.

2.1.10 Responsibility for keeping records

The responsibility for keeping records concerning committee work and the background to the publication of International Standards and other ISO deliverables is divided between committee secretariats and the ISO Central Secretariat. The maintenance of such records is of particular importance in the context of changes of secretariat responsibility from one member body to another. It is also important that information on key decisions and important correspondence pertaining to the preparation of International Standards and other ISO deliverables should be readily retrievable in the event of any dispute arising out of the provenance of the technical content of the publications.

The secretariats of committees shall establish and maintain records of all official transactions concerning their committees, in particular reference copies of approved minutes of meetings and resolutions. Copies of working documents, results of ballots etc. shall be kept at least until such time as the publications to which they refer have been revised or have completed their next systematic review, but in any case for a minimum of five years after the publication of the related International Standards or other ISO deliverable.

The ISO Central Secretariat shall keep reference copies of all International Standards and other ISO deliverables, including withdrawn editions, and shall keep up-to-date records of member body votes in respect of these publications. Copies of draft International Standards (DIS) and of final draft International Standards (FDIS), including associated reports of voting, and final proofs shall be kept at least until such time as the publications to which they refer have been revised or have completed their next systematic review, but in any case for a minimum of five years after publication.

2.2 Preliminary stage

2.2.1 Technical committees or subcommittees may introduce into their work programmes, by a simple majority vote of their P-members, preliminary work items (for example, corresponding to subjects dealing with emerging technologies), which are not yet sufficiently mature for processing to further stages and for which no target dates can be established.

Such items may include, for example, those listed in the strategic business plan, particularly as given under 2.1.2 d) giving a prospective view on emerging needs.

2.2.2 All preliminary work items shall be registered into the programme of work.

2.2.3 All preliminary work items shall be subject to regular review by the committee. The committee shall evaluate the market relevance and resources required for all such items.

All preliminary work items that have not progressed to the proposal stage in the IEC by the expiration date given by the TC/SC and in ISO within 3 years will be automatically cancelled.

2.2.4 This stage can be used for the elaboration of a new work item proposal (see 2.3) and the development of an initial draft.

2.2.5 Before progressing to the preparatory stage, all such items shall be subject to approval in accordance with the procedures described in 2.3.

2.3 Proposal stage

In the case of proposals to prepare management system deliverables, see Annex L.

2.3.1 A new work item proposal (NP) is a proposal for:

— a new standard;

— a new part of an existing standard;

— a Technical Specification (see 3.1) or a Publicly Available Specification (see 3.2).

The NP stage (clause 2.3) is not required for:

— the revision or amendment of an existing standard, a TS or a PAS (if within its 6 year lifespan),

— the conversion of a TS or PAS to an IS.

However, the committee shall pass a resolution containing the following elements: 1) target dates, 2) confirmation that the scope will not be expanded, and 3) the convenor or project leader. The committee shall also launch a call for experts (Form 4 is not required).

For the conversion of a TS or a PAS to an IS, a two-thirds majority resolution is required.

If the revision or the amendment results in an expanded scope, 2.3 applies (NP ballot shall be initiated and Form 4 is required).

2.3.2 A new work item proposal within the scope of an existing technical committee or subcommittee may be made in the respective organization by

— a National Body;

— the secretariat of that technical committee or subcommittee;

— another technical committee or subcommittee;

— an organization in category A liaison;

— the technical management board or one of its advisory groups;

— the Chief Executive Officer.

2.3.3 Where both an ISO and an IEC technical committee are concerned, the Chief Executive Officers shall arrange for the necessary coordination. (See also Annex B.)

2.3.4 Each new work item proposal shall be presented using the appropriate form, and shall be fully justified and properly documented (see Annex C).

The proposers of the new work item proposal shall

— make every effort to provide a first working draft for discussion, or shall at least provide an outline of such a working draft;

— nominate a project leader;

— discuss the proposal with the committee leadership prior to submitting the appropriate form, in order to decide on an appropriate development track (based on market needs) and draft a project plan including key milestones and the proposed date of the first meeting.

The form shall be submitted to the office of the CEO or to the secretariat of the relevant committee for proposals within the scope of an existing committee.

The office of the CEO or the relevant committee chair and secretariat shall ensure that the proposal is properly developed in accordance with ISO and IEC requirements (see Annex C) and provides sufficient information to support informed decision making by National Bodies.

The office of the CEO or the relevant committee chair and secretariat shall also assess the relationship of the proposal to existing work, and may consult interested parties, including the technical management board or committees conducting related existing work. If necessary, an ad hoc group may be established to examine the proposal. Any review of proposals should not exceed 2 weeks.

In all cases, the office of the CEO or the relevant committee chair and secretariat may also add comments and recommendations to the proposal form.

See Annex K for new work item proposals for project committees.

Copies of the completed form shall be circulated to the members of the technical committee or subcommittee for P-member ballot and to the O-members and liaison members for information.

The proposed date of availability of the publication shall be indicated on the form.

A decision upon a new work item proposal shall be taken by correspondence.

Votes shall be returned within 12 weeks.

The committee may decide on a case-by-case basis by way of a resolution to shorten the voting period for new work item proposals to 8 weeks.

When completing the ballot form, National Bodies shall provide a statement justifying their decision for negative votes ("justification statement"). If no such statement is provided, the negative vote of a National Body will not be registered and considered.

2.3.5 Acceptance requires

a) approval of the work item by a 2/3 majority of the P-members of the technical committees or subcommittees voting — abstentions are excluded when the votes are counted; and

b) a commitment to participate actively in the development of the project, i.e. to make an effective contribution at the preparatory stage, by nominating technical experts and by commenting on working drafts, by at least 4 P-members in committees with 16 or fewer P-members, and at least 5 P-members in committees with 17 or more P-members; only P-members having also approved the inclusion of the work item in the programme of work [see a)] will be taken into account when making this tally. If experts are not nominated on the form accompanying an approval vote, then the National Body's commitment to active participation will not be registered and considered when determining if the approval criteria have been met on this ballot.

If in the context of an NP, a member body does not provide a clear justification statement for why it voted "no", the committee secretariat should go back to the member body and give it two (2) weeks to provide an explanation.

If the member body does not provide a response within that 2-week period, the vote will not be counted in the result.

Secretariats shall not make value judgments about the justification and shall ask the member body in case of doubt.

If member bodies do not name an expert in the Form, they have two (2) weeks following the result of the vote to name their expert. If this delay is not respected, the member body's participation will not be counted, thereby affecting the approval requirement for b) above.

Individual committees may increase this minimum requirement of nominated experts.

In cases, where it can be documented that the industry and/or technical knowledge exists only with a very small number of P-members, then the committee may request permission from the technical management board to proceed with fewer than 4 or 5 nominated technical experts.

2.3.6 Once a new work item proposal is accepted, it shall be registered in the programme of work of the relevant technical committee or subcommittee as a new project with the appropriate priority. The agreed target dates (see 2.1.6) shall be indicated on the appropriate form.

The voting results will be reported to the ISO Central Secretariat (using Form 6) or the IEC Central Office (using Form RVN) within 4 weeks after the close of the ballot.

2.3.7 The inclusion of the project in the programme of work concludes the proposal stage.

2.4 Preparatory stage

2.4.1 The preparatory stage covers the preparation of a working draft (WD) conforming to the ISO/IEC Directives, Part 2.

2.4.2 When a new project is accepted the project leader shall work with the experts nominated by the P-members during the approval [see 2.3.5 a)].

2.4.3 The secretariat may propose to the technical committee or subcommittee, either at a meeting or by correspondence, to create a working group the convenor of which will normally be the project leader.

Such a working group shall be set up by the technical committee or subcommittee, which shall define the task(s) and set the target date(s) for submission of draft(s) to the technical committee or subcommittee (see also 1.12). The working group convenor shall ensure that the work undertaken remains within the scope of the balloted work item.

2.4.4 In responding to the proposal to set up a working group those P-members having agreed to participate actively [see 2.3.5 a)] shall each confirm their technical expert(s). Other P-members or A- or C- liaison organizations may also nominate expert(s).

2.4.5 The project leader is responsible for the development of the project and will normally convene and chair any meetings of the working group. S/he may invite a member of the working group to act as its secretary.

2.4.6 Every possible effort shall be made to prepare both a French and an English version of the text in order to avoid delays in the later stages of the development of the project.

If a trilingual (English — French — Russian) standard is to be prepared, this provision should include the Russian version.

2.4.7 For time limits relating to this stage, see 2.1.6.

2.4.8 The preparatory stage ends when a working draft is available for circulation to the members of the technical committee or subcommittee as a first committee draft (CD) and is registered by the office of the CEO. The committee may also decide to publish the final working draft as a PAS (see 3.2) to respond particular market needs.

If the committee has opted to skip the CD (see 2.5), the preparatory stage ends when the enquiry draft (DIS) is available for circulation (see 2.6).

2.5 Committee stage

2.5.1 The committee stage is the principal stage at which comments from National Bodies are taken into consideration, with a view to reaching consensus on the technical content. National Bodies shall therefore carefully study the texts of committee drafts and submit all pertinent comments at this stage.

Committees may decide to skip the CD stage in accordance with Annex SS.

Any graphical symbol shall be submitted to the relevant ISO committee responsible for the registration of graphical symbols (see Annex SH).

2.5.2 As soon as it is available, a committee draft shall be circulated to all P-members and O-members of the technical committee or subcommittee for consideration, with a clear indication of the latest date for submission of replies.

A period of 8, 12 or 16 weeks as agreed by the technical committee or subcommittee shall be available for National Bodies to comment.

The default for CD circulation is 8 weeks.

Comments shall be sent for preparation of the compilation of comments, in accordance with the instructions given.

National Bodies shall fully brief their delegates on the national position before meetings.

2.5.3 No more than 4 weeks after the closing date for submission of replies, the secretariat shall prepare the compilation of comments and arrange for its circulation to all P-members and O-members of the technical committee or subcommittee. When preparing this compilation, the secretariat shall indicate its proposal, made in consultation with the chair of the technical committee or subcommittee and, if necessary, the project leader, for proceeding with the project, either

a) to discuss the committee draft and comments at the next meeting, or

b) to circulate a revised committee draft for consideration, or

c) to register the committee draft for the enquiry stage (see 2.6).

In the case of b) and c), the secretariat shall indicate in the compilation of comments the action taken on each of the comments received. This shall be made available to all P-members, if necessary by the circulation of a revised compilation of comments, no later than in parallel with the submission of a revised CD for consideration by the committee (case b) or simultaneously with the submission of the finalized version of the draft to the office of the CEO for registration for the enquiry stage (case c).

Committees are required to respond to all comments received.

If, within 8 weeks from the date of dispatch, 2 or more P-members disagree with proposal b) or c) of the secretariat, the committee draft shall be discussed at a meeting (see 4.2.1.3).

2.5.4 If a committee draft is considered at a meeting but agreement on it is not reached on that occasion, a further committee draft incorporating decisions taken at the meeting shall be distributed within 12 weeks for consideration. A period of 8, 12 or 16 weeks as agreed by the technical committee or subcommittee shall be available to National Bodies to comment on the draft and on any subsequent versions.

2.5.5 Consideration of successive drafts shall continue until consensus of the P-members of the technical committee or subcommittee has been obtained or a decision to abandon or defer the project has been made.

2.5.6 The decision to circulate an enquiry draft (see 2.6.1) shall be taken on the basis of the consensus principle.

It is the responsibility of the chair of the technical committee or subcommittee, in consultation with the secretary of his/her committee and, if necessary, the project leader, to judge whether there is sufficient support bearing in mind the definition of consensus given in ISO/IEC Guide 2:2004.

> "**consensus**: General agreement, characterized by the absence of sustained opposition to substantial issues by any important part of the concerned interests and by a process that involves seeking to take into account the views of all parties concerned and to reconcile any conflicting arguments.
>
> NOTE   Consensus need not imply unanimity."

The following applies to the definition of consensus:

In the process of reaching consensus, many different points of views will be expressed and addressed as the document evolves. However, "sustained oppositions" are views expressed at minuted meetings of committee, working group (WG) or other groups (e.g. task forces, advisory groups, etc.) and which are maintained by an important part of the concerned interest and which are incompatible with the committee consensus. The notion of "concerned interest(s)" will vary depending on the dynamics of the committee and shall therefore be determined by the committee leadership on a case by case basis. The concept of sustained opposition is not applicable in the context of member body votes on CD, DIS or FDIS since these are subject to the applicable voting rules.

Those expressing sustained oppositions have a right to be heard and the following approach is recommended when a sustained opposition is declared:

— The leadership shall first assess whether the opposition can be considered a "sustained opposition", i.e. whether it has been sustained by an important part of the concerned interest. If this is not the case, the leadership will register the opposition (i.e. in the minutes, records, etc.) and continue to lead the work on the document.

— If the leadership determines that there is a sustained opposition, it is required to try and resolve it in good faith. However, a sustained opposition is not akin to a right to veto. The obligation to address the sustained oppositions does not imply an obligation to successfully resolve them.

The responsibility for assessing whether or not consensus has been reached rests entirely with the leadership. This includes assessing whether there is sustained opposition or whether any sustained opposition can be resolved without compromising the existing level of consensus on the rest of the document. In such cases, the leadership will register the opposition and continue the work.

Those parties with sustained oppositions may avail themselves of appeals mechanisms as detailed in Clause 5.

In case of doubt concerning consensus, approval by a two-thirds majority of the P-members of the technical committee or subcommittee voting may be deemed to be sufficient for the committee draft to be accepted for registration as an enquiry draft; however every attempt shall be made to resolve negative votes.

Abstentions are excluded when the votes are counted, as well as negative votes not accompanied by technical reasons.

The secretariat of the technical committee or subcommittee responsible for the committee draft shall ensure that the enquiry draft fully embodies decisions taken either at meetings or by correspondence.

2.5.7   When consensus has been reached in a technical committee or subcommittee, its secretariat shall submit the finalized version of the draft in electronic form suitable for distribution to the national members for enquiry (2.6.1), to the office of the CEO (with a copy to the technical committee secretariat in the case of a subcommittee) within a maximum of 16 weeks.

The secretariat shall submit the proposed draft International Standard (DIS) to the ISO Central Secretariat in electronic format together with a completed explanatory report (ISO form 8A) and the compilation of comments and actions taken in response to comments on the final CD.

2.5.8 For time limits relating to this stage, see 2.1.6.

2.5.9 The committee stage ends when all technical issues have been resolved and a committee draft is accepted for circulation as an enquiry draft and is registered by the office of the CEO. Texts that do not conform to the ISO/IEC Directives, Part 2 shall be returned to the secretariat with a request for correction before they are registered.

2.5.10 If the technical issues cannot all be resolved within the appropriate time limits, technical committees and subcommittees may wish to consider publishing an intermediate deliverable in the form of a Technical Specification (see 3.1) pending agreement on an International Standard.

2.6 Enquiry stage

2.6.1 At the enquiry stage, the enquiry draft (DIS in ISO, CDV in IEC) shall be circulated by the office of the CEO to all National Bodies for a 12-week vote.

For policy on the use of languages, see Annex E.

National Bodies shall be advised of the date by which completed ballots are to be received by the office of the CEO.

At the end of the voting period, the Chief Executive Officer shall send within 4 weeks to the chair and secretariat of the technical committee or subcommittee the results of the voting together with any comments received, for further speedy action.

2.6.2 Votes submitted by National Bodies shall be explicit: positive, negative, or abstention.

A positive vote may be accompanied by editorial or technical comments, on the understanding that the secretary, in consultation with the chair of the technical committee or subcommittee and project leader, will decide how to deal with them.

If a National Body finds an enquiry draft unacceptable, it shall vote negatively and state the technical reasons. It may indicate that the acceptance of specified technical modifications will change its negative vote to one of approval, but it shall not cast an affirmative vote which is conditional on the acceptance of modifications.

In the case where a member body has voted negatively without submitting a justification, the vote will not be counted.

In the case where a Member Body has voted negatively and has submitted comments that are not clearly of a technical nature, the committee secretary shall contact the ISO/CS Technical Programme Manager within 2 weeks of the ballot closure.

2.6.3 An enquiry draft is approved if

a) a two-thirds majority of the votes cast by the P-members of the technical committee or subcommittee are in favour, and

b) not more than one-quarter of the total number of votes cast are negative.

Abstentions are excluded when the votes are counted, as well as negative votes not accompanied by technical reasons.

Comments received after the normal voting period are submitted to the technical committee or subcommittee secretariat for consideration at the time of the next review of the International Standard.

2.6.4 On receipt of the results of the voting and any comments, the chair of the technical committee or subcommittee, in cooperation with its secretariat and the project leader, and in consultation with the office of the CEO, shall take one of the following courses of action:

a) when the approval criteria of 2.6.3 are met and no technical changes are to be included, to proceed directly to publication (see 2.8);

b) when the approval criteria of 2.6.3 are met, but technical changes are to be included, to register the enquiry draft, as modified, as a final draft international standard;

c) when the approval criteria of 2.6.3 are not met:

    1) to circulate a revised enquiry draft for voting (see 2.6.1), or

    NOTE   A revised enquiry draft will be circulated for a voting period of 8 weeks, which may be extended up to 12 weeks at the request of one or more P-members of the committee concerned.

    2) to circulate a revised committee draft for comments, or

    3) to discuss the enquiry draft and comments at the next meeting.

2.6.5 Not later than 12 weeks after the end of the voting period, a full report shall be prepared by the secretariat of the technical committee or subcommittee and circulated by the office of the CEO to the National Bodies. The report shall

a) show the result of the voting;

b) state the decision of the chair of the technical committee or subcommittee;

c) reproduce the text of the comments received; and

d) include the observations of the secretariat of the technical committee or subcommittee on each of the comments submitted.

Every attempt shall be made to resolve negative votes.

If, within 8 weeks from the date of dispatch, two or more P-members disagree with decision 2.6.4 c)1) or 2.6.4 c)2) of the chair, the draft shall be discussed at a meeting (see 4.2.1.3).

Committees are required to respond to all comments received.

2.6.6 When the chair has taken the decision to proceed to the approval stage (see 2.7) or publication stage (see 2.8), the secretariat of the technical committee or subcommittee shall prepare, within a maximum of 16 weeks after the end of the voting period and with the assistance of its editing committee, a final text and send it to the office of the CEO for preparation and circulation of the final draft International Standard.

The secretariat shall provide the office of the CEO with the text in a revisable electronic format and also in a format which permits validation of the revisable form.

Texts that do not conform to the ISO/IEC Directives, Part 2 shall be returned to the secretariat with a request for correction before they are registered.

The revised text shall be submitted to ISO Central Secretariat in electronic format together with the decision of the chair taken as a result of the voting, using ISO Form 13, and including a detailed indication of the decisions taken for each comment as Annex B to the ISO Form 13.

2.6.7 For time limits relating to this stage, see 2.1.6.

2.6.8 The enquiry stage ends with the registration, by the office of the CEO, of the text for circulation as a final draft International Standard or publication as an International Standard, in the case of 2.6.4 a) and b).

2.7 Approval stage

2.7.1 At the approval stage, the final draft International Standard (FDIS) shall be distributed by the office of the CEO within 12 weeks to all National Bodies for a 8-week vote (6 weeks in IEC).

National Bodies shall be advised of the date by which ballots are to be received by the office of the CEO.

2.7.2 Votes submitted by National Bodies shall be explicit: positive, negative, or abstention.

A National Body may submit comments on any FDIS vote.

If a National Body finds a final draft International Standard unacceptable, it shall vote negatively and state the technical reasons. It shall not cast an affirmative vote that is conditional on the acceptance of modifications.

In the case where a member body has voted negatively without submitting a justification, the vote will not be counted.

In the case where a Member Body has voted negatively and has submitted comments that are not clearly of a technical nature, the committee secretary shall contact the ISO/CS Technical Programme Manager within 2 weeks of the ballot closure.

2.7.3 A final draft International Standard having been circulated for voting is approved if

a) a two-thirds majority of the votes cast by the P-members of the technical committee or subcommittee are in favour, and

b) not more than one-quarter of the total number of votes cast are negative.

Abstentions are excluded when the votes are counted, as well as negative votes not accompanied by technical reasons.

2.7.4 The secretariat of the technical committee or subcommittee has the responsibility of bringing any errors that may have been introduced in the preparation of the draft to the attention of the office of the CEO by the end of the voting period; further editorial or technical amendments are not acceptable at this stage.

2.7.5 All comments received will be retained for the next review and will be recorded on the voting form as "noted for future consideration". However, the Secretary along with the office of the CEO may seek to resolve obvious editorial errors. Technical changes to an approved FDIS are not allowed.

Within 2 weeks after the end of the voting period, the office of the CEO shall circulate to all National Bodies a report showing the result of voting and indicating either the formal approval by National Bodies to issue the International Standard or formal rejection of the final draft International Standard.

2.7.6 If the final draft International Standard has been approved in accordance with the conditions of 2.7.3, it shall proceed to the publication stage (see 2.8).

2.7.7 If the final draft International Standard is not approved in accordance with the conditions in 2.7.3, the document shall be referred back to the technical committee or subcommittee concerned for reconsideration in the light of the technical reasons submitted in support of the negative votes.

The committee may decide to:

— resubmit a modified draft as a committee draft, enquiry draft or, in ISO and JTC 1, final draft International Standard;

— publish a Technical Specification (see 3.1);

— publish a Publicly Available Specification (see 3.2);

— cancel the project.

2.7.8 The approval stage ends with the circulation of the voting report (see 2.7.5) stating that the FDIS has been approved for publication as an International Standard, with the publication of a Technical Specification (see 3.1.1.2), or with the document being referred back to the committee.

2.8 Publication stage

2.8.1 Within 6 weeks, the office of the CEO shall correct any errors indicated by the secretariat of the technical committee or subcommittee, and print and distribute the International Standard.

2.8.2 The publication stage ends with the publication of the International Standard.

2.9 Maintenance of deliverables

The procedures for the maintenance of deliverables are given in the respective Supplements to the ISO/IEC Directives.

2.9.1 Introduction

Every International Standard and other deliverable published by ISO or jointly with IEC shall be subject to systematic review in order to determine whether it should be confirmed, revised/amended, converted to another form of deliverable, or withdrawn, according to Table S1.

A committee may at any time between systematic reviews pass a resolution initiating a revision or amendment of a standard.

See clause 2.3.1 for the process for initiating a revision or amendment of an existing standard.

For minor changes, e.g. updating and editorial changes, that do not impact the technical content, a shortened procedure called "minor revision" may be applied. This is comprised only of the proposal for a minor revision by the committee (through a resolution and completion of ISO form 8B), approval and publication stages (see 2.7 and 2.8). Subsequent to the resolution of the responsible technical committee and consultation of the responsible ISO Technical Programme Manager, a final draft of the revised deliverable shall be circulated for an 8 week FDIS vote, and 12 weeks in the case of Vienna Agreement documents. The Foreword of the next edition of the deliverable shall indicate that it is a minor revision and list the updates and editorial changes made.

Table S1 — Timing of systematic reviews

Deliverable	Max. elapsed time before systematic review	Max. number of times deliverable may be confirmed	Max. life
International Standard	5 years	Not limited	Not limited
Technical Specification (see 3.1.3)	3 years	Once recommended	6 years recommended
Publicly Available Specification (see 3.2.4)	3 years	Once	6 years (If not converted after this period, the deliverable is proposed for withdrawal)
Technical Report (see 3.3.3)	Not specified	Not specified	Not limited

A systematic review will typically be initiated in the following circumstances:

— (all deliverables) on the initiative and as a responsibility of the secretariat of the responsible committee, typically as the result of the elapse of the specified period since publication or the last confirmation of the document, or

— (for International Standards and Technical Specifications) a default action by ISO Central Secretariat if a systematic review of the International Standard or Technical Specification concerned has not been initiated by the secretariat of the responsible committee, or

— (all deliverables) at the request of one or more National Body, or

— (all deliverables) at the request of the CEO.

The timing of a systematic review is normally based either on the year of publication or, where a document has already been confirmed, on the year in which it was last confirmed. However, it is not necessary to wait for the maximum period to elapse before a document is reviewed.

2.9.2 Review

The review period is 20 weeks.

NOTE   Systematic reviews are administered electronically by the ISO Central Secretariat and all ISO member bodies are invited to respond to such reviews. P-members of a given committee have an obligation to vote on all systematic review ballots for deliverables under the responsibility of that committee. The purpose of the reviews has been extended to include obtaining information when member bodies have needed to make modifications in order to make ISO standards suitable for national adoption. Such modifications need to be considered by committees in order to determine whether they need to be taken into account to improve the global relevance of a standard. The final decision, to revise, confirm or withdraw a standard, remains with the P-members of the responsible committee.

After the closing of the vote, the secretariat's proposal reflecting the voting results shall be circulated to the members of the technical committee or subcommittee using Form 21. No more than 6 months after the closing of the vote, the committee shall take a final decision as to whether to revise, confirm or withdraw the standard, following which the secretariat shall submit the committee's decision to the ISO Central Secretariat.

2.9.3 Interpretation of ballot results

2.9.3.1 General

Typically, a decision as to the appropriate action to take following a systematic review shall be based on a simple majority of P-members voting for a specific action. However, in some cases a more detailed analysis of the results may indicate that another interpretation may be more appropriate.

NOTE   It is not feasible to provide concrete rules for all cases when interpreting the ballot results due to the variety of possible responses, degrees of implementation, and the relative importance of comments, etc.

Where voting results are not definitive and/or a decision is based on interpretation of responses the secretariat shall invite approval of a proposed course of action within a specified time delay, for example within two months.

In proposing future action, due account shall be taken of the maximum possible number of confirmations and specified maximum life of the deliverable concerned (see Table S1).

2.9.3.2 Interpretation of ballot results for International Standards

Option 1: Confirmation (retention without technical change)

Where it has been verified that a document is used, that it should continue to be made available, and that no technical changes are needed, a deliverable may be confirmed. The criteria is as follows:

— the standard has been adopted with or without change or is used in at least five countries (when this criteria is not met, the standard should be withdrawn); and

— a simple majority of the P-members of the committee voting propose confirmation.

Option 2: Amendment or revision (Retention, with change/s)

Where it has been verified that a document is used, that it should continue to be made available, but that technical changes are needed, a deliverable may be proposed for amendment or revision. The criteria is as follows:

— the standard has been adopted with or without change or is used in at least five countries (when this criteria is not met, the standard should be withdrawn); and

— a simple majority of the P-members of the committee voting considers there is a need for amendment or revision.

In that case, an item may be registered as an approved work item (stage 10.99).

A call for experts shall be launched. However, there is no minimum number of active P-members required.

Where an amendment or revision is not immediately started following approval by the committee, it is recommended that the project is first registered as a preliminary work item and that the standard is registered as confirmed. When it is eventually proposed for registration at stage 10.99, reference shall be made to the results of the preceding systematic review and the committee shall pass a resolution (see clause 2.3.1 for the process for initiating a revision or amendment of an existing standard).

Where it is decided that the International Standard needs to be revised or amended, it becomes a new project and shall be added to the programme of work of the technical committee or subcommittee. The steps for revision or amendment are the same as those for preparation of a new standard (see ISO/IEC Directives, Part 1, clauses 2.3 to 2.8), and include the establishment of target dates for the completion of the relevant stages.

Option 3: Withdrawal

When the standard has not been adopted with or without change or is not used in at least five countries, the standard should be withdrawn (see Options 1 and 2 above).

In the case of the proposed withdrawal of an International Standard, the National Bodies shall be informed by the CEO of the decision of the technical committee or subcommittee, with an invitation to inform the office of the CEO within 8 weeks if they object to that decision.

Any objection received shall be referred to the ISO Technical Management Board for consideration and decision.

2.9.3.3 Conversion to an International Standard (Technical Specifications and Publicly Available Specifications only)

In addition to the three basic options of confirmation, amendment or revision, or withdrawal, in the cases of the systematic review of Technical Specifications and Publicly Available Specifications a fourth option is their conversion to an International Standard.

To initiate conversion to an International Standard, a text, up-dated as appropriate, is submitted to the normal development procedures as specified for an International Standard (see clause 2.3.1).

The conversion procedure will typically start with a DIS vote. Where changes considered to be required are judged as being so significant as to require a full review in the committee prior to DIS ballot, a revised version of the document shall be submitted for review and ballot as a CD.

2.9.4 Reinstatement of withdrawn standards

If, following withdrawal of an International Standard, a committee determines that it is still needed, it may propose that the standard be reinstated. The standard shall be issued either as a draft International Standard or as a final draft International Standard, as decided by the committee, for voting by the member bodies. The usual approval criteria shall apply. If approved, the standard shall be published as a new edition with a new date of publication. The foreword shall explain that the standard results from the reinstatement of the previous edition.

2.10 Corrections and amendments

2.10.1 General

A published International Standard may subsequently be modified by the publication of

— a technical corrigendum (in IEC only);

— a corrected version;

— an amendment; or

— a revision (as part of the maintenance procedure in 2.9).

NOTE   In case of revision a new edition of the International Standard will be issued.

2.10.2 Corrections

A correction is only issued to correct an error or ambiguity, inadvertently introduced either in drafting or in publishing and which could lead to incorrect or unsafe application of the publication.

Corrections are not issued to update information that has become outdated since publication.

Suspected errors shall be brought to the attention of the secretariat of the technical committee or subcommittee concerned. After confirmation by the secretariat and chair, if necessary in consultation with the project leader and P-members of the technical committee or subcommittee, the secretariat shall submit to the office of the CEO a proposal for correction, with an explanation of the need to do so.

The Chief Executive Officer shall decide, in consultation with the secretariat of the technical committee or subcommittee, and bearing in mind both the financial consequences to the organization and the interests of users of the publication, whether to publish a technical corrigendum (in IEC only) and/or a corrected version of the existing edition of the publication (see also 2.10.4). The secretariat of the committee will then inform the members of the committee of the outcome.

The corrections are mentioned in the Foreword of the corrected version.

In general, a correction will not be issued for a publication that is older than 3 years.

2.10.3 Amendments

An amendment alters and/or adds to previously agreed technical provisions in an existing International Standard. An amendment is considered as a partial revision: the rest of the International Standard is not open for comments.

An amendment is normally published as a separate document, the edition of the International Standard affected remaining in use.

The procedure for developing and publishing an amendment shall be as described in 2.3 (ISO and JTC 1), or the review and maintenance procedures (see IEC Supplement) and 2.4, 2.5, 2.6 (draft amendment, DAM), 2.7 (final draft amendment, FDAM), and 2.8.

At the approval stage (2.7), the Chief Executive Officer shall decide, in consultation with the secretariat of the technical committee or subcommittee, and bearing in mind both the financial consequences to the organization and the interests of users of the International Standard, whether to publish an amendment or a new edition of the International Standard, incorporating the amendment. (See also 2.10.4.)

NOTE   Where it is foreseen that there will be frequent *additions* to the provisions of an International Standard, the possibility should be borne in mind at the outset of developing these additions as a series of parts (see ISO/IEC Directives, Part 2).

2.10.4   Avoidance of proliferation of modifications

No more than 2 separate documents in the form of technical corrigenda (in IEC only) or amendments shall be published modifying a current International Standard. The development of a third such document shall result in publication of a new edition of the International Standard.

2.11   Maintenance agencies

When a technical committee or subcommittee has developed a standard that will require frequent modification, it may decide that a maintenance agency is required. Rules concerning the designation of maintenance agencies are given in Annex G.

2.12   Registration authorities

When a technical committee or subcommittee has developed a standard that includes registration provisions, a registration authority is required. Rules concerning the designation of registration authorities are given in Annex H.

In ISO, see also Annex SN, Registration Authority Policy.

2.13   Copyright

The copyright for all drafts and International Standards and other publications belongs to ISO, IEC or ISO and IEC, respectively as represented by the office of the CEO.

The content of, for example, an International Standard may originate from a number of sources, including existing national standards, articles published in scientific or trade journals, original research and development work, descriptions of commercialized products, etc. These sources may be subject to one or more rights.

In ISO and IEC, there is an understanding that original material contributed to become a part of an ISO, IEC or ISO/IEC publication can be copied and distributed within the ISO and/or IEC systems (as relevant) as part of the consensus building process, this being without prejudice to the rights of the original copyright owner to exploit the original text elsewhere. Where material is already subject to copyright, the right should be granted to ISO and/or IEC to reproduce and circulate the material. This is frequently done without recourse to a written agreement, or at most to a simple written statement of acceptance. Where contributors wish a formal signed agreement concerning copyright of any submissions they make to ISO and/or IEC, such requests shall be addressed to ISO Central Secretariat or the IEC Central Office, respectively.

Attention is drawn to the fact that the respective members of ISO and IEC have the right to adopt and re-publish any respective ISO and/or IEC standard as their national standard. Similar forms of endorsement do or may exist (for example, with regional standardization organizations).

2.14 Reference to patented items (see also Annex I)

2.14.1 If, in exceptional situations, technical reasons justify such a step, there is no objection in principle to preparing an International Standard in terms which include the use of items covered by patent rights — defined as patents, utility models and other statutory rights based on inventions, including any published applications for any of the foregoing — even if the terms of the standard are such that there are no alternative means of compliance. The rules given below shall be applied.

2.14.2 If technical reasons justify the preparation of a document in terms which include the use of items covered by patent rights, the following procedures shall be complied with:

a) The proposer of a proposal for a document shall draw the attention of the committee to any patent rights of which the proposer is aware and considers to cover any item of the proposal. Any party involved in the preparation of a document shall draw the attention of the committee to any patent rights of which it becomes aware during any stage in the development of the document.

b) If the proposal is accepted on technical grounds, the proposer shall ask any holder of such identified patent rights for a statement that the holder would be willing to negotiate worldwide licences under his/her rights with applicants throughout the world on reasonable and non-discriminatory terms and conditions. Such negotiations are left to the parties concerned and are performed outside ISO and/or IEC. A record of the right holder's statement shall be placed in the registry of the ISO Central Secretariat or IEC Central Office as appropriate, and shall be referred to in the introduction to the relevant document. If the right holder does not provide such a statement, the committee concerned shall not proceed with inclusion of an item covered by a patent right in the document without authorization from ISO Council or IEC Council Board as appropriate.

c) A document shall not be published until the statements of the holders of all identified patent rights have been received, unless the council board concerned gives authorization.

2.14.3 Should it be revealed after publication of a document that licences under patent rights, which appear to cover items included in the document, cannot be obtained under reasonable and non-discriminatory terms and conditions, the document shall be referred back to the relevant committee for further consideration.

3 Development of other deliverables

3.1 Technical Specifications

Technical Specifications may be prepared and published under the following circumstances and conditions.

3.1.1 When the subject in question is still under development or where for any other reason there is the future but not immediate possibility of an agreement to publish an International Standard, the technical committee or subcommittee may decide, by following the procedure set out in 2.3, that the publication of a Technical Specification would be appropriate. The procedure for preparation of such a Technical Specification shall be as set out in 2.4 and 2.5. The decision to publish the resulting document as a Technical Specification shall require a two-thirds majority vote of the P-members voting of the technical committee or subcommittee. A Technical Specification is a normative document.

In ISO, when the required support cannot be obtained for a final draft International Standard to pass the approval stage (see 2.7), or in case of doubt concerning consensus, the technical committee or subcommittee may decide, by a two-thirds majority vote of P-members voting, that the document should be published in the form of a Technical Specification.

3.1.2 When the P-members of a technical committee or subcommittee have agreed upon the publication of a Technical Specification, the draft specification shall be submitted electronically by the secretariat of the technical committee or subcommittee to the office of the CEO within 16 weeks for publication. Competing technical specifications offering different technical solutions are possible provided that they do not conflict with existing International Standards.

3.1.3 Technical Specifications shall be subject to review by the technical committee or subcommittee not later than 3 years after their publication. The aim of such review shall be to re-examine the situation which resulted in the publication of a Technical Specification and if possible to achieve the agreement necessary for the publication of an International Standard to replace the Technical Specification. In IEC, the date for this review is based on the stability date which shall be agreed in advance of the publication of the Technical Specification (review date).

3.2 Publicly Available Specifications (PAS)

The procedures set out in 2.3 apply to the development of a PAS.

3.2.1 A PAS may be an intermediate specification, published prior to the development of a full International Standard, or, in IEC may be a "dual logo" publication published in collaboration with an external organization. It is a document not fulfilling the requirements for a standard. A PAS is a normative document.

Normally, the decision to develop a PAS should be agreed at the outset, i.e. simultaneously with the approval of the new work item proposal.

3.2.2 A proposal for submission of a PAS may be made by the Secretariat, an A-liaison or by any P-member of the committee. In IEC, a C-liaison may also submit a PAS (see 1.17).

3.2.3 The PAS is published after verification of the presentation and checking that there is no conflict with existing International Standards by the committee concerned and following simple majority approval of the P-members voting of the committee concerned. Competing PAS offering different technical solutions are possible provided that they do not conflict with existing International Standards.

3.2.4 A PAS shall remain valid for an initial maximum period of 3 years in ISO and 2 years in IEC. The validity may be extended for a single period up to a maximum of 3 years in ISO and 2 years in IEC, at the end of which it shall be transformed with or without change into another type of normative document, or shall be withdrawn.

3.3 Technical Reports

3.3.1 When a technical committee or subcommittee has collected data of a different kind from that which is normally published as an International Standard (this may include, for example, data obtained from a survey carried out among the National Bodies, data on work in other international organizations or data on the "state of the art" in relation to standards of National Bodies on a particular subject), the technical committee or subcommittee may decide, by a simple majority vote of P-members voting, to request the Chief Executive Officer to publish such data in the form of a Technical Report. The document shall be entirely informative in nature and shall not contain matter implying that it is normative. It shall clearly explain its relationship to normative aspects of the subject which are, or will be, dealt with in International Standards related to the subject. The Chief Executive Officer, if necessary in consultation with the technical management board, shall decide whether to publish the document as a Technical Report.

3.3.2 When the P-members of a technical committee or subcommittee have agreed upon the publication of a Technical Report, the draft report shall be submitted electronically by the secretariat of the technical committee or subcommittee to the Chief Executive Officer within 16 weeks for publication.

3.3.3 It is recommended that Technical Reports are regularly reviewed by the committee responsible, to ensure that they remain valid. Withdrawal of a Technical Report is decided by the technical committee or subcommittee responsible.

Technical Reports are not subject to systematic revision.

4 Meetings

4.1 General

National Bodies are reminded that they are not permitted to charge delegates/experts any sort of participation fee, nor require accommodations at specific hotels or hotel rates for any meetings of technical committees, subcommittees, working groups, maintenance and project teams. The basic meeting facilities shall be funded entirely by resources from a National Body and/or voluntary sponsors. For more information in IEC, see Meeting Guide (http://www.iec.ch/members_experts/refdocs/iec/IEC_Meeting_Gu ide_2012.pdf) and for ISO, see Annex SF for further details.

4.1.1 Technical committees and subcommittees shall use current electronic means to carry out their work (for example, e-mail, groupware and teleconferencing) wherever possible. A meeting of a technical committee or subcommittee should be convened only when it is necessary to discuss committee drafts (CD) or other matters of substance which cannot be settled by other means.

4.1.2 The technical committee secretariat should look ahead with a view to drawing up, in consultation with the office of the CEO, a planned minimum 2-year programme of meetings of the technical committee and its subcommittees and, if possible, its working groups, taking account of the programme of work.

4.1.3 In planning meetings, account should be taken of the possible advantage of grouping meetings of technical committees and subcommittees dealing with related subjects, in order to improve communication and to limit the burden of attendance at meetings by delegates who participate in several technical committees or subcommittees.

4.1.4 In planning meetings, account should also be taken of the advantages for the speedy preparation of drafts of holding a meeting of the editing committee immediately after the meeting of the technical committee or subcommittee and at the same place.

4.2 Procedure for calling a meeting

For information relating to hosting meetings, refer to Annex SF.

4.2.1 Technical committee and subcommittee meetings

4.2.1.1 The date and place of a meeting shall be subject to an agreement between the chair and the secretariat of the technical committee or subcommittee concerned, the Chief Executive Officer and the National Body acting as host. In the case of a subcommittee meeting, the subcommittee secretariat shall firse consult with the secretariat of the parent technical committee in order to ensure coordination of meetings (see also 4.1.3).

4.2.1.2 A National Body wishing to act as host for a particular meeting shall contact the Chief Executive Officer and the technical committee or subcommittee secretariat concerned.

The National Body shall first ascertain that there are no restrictions imposed by its country to the entry of representatives of any P-member of the technical committee or subcommittee for the purpose of attending the meeting.

In accrediting delegates to attend meetings, national standards bodies from P- and O-member countries shall register them in the ISO Meetings application.

The hosting national standards body can access the list of delegates through the ISO Meetings application so that it can make appropriate arrangements for the meeting. It is the responsibility of the national standards bodies in countries with participants who need invitation letters to send the names of these participants directly to the hosting national standards body.

The hosting organizations are advised to verify and provide information on access means to meeting facilities. As per clause 4.2.1.3, a document describing logistics for the meeting shall be circulated. As well as location and transport information, it should provide details of the accessibility of meeting facilities.

During the planning process, there should be a request for notification of specific accessibility requirements. The hosting body should make its best efforts to satisfy these requirements.

4.2.1.3 The secretariat shall ensure that arrangements are made for the agenda and logistical information to be circulated by the office of the CEO (in the IEC) or by the secretariat with a copy to the office of the CEO (in ISO) at the latest 16 weeks before the date of the meeting.

NOTE   All new work item proposals must be approved by correspondence (committee internal ballot – CIB) see 2.3.4.

Only those committee drafts for which the compilation of comments will be available at least 6 weeks before the meeting shall be included on the agenda and be eligible for discussion at the meeting.

Any other working documents, including compilations of comments on drafts to be discussed at the meeting, shall be distributed not less than 6 weeks in advance of the meeting.

The agenda shall clearly state the starting and estimated finishing times.

In the event of meetings over running the estimated finishing time, the Chair shall ensure that the P-members are willing to take voting decisions. However if P-members leave, they may request the Chair not to take any further voting decisions.

Any decisions made after the estimated finishing time of the meeting and after any P-members have left shall be confirmed by correspondence after the meeting.

NOTE   Attendees should take the estimated meeting time into consideration when booking their travel.

See Annex SK for the issuance of notices, agendas and documents before committee and working group meetings.

4.2.2 Working group meetings

4.2.2.1 Working groups shall use current electronic means to carry out their work (for example, e-mail, groupware and teleconferencing) wherever possible. When a meeting needs to be held, notification by the convenor of the meetings of a working group shall be sent to its members and to the secretariat of the parent committee, at least 6 weeks in advance of the meeting.

Arrangements for meetings shall be made between the convenor and the member of the working group in whose country the meeting is to be held. The latter member shall be responsible for all practical working arrangements.

4.2.2.2 If a working group meeting is to be held in conjunction with a meeting of the parent committee, the convenor shall coordinate arrangements with the secretariat of the parent committee. In particular it shall be ensured that the working group members receive all general information for the meeting, which is sent to delegates to the meeting of the parent committee.

4.2.2.3 Either the WG (or PT/MT/AC in IEC) leader or the Secretary of the relevant committee shall notify National Body Secretariats of any WG (or PT/MT/AC in IEC) meeting held in their country.

4.3 Languages at meetings

While the official languages are English, French and Russian, meetings are conducted in English by default.

The National Body for the Russian Federation provides all interpretation and translation into or from the Russian language.

The chair and secretariat are responsible for dealing with the question of language at a meeting in a manner acceptable to the participants following the general rules of ISO or IEC, as appropriate. (See also Annex E.)

4.4 Cancellation of meetings

Every possible effort shall be made to avoid cancellation or postponement of a meeting once it has been convened. Nevertheless, if the agenda and basic documents are not available within the time required by 4.2.1.3, then the Chief Executive Officer has the right to cancel the meeting.

4.5 Distribution of documents

For requirements relating to document distribution, refer to Annex SB. A copy of the agenda and calling notice for a committee meeting shall be made available to the ISO Central Secretariat for information.

4.6 Remote participation at committee meetings

Remote participation at committee meetings is supported in ISO to enable the goals of increasing stakeholder engagement and better coordination of the committee work. Remote participation is allowed for all ISO meetings and ISO committees, provided that:

— The committee secretary checks with the host prior to meeting and follows the 'Guidelines on remote participation at committee meetings'.

— The host agrees and can provide the necessary technology and support.

— The same registration and accreditation rules apply to remote and physical participants.

— The 'Guidelines on remote participation at committee meetings' are provided to all participants in advance of the meeting.

5 Appeals

5.1 General

5.1.1 National Bodies have the right of appeal

a) to the parent technical committee on a decision of a subcommittee;

b) to the technical management board on a decision of a technical committee;

c) to the council board on a decision of the technical management board,

within 12 weeks in ISO and 8 weeks in IEC of the decision in question.

The decision of the council board on any case of appeal is final.

5.1.2 A P-member of a technical committee or subcommittee may appeal against any action, or inaction, on the part of the technical committee or subcommittee, when the P-member considers that such action or inaction is

a) not in accordance with

— the Statutes and Rules of Procedure;

— the ISO/IEC Directives; or

b) not in the best interests of international trade and commerce, or such public factors as safety, health or environment.

5.1.3 Matters under appeal may be either technical or administrative in nature.

Appeals on decisions concerning new work item proposals, committee drafts, enquiry drafts and final draft International Standards are only eligible for consideration if

— questions of principle are involved, or

— the contents of a draft may be detrimental to the reputation of ISO or IEC.

5.1.4 All appeals shall be fully documented to support the P-member's concern.

5.2 Appeal against a subcommittee decision

5.2.1 The documented appeal shall be submitted by the P-member to the secretariat of the parent technical committee, with a copy to the Chief Executive Officer.

5.2.2 Upon receipt, the secretariat of the parent technical committee shall advise all its P-members of the appeal and take immediate action, by correspondence or at a meeting, to consider and decide on the appeal, consulting the Chief Executive Officer in the process.

5.2.3 If the technical committee supports its subcommittee, then the P-member which initiated the appeal may either

— accept the technical committee decision, or

— appeal against it.

5.3 Appeal against a technical committee decision

5.3.1 Appeals against a technical committee decision may be of 2 kinds:

— an appeal arising out of 5.2.3 above, or

— an appeal against an original decision of a technical committee.

5.3.2 The documented appeal shall, in all cases, be submitted to the Chief Executive Officer, with a copy to the chair and secretariat of the technical committee.

5.3.3 The Chief Executive Officer shall, following whatever consultations s/he deems appropriate, refer the appeal together with his/her comments to the technical management board within 4 weeks after receipt of the appeal.

5.3.4 The technical management board shall decide whether an appeal shall be further processed or not. If the decision is in favour of proceeding, the chair of the technical management board shall form a conciliation panel.

The conciliation panel shall hear the appeal within 12 weeks and attempt to resolve the difference of opinion as soon as practicable. The conciliation panel shall give a final report within 12 weeks. If the conciliation panel is unsuccessful in resolving the difference of opinion, this shall be reported to the Chief Executive Officer, together with recommendations on how the matter should be settled.

5.3.5 The Chief Executive Officer, on receipt of the report of the conciliation panel, shall inform the technical management board, which will make its decision.

5.4 Appeal against a technical management board decision

An appeal against a decision of the technical management board shall be submitted to the Chief Executive Officer with full documentation on all stages of the case.

The Chief Executive Officer shall refer the appeal together with his/her comments to the members of the council board within 4 weeks after receipt of the appeal.

The council board shall make its decision within 12 weeks.

5.5 Progress of work during an appeal process

When an appeal is against a decision respecting work in progress, the work shall be continued, up to and including the approval stage (see 2.7).

Annex A
(normative)
Guides

A.1 Introduction

In addition to International Standards, Technical Specifications, Publicly Available Specifications and Technical Reports prepared by technical committees, ISO and IEC publish Guides on matters related to international standardization. Guides shall be drafted in accordance with the ISO/IEC Directives, Part 2.

Guides shall not be prepared by technical committees and subcommittees. They may be prepared by an ISO Policy Development Committee, an IEC Advisory Committee or Strategic Group, an ISO group reporting to the ISO technical management board, or an ISO/IEC Joint Coordination Group. These bodies are referred to below as the "Committee or Group responsible for the project".

The procedure for preparation and publication of a Guide is as described below.

A.2 Proposal stage

The ISO and/or IEC technical management board will approve proposals for new Guides or revisions of Guides and decide on the secretariat and composition of the Committee or Group responsible for the project.

Once a project is approved by the ISO and/or IEC technical management board, the secretariat of the Committee or Group responsible for the project shall ensure that the appropriate interests in ISO and IEC are informed.

A.3 Preparatory stage

The Committee or Group responsible for the project shall ensure that the appropriate interests in ISO and IEC have the opportunity to be represented during the preparation of the working draft.

A.4 Committee stage

Once a working draft is available for circulation as a committee draft, the secretariat of the Committee or Group responsible for the project shall send it to the parent committee or ISO and/or IEC technical management board for vote, comments and to approve its advancement to the Enquiry stage.

A.5 Enquiry stage

A.5.1 The office of the CEOs shall circulate both the English and French texts of the revised draft Guide to all National Bodies for a 16-week vote.

A.5.2 The draft Guide is approved for publication as a Guide if not more than one-quarter of the votes cast are negative, abstentions being excluded when the votes are counted.

In the case of ISO/IEC Guides, the draft shall be submitted for approval to the National Bodies of both ISO and IEC. The National Bodies of both organizations need to approve the document if it is to be published as an ISO/IEC Guide.

If this condition is satisfied for only one of the organizations, ISO or IEC, the Guide may be published under the name of the approving organization only, unless the Committee or Group responsible for the project decides to apply the procedure set out in A.5.3.

A.5.3 If a draft Guide is not approved, or if it is approved with comments the acceptance of which would improve consensus, the chair of the Committee or Group responsible for the project may decide to submit an amended draft for a 8-week vote. The conditions for acceptance of the amended draft are the same as in A.5.2.

A.6 Publication stage

The publication stage shall be the responsibility of the office of the CEO of the organization to which the Committee or Group responsible for the project belongs.

In the case of a Joint ISO/IEC Group, the responsibility shall be decided by agreement between the Chief Executive Officers.

A.7 Withdrawal of a Guide

The Committee or Group responsible for the Guide shall be responsible for deciding if the Guide shall be withdrawn. The formal withdrawal shall be ratified by the technical management board (TMB) in accordance with its normal procedures.

A.8 Procedure for transforming a guide into an ISO International Standard

1. Submission of a proposal for transformation of a Guide into an International Standard

Any member body, organization in liaison or group reporting to the TMB ("Authorized bodies"), or the ISO CEO, can propose that an approved Guide be transformed into an ISO International Standard. All proposals shall be accompanied by justification for the transformation into an International Standard.

2. Circulation of the proposal

The offices of the CEO's circulate the proposal to all member bodies for approval.

3. Member body ballot

3.1 The ballot will ask the question: Do you approve the transformation of Guide XX into an ISO International Standard?

3.2 Members will have the option of responding yes or no and the reasons for their negative vote should be provided.

3.3 Because the Guide in question has been developed and approved according to the consensus building procedures of ISO and/or IEC, technical or editorial comments will not be considered. Consideration of such comments will be held over to the next revision of the International Standard if the transformation of the Guide is approved.

3.4 The voting period and voting rules are the same as for the normal DIS vote.

4. Approval criteria

4.1 If more than one quarter of all votes received is negative, the transformation into an International Standard is not approved and the approval process ends.

4.2 If the transformation of a Guide into a standard is approved, the International Standard will be published, and the original Guide withdrawn.

5. Maintenance of the International Standard resulting from a transformed guide

The ISO technical management board will decide on a case-by-case basis how to handle the maintenance of transformed Guide. Preference will be given to assigning the maintenance responsibility to an appropriate TC, but, in the absence of an appropriate TC, the TMB will decide where to assign maintenance responsibility, including through the establishment of a new TC or PC.

Annex B
(normative)

ISO/IEC procedures for liaison and work allocation

B.1 Introduction

By the ISO/IEC Agreement of 1976 [1], ISO and IEC together form a system for international standardization as a whole. For this system to operate efficiently, the following procedures are agreed for coordination and allocation of work between the technical committees and subcommittees of both organizations.

B.2 General considerations

The allocation of work between ISO and IEC is based on the agreed principle that all questions relating to international standardization in the electrical and electronic engineering fields are reserved to IEC, the other fields being reserved to ISO and that allocation of responsibility for matters of international standardization where the relative contribution of electrical and non-electrical technologies is not immediately evident will be settled by mutual agreement between the organizations.

Questions of coordination and work allocation may arise when establishing a new ISO or IEC technical committee, or as a result of the activities of an existing technical committee.

The following levels of coordination and work allocation agreement are available. Matters should be raised at the next higher level only after all attempts to resolve them at the lower levels have failed.

a) **Formal liaisons** between ISO and IEC committees for normal inter-committee cooperation.

b) **Organizational consultations**, including technical experts and representatives of the Chief Executive Of ficers, for cases where technical coordination may have an effect on the future activities of the organizations in a larger sense than the point under consideration.

c) Decisions on work allocation

   — by the technical management boards or, if necessary,

   — the ISO/IEC Joint Technical Advisory Board (JTAB).

Questions affecting both ISO and IEC, on which it has not proved possible to obtain a common decision by the ISO Technical Management Board and the IEC Standardization Management Board, are referred to the ISO/IEC Joint Technical Advisory Board (JTAB) for decision (see 1.3.1).

B.3 Establishing new technical committees

Whenever a proposal to establish a new technical committee is made to the National Bodies of ISO or of IEC respectively, the proposal shall also be submitted to the other organization requesting comment and/or agreement. As a result of these consultations, two cases may arise:

a) the opinion is unanimous that the work should be carried out in one of the organizations;

b) opinions are divided.

---

1) ISO Council resolutions 49/1976 and 50/1976 and IEC Administrative Circular No. 13/1977.

In case a), formal action may then be taken to establish the new technical committee according to the unanimous opinion.

In case b), a meeting of experts in the field concerned shall be arranged with representatives of the Chief Executive Officers with a view to reaching a satisfactory agreement for allocation of the work (i.e., organizational level). If agreement is reached at this level, formal action may be taken by the appropriate organization to implement the agreement.

In the case of disagreement after these consultations, the matter may be referred by either organization to the ISO/IEC Joint Technical Advisory Board (JTAB).

B.4 Coordinating and allocating work between ISO and IEC technical committees

B.4.1 Formal liaison at TC level

Most coordination needs arising between individual ISO and IEC committees are successfully dealt with through formal technical liaison arrangements. These arrangements, when requested by either organization, shall be honoured by the other organization. Requests for formal liaison arrangements are controlled by the offices of the CEOs. The requesting organization shall specify the type of liaison required, such as:

a) full or selective exchange of committee documents;

b) regular or selective attendance of liaison representatives at meetings;

c) participation in a standing coordination (or steering) committee for selected ISO and IEC technical committees;

d) setting up of a Joint Working Group (JWG).

B.4.2 Details of agreement

B.4.2.1 Continual efforts shall be made to minimize the overlap areas between IEC and ISO by entrusting areas of work to one of the two organizations.

For areas of work so entrusted, IEC and ISO shall agree through the JTAB on how the views and interests of the other organization are to be fully taken into account.

B.4.2.2 Five working modes of cooperation have been established, as follows:

Mode 1 — Informative relation

One organization is fully entrusted with a specific work area and keeps the other fully informed of all progress.

Mode 2 — Contributive relation

One organization should take the lead of the work and the other should make written contributions where considered appropriate during the progress of this work. This relation also includes the exchange of full information.

Mode 3 — Subcontracting relation

One organization is fully entrusted with the realization of the work on an identified item, but due to specialization of the other, a part of the work is subcontracted and that part is prepared under the responsibility of the second organization. Necessary arrangements shall be made to guarantee the correct integration of the resulting subcontracted work into the main part of the programme. The enquiry and approval stages are handled by the organization being the main contractor for the standardization task.

Mode 4 — Collaborative relation

One organization takes the lead in the activities, but the work sessions and meetings receive liaison representatives from the other. Such liaison representatives should have the right to intervene in the debate but have no right to vote. The full flow of information is oriented through this liaison.

Mode 5 — Integrated liaison

Joint Working Groups and Joint Technical Committees ensure integrated meetings for handling together the realization of standards under a principle of total equality of participation.

Joint Working Groups between technical committees of the two organizations shall operate in accordance with 1.12.6.

B.4.2.3 The allocation of work between IEC and ISO for potentially overlapping areas will be set out as required in schedules or programmes which, when agreed by the relevant parties, will form addenda to this agreement.

A consequence of this agreement is that the parties agree to cross-refer to the relevant standards of the other in the respective competent fields of interest.

When the standard being referred to is updated, it is the responsibility of the body making the reference to take care of the updating of the reference where appropriate.

B.4.2.4 For work for which one organization has assumed responsibility and for which there will be subcontracting of work to the other, the fullest account shall be taken of the interests participating in the subcontracted work in defining the objectives of that work.

B.4.2.5 The necessary procedures for enquiry and approval shall be realized by the organization entrusted with a particular standardization task, except as otherwise agreed by the two technical management boards.

B.4.2.6 For standards developed under the Mode 5 — Integrated liaison, the committee, enquiry and approval stages shall be carried out in parallel in both ISO and IEC in accordance with the rules of the organization with the administrative lead. The committee/organization with the administrative responsibility for the project shall submit drafts for the committee, enquiry and approval stages to the other organization two weeks prior to the circulation date.

B.4.2.7 When the enquiry draft has not fulfilled the approval criteria (see 2.6.3) in one of the organizations, then:

— the officers of the committees involved in the joint working group may select one of options given in 2.6.4 c) or

— in exceptional circumstances, if agreed between the officers of the ISO and IEC committees involved in the joint working group and the offices of the CEO, the project may proceed as a single logo standard of the organization in which the enquiry draft was approved. The joint working group is automatically disbanded.

B.4.2.8 If the final draft International Standard is not approved in accordance of the conditions in 2.7.3 then:

— the committees involved in the joint working group may select one of the options given in 2.7.7, noting that in IEC the circulation of a second final draft International Standard is not allowed and will require a derogation of the TMB or

— in exceptional circumstances, if agreed between the officers of the ISO and IEC committees involved in the joint working group and the offices of the CEO, the standard may be published as a single logo standard of the organization in which the final draft International Standard was approved. The joint working group is automatically disbanded.

B.4.2.9 Standards developed under the Mode 5 — Integrated liaison via a joint working group between ISO and IEC are published by the organization of the committee having administrative responsibility. That organization assigns the reference number of the standard and owns the copyright of the standard. The standard carries the logo of the other organization and may be sold by both organizations. The foreword of the International Standard will identify all the committees responsible for the development. For those standards where the committee with the administrative responsibility is in the IEC, then the foreword will also give the ISO voting results. ISO-lead documents are assigned numbers from 1 to 59999. IEC-lead documents are assigned numbers from 60000 to 79999. In the case of multi-part standards, some parts being under ISO responsibility and some being under IEC responsibility, a number in the 80000 series is assigned (e.g. ISO 80000-1, IEC 80000-6).

B.4.2.10 The maintenance procedures to be used for standards developed under the Mode 5 — Integrated liaison will be those currently applied in the organization which has the committee with the administrative responsibility.

B.4.2.11 If there is a reason, during the development of the project, to change from one mode of operation to another, a recommendation shall be made by both technical committees concerned and submitted to the two technical management boards for information.

B.4.3 Cooperation of secretariats

The secretariats of the technical committees/subcommittees from the two organizations concerned shall cooperate on the implementation of this agreement. There shall be a complete information flow on on-going work and availability on demand to each other of working documents, in accordance with normal procedures.

Annex C
(normative)

Justification of proposals for the establishment of standards

C.1 General

C.1.1 Because of the large financial resources and manpower involved and the necessity to allocate these according to the needs, it is important that any standardization activity begin by identifying the needs, determining the aims of the standard(s) to be prepared and the interests that may be affected. This will, moreover, help to ensure that the standards produced will cover appropriately the aspects required and be market relevant for the affected sectors. Any new activity shall therefore be reasonably justified before it is begun.

C.1.2 It is understood that, whatever conclusions may be drawn on the basis of the annex, a prerequisite of any new work to be commenced would be a clear indication of the readiness of a sufficient number of relevant interested parties to allocate necessary manpower, funds and to take an active part in the work.

C.1.3 This annex sets out rules for proposing and justifying new work, so that proposals will offer to others the clearest possible idea of the purposes and extent of the work, in order to ensure that standardization resources are really allocated by the parties concerned and are used to the best effect.

C.1.4 This annex does not contain rules of procedure for implementing and monitoring the guidelines contained in it, nor does it deal with the administrative mechanism which should be established to this effect.

C.1.5 This annex is addressed primarily to the proposer of any kind of new work to be started but may serve as a tool for those who will analyse such a proposal or comment on it, as well as for the body responsible for taking a decision on the proposal.

C.2 Terms and definitions

C.2.1
proposal for new work
proposal for a new field of technical activity or for a new work item

C.2.2
proposal for a new field of technical activity
proposal for the preparation of (a) standard(s) in a field that is not covered by an existing committee (such as a technical committee, subcommittee or project committee) of the organization to which the proposal is made

C.2.3
proposal for a new work item
proposal for the preparation of a standard or a series of related standards in the field covered by an existing committee (such as a technical committee) of the organization to which the proposal is made

C.3 General principles

C.3.1 Any proposal for new work shall lie within the scope of the organization to which it is submitted.

NOTE   For example, the objects of ISO are laid down in its Statutes and of IEC in Article 2 of its Statutes.

C.3.2 The documentation justifying new work in ISO and IEC shall make a substantial case for the market relevance of the proposal.

C.3.3 The documentation justifying new work in ISO and IEC shall provide solid information as a foundation for informed ISO or IEC National Body voting.

C.3.4 Within the ISO and IEC systems, the onus is considered to be placed on the proposer to provide the proper documentation to support principles C.3.2 and C.3.3 stated above.

C.4 Elements to be clarified when proposing a new field of technical activity or a new work item

C.4.1 Proposals for new fields of technical activity and new work items shall include the following fields of information (C.4.2 to C.4.13).

C.4.2 Title

The title shall indicate clearly yet concisely the new field of technical activity or the new work item which the proposal is intended to cover.

EXAMPLE 1   (proposal for a new technical activity) "Machine tools".

EXAMPLE 2   (proposal for a new work item) "Electrotechnical products — Basic environmental testing procedures".

C.4.3 Scope

C.4.3.1 For new fields of technical activity

The scope shall precisely define the limits of the field of activity. Scopes shall not repeat general aims and principles governing the work of the organization but shall indicate the specific area concerned.

EXAMPLE   "Standardization of all machine tools for the working of metal, wood and plastics, operating by removal of material or by pressure".

C.4.3.2 For new work items

The scope shall give a clear indication of the coverage of the proposed new work item and, if necessary for clarity, exclusions shall be stated.

EXAMPLE 1

This standard lists a series of environmental test procedures, and their severities, designed to assess the ability of electrotechnical products to perform under expected conditions of service.

Although primarily intended for such applications, this standard may be used in other fields where desired.

Other environmental tests, specific to the individual types of specimen, may be included in the relevant specifications.

EXAMPLE 2

Standardization in the field of fisheries and aquaculture, including, but not limited to, terminology, technical specifications for equipment and for their operation, characterization of aquaculture sites and maintenance of appropriate physical, chemical and biological conditions, environmental monitoring, data reporting, traceability and waste disposal.

Excluded:

— methods of analysis of food products (covered by ISO/TC 34);

— personal protective clothing (covered by ISO/TC 94);

— environmental monitoring (covered by ISO/TC 207).

C.4.4 Proposed initial programme of work (for proposals for new fields of technical activity only)

C.4.4.1 The proposed programme of work shall correspond to and clearly reflect the aims of the standardization activities and shall, therefore, show the relationship between the subjects proposed.

C.4.4.2 Each item on the programme of work shall be defined by both the subject and aspect(s) to be standardized (for products, for example, the items would be the types of products, characteristics, other requirements, data to be supplied, test methods, etc.).

C.4.4.3 Supplementary justification may be combined with particular items in the programme of work.

C.4.4.4 The proposed programme of work shall also suggest priorities and target dates for new work items (when a series of standards is proposed, priorities shall be suggested).

C.4.5 Indication(s) of the preferred type or types of deliverable(s) to be produced

In the case of proposals for new fields of technical activity, this may be provided under C.4.4.

C.4.6 A listing of relevant existing documents at the international, regional and national levels

Any known relevant documents (such as standards and regulations) shall be listed, regardless of their source and should be accompanied by an indication of their significance.

C.4.7 Relation to and impact on existing work

C.4.7.1 A statement shall be provided regarding any relation or impact the proposed work may have on existing work, especially existing ISO and IEC deliverables. The proposer should explain how the work differs from apparently similar work, or explain how duplication and conflict will be minimized.

C.4.7.2 If seemingly similar or related work is already in the scope of other committees of the organization or in other organizations, the proposed scope shall distinguish between the proposed work and the other work.

C.4.7.3 The proposer shall indicate whether his or her proposal could be dealt with by widening the scope of an existing committee or by establishing a new committee.

C.4.8 Relevant country participation

C.4.8.1 For proposals for new fields of technical activity, a listing of relevant countries should be provided where the subject of the proposal is important to their national commercial interests.

C.4.8.2 For proposals for new work item within existing committees, a listing of relevant countries should be provided which are not already P-members of the committee, but for whom the subject of the proposal is important to their national commercial interests.

C.4.9 Cooperation and liaison

C.4.9.1 A list of relevant external international organizations or internal parties (other than ISO and/or IEC committees) to be engaged as liaisons in the development of the deliverable(s) shall be provided.

C.4.9.2 In order to avoid conflict with, or duplication of efforts of, other bodies, it is important to indicate all points of possible conflict or overlap.

C.4.9.3 The result of any communication with other interested bodies shall also be included.

C.4.10 Affected stakeholders

A simple and concise statement shall be provided identifying and describing relevant affected stakeholder categories (including small and medium sized enterprises) and how they will each benefit from or be impacted by the proposed deliverable(s).

C.4.11 Base document (for proposals for new work items only)

C.4.11.1 When the proposer considers that an existing well-established document may be acceptable as a standard (with or without amendments) this shall be indicated with appropriate justification and a copy attached to the proposal.

C.4.11.2 All proposals for new work items shall include an attached existing document to serve as an initial basis for the ISO or IEC deliverable or a proposed outline or table of contents.

C.4.11.3 If an existing document is attached that is copyrighted or includes copyrighted content, the proposer shall ensure that appropriate permissions have been granted in writing for ISO or IEC to use that copyrighted content.

C.4.12 Leadership commitment

C.4.12.1 In the case of a proposal for a new field of technical activity, the proposer shall indicate whether his/her organization is prepared to undertake the secretariat work required.

C.4.12.2 In the case of a proposal for new work item, the proposer shall also nominate a project leader.

C.4.13 Purpose and justification

C.4.13.1 The purpose and justification of the standard to be prepared shall be made clear and the need for standardization of each aspect (such as characteristics) to be included in the standard shall be justified.

C.4.13.2 If a series of new work items is proposed the purpose and the justification of which is common, a common proposal may be drafted including all elements to be clarified and enumerating the titles and scopes of each individual item.

C.4.13.3 Please note that the items listed in the bullet points below represent a menu of suggestions or ideas for possible documentation to support the purpose and justification of proposals. Proposers should consider these suggestions, but they are not limited to them, nor are they required to comply strictly with them. What is most important is that proposers develop and provide purpose and justification information that is most relevant to their proposals and that makes a substantial business case for the market relevance and need of their proposals. Thorough, well-developed and robust purpose and justification documentation will lead to more informed consideration of proposals and ultimately their possible success in the ISO and IEC systems.

— A simple and concise statement describing the business, technological, societal or environmental issue that the proposal seeks to address, preferably linked to the Strategic Business Plan of the concerned ISO or IEC committee.

— Documentation on relevant global metrics that demonstrate the extent or magnitude of the economic, technological, societal or environmental issue, or the new market. This may include an estimate of the potential sales of the resulting standard(s) as an indicator of potential usage and global relevance.

— Technological benefit — a simple and concise statement describing the technological impact of the proposal to support coherence in systems and emerging technologies, convergence of merging technologies, interoperability, resolution of competing technologies, future innovation, etc.

— Economic benefit — a simple and concise statement describing the potential of the proposal to remove barriers to trade, improve international market access, support public procurement, improve business efficiency for a broad range of enterprises including small and medium sized ones, and/or result in a flexible, cost-effective means of complying with international and regional rules/conventions, etc. A simple cost/benefit analysis relating the cost of producing the deliverable(s) to the expected economic benefit to businesses worldwide may also be helpful.

— Societal benefit(s) — a simple and concise statement describing any societal benefits expected from the proposed deliverable(s).

— Environmental benefit(s) — a simple and concise statement describing any environmental or wider sustainability benefits expected from the proposed deliverable(s).

— A simple and concise statement clearly describing the intended use(s) of the proposed deliverable(s), for example, whether the deliverable is intended as requirements to support conformity assessment or only as guidance or recommended best practices; whether the deliverable is a management system standard; whether the deliverable is intended for use or reference in technical regulation; whether the deliverable is intended to be used to support legal cases in relation to international treaties and agreements.

— A simple and concise statement of metrics for the committee to track in order to assess the impact of the published standard over time to achieve the benefits to stakeholders documented under C.4.10 above.

— A statement assessing the prospect of the resulting deliverable(s) being compliant with, for the IEC, the IEC Global Relevance Policy: http://www.iec.ch/members_experts/refdocs/ac_cl/AC_200817e_AC.pdf and for ISO, with ISO's Global Relevance Policy http://www.iso.org/iso/home/standards_development/governance_of_technical_work.htm and the ISO/TMB recommendations (see NOTE below) regarding sustainable development and sustainability, where relevant.

NOTE   For ISO, the ISO/TMB confirmed the following recommendations: 1) When a committee (in any sector) develops a standard dealing with sustainability/sustainable development the standard must remain within the context of the committee's scope of work; 2) The committee should also notify the TMB with the title and scope as early as possible; 3) The committee undertaking such work should clarify its intentions in the Introduction of the specific standard(s); 4) The most widely used definition of sustainable development is the one from the UN Brundtland committee on sustainable development: development that meets the needs of the present without compromising the ability of future generations to meet their own needs.

— A statement assessing the proposal's compliance with the Principles for developing ISO and IEC Standards related to or supporting public policy initiatives (for ISO see Annex SO in the Consolidated ISO Supplement and for IEC and ISO see Using and referencing ISO and IEC standards to support public policy: http://www.iso.org/sites/policy/) and the possible relation of the resulting deliverable(s) to public policy, including a statement regarding the potential for easier market access due to conformity with appropriate legislation.

Annex D
(normative)

Resources of secretariats and qualifications of secretaries

D.1 Terms and definitions

D.1.1
secretariat
National Body to which has been assigned, by mutual agreement, the responsibility for providing technical and administrative services to a technical committee or subcommittee

D.1.2
secretary
individual appointed by the secretariat to manage the technical and administrative services provided

D.2 Resources of a secretariat

A National Body to which a secretariat has been assigned shall recognize that, no matter what arrangements it makes in its country to provide the required services, it is the National Body itself that is ultimately responsible for the proper functioning of the secretariat. National Bodies undertaking secretariat functions shall become party to the ISO Service Agreement or IEC Basic Agreement, as appropriate.

The secretariat shall therefore have adequate administrative and financial means or backing to ensure:

a) facilities for word-processing in English and/or French, for providing texts electronically, and for any necessary reproduction of documents;

b) preparation of adequate technical illustrations;

c) identification and use, with translation where necessary, of documents received in the official languages;

d) updating and continuous supervision of the structure of the committee and its subsidiary bodies, if any;

e) reception and prompt dispatch of correspondence and documents;

f) adequate communication facilities by telephone, telefax and electronic mail;

g) access to the Internet;

h) arrangements and facilities for translation, interpretation and services during meetings, in collaboration with the host National Body, as required;

i) attendance of the secretary at any meetings requiring his/her presence, including technical committee and/or subcommittee meetings, editing committee meetings, working group meetings, and consultations with the chair when necessary;

j) access by the secretary to basic International Standards (see the ISO/IEC Directives, Part 2 on "Referencing Documents and sources for drafting") and to International Standards, national standards and/or related documents in the field under consideration;

k) access by the secretary, when necessary, to experts capable of advising on technical issues in the field of the committee.

Whilst the Chief Executive Officer endeavours to send his/her representative to the first meeting of a technical committee, to meetings of technical committees with new secretariats, and to any technical committee or subcommittee meeting where such presence is desirable for solving problems, the office of the CEO cannot undertake to carry out the work for a secretariat, on a permanent or temporary basis.

D.3 Requirements of a secretary

The individual appointed as secretary shall

a) have sufficient knowledge of English and/or French;

b) be familiar with the *Statutes and rules of procedure*, as appropriate, and with the ISO/IEC Directives (see the respective Supplements to the ISO/IEC Directives);

c) be in a position to advise the committee and any subsidiary bodies on any point of procedure or drafting, after consultation with the office of the CEO if necessary;

d) be aware of any council board or technical management board decision regarding the activities of the technical committees in general and of the committee for which s/he is responsible in particular;

e) be a good organizer and have training in and ability for technical and administrative work, in order to organize and conduct the work of the committee and to promote active participation on the part of committee members and subsidiary bodies, if any;

f) be familiar with the documentation supplied by the offices of CEO, in particular the use of electronic tools and services.

It is recommended that newly appointed secretaries of technical committees should make an early visit to the office of the CEO in Geneva in order to discuss procedures and working methods with the staff concerned.

Annex E
(normative)

General policy on the use of languages

E.1 Expressing and communicating ideas in an international environment

At the international level, it is common practice to publish deliverables in at least two languages. There are a number of reasons why it is advantageous to use two languages, for example:

— greater clarity and accuracy of meaning can be achieved by expressing a given concept in two languages which have different grammar and syntax;

— if consensus is reached on the basis of a text drafted in only one language, difficulties may arise when it comes to putting that text into another language. Some questions may have to be rediscussed, and this can cause delay if the text originally agreed upon has to be altered. Subsequent drafting into a second language of a text already approved in the first language often brings to light difficulties of expression that could have been avoided if both versions had been prepared at the same time and then amended together;

— to ensure that international meetings will be as productive as possible, it is important for the agreements reached to be utterly devoid of ambiguity, and there has to be no risk that these agreements can be called back into question because of misunderstandings of a linguistic nature;

— the use of two languages chosen from two linguistic groups widens the number of prospective delegates who might be appointed to attend the meetings;

— it becomes easier to express a concept properly in other languages if there are already two perfectly harmonized versions.

E.2 The use of languages in the technical work

The official languages are English, French and Russian.

The work of the technical committees and the correspondence are in English by default.

For the purposes of the above, the National Body for the Russian Federation provides all interpretation and translation into and from the Russian language.

In IEC, a definitive language of development for each deliverable shall be designated in the Foreword. Specific exceptions apply to the IEV and/or database standards.

E.3 International Standards

International Standards are published by the ISO and IEC in English and in French (and sometimes in multilingual editions also including Russian and other languages, especially in cases of terminology). These versions of a given International Standard are equivalent, and each is regarded as being an original-language version.

It is advantageous for the technical content of a standard to be expressed in both English and French from the outset of the drafting procedure, so that these two versions will be studied, amended and adopted at the same time and their linguistic equivalence will be ensured at all times. (See also the ISO/IEC Directives, Part 2, clause on "Language versions").

This may be done

— by the secretariat or, under the latter's responsibility, with outside assistance, or

— by the editing committee of the responsible technical committee or subcommittee, or

— by National Bodies whose national language is English or French and under an agreement concluded between those National Bodies and the secretariat concerned.

When it is decided to publish a multilingual International Standard (a vocabulary, for example), the National Body for the Russian Federation takes charge of the Russian portion of the text; similarly, when it is decided to publish an International Standard containing terms or material in languages other than the official languages, the National Bodies whose national languages are involved are responsible for selecting the terms or for drafting the portions of text which are to be in those languages.

E.4 Other publications developed by technical committees

Other publications may be issued in one official language only.

E.5 Documents for technical committee and subcommittee meetings

E.5.1 Drafts and documents referred to the agenda

The documents prepared and circulated prior to a meeting are the following.

a) **Draft agendas**

Draft agendas are prepared and distributed in the language(s) of the meeting (English by default) by the responsible secretariats.

b) **Committee drafts referred to in the agenda**

It is desirable that versions of the committee drafts referred to in the agenda will be available for the meeting in the language(s) of the meeting (English by default).

Enquiry drafts shall be available in English and French. The ISO Council or IEC Standardization Management Board guidelines shall be applied where one of the language versions is not available in due time.

Other documents (sundry proposals, comments, etc.) relating to agenda items may be prepared in only one language (English or French).

E.5.2 Documents prepared and circulated during a meeting

The documents prepared and circulated during a meeting are the following.

a) **Resolutions adopted during the meeting**

An ad hoc drafting committee, formed at the beginning of each meeting and comprising the secretary and, whenever possible, one or more delegates of English and/or French mother tongue, edits each of the proposed resolutions.

b) **Brief minutes, if any, prepared after each session**

If such minutes are prepared, they shall be drafted in English or French and preferably in both with, if necessary, the assistance of the ad hoc drafting committee.

E.5.3 Documents prepared and circulated after a meeting

After each technical committee or subcommittee meeting, the secretariat concerned shall draft a report of the meeting, which may be in only one language (English or French) and which includes, as annex, the full text of the resolutions adopted, preferably in both English and French.

E.6 Documents prepared in languages other than English or French

National Bodies whose national language is neither English nor French may translate any documents circulated by secretariats into their own national language in order to facilitate the study of those documents by the experts of their country or to assist the delegates they have appointed to attend the meetings of the technical committees and subcommittees.

If one language is common to two or more National Bodies, one of them may at any time take the initiative of translating technical documents into that language and of providing copies to other National Bodies in the same linguistic group.

The terms of the above two paragraphs may be applied by the secretariats for their own needs.

E.7 Technical meetings

E.7.1 Purpose

The purpose of technical meetings is to achieve as full agreement as possible on the various agenda items and every effort shall be made to ensure that all delegates understand one another.

E.7.2 Interpretation of debates into English and French

Although the basic documents may be available in both English and French, it has to be determined according to the case whether interpretation of statements expressed in one language should be given in the other language

— by a volunteer delegate,

— by a staff member from the secretariat or host National Body, or

— by an adequately qualified interpreter.

Care should also be taken that delegates who have neither English nor French as mother tongue can follow the meeting to a sufficient extent.

It is impractical to specify rules concerning the necessity of interpreting the debates at technical meetings. It is essential, of course that all delegates should be able to follow the discussions, but it may not be altogether essential to have a word-for-word interpretation of each statement made.

In view of the foregoing, and except in special cases where interpretation may not be necessary, the following practice is considered appropriate:

a) for meetings where procedural decisions are expected to be taken, brief interpretation may be provided by a member of the secretariat or a volunteer delegate;

b) at working group meetings, the members should, whenever possible, arrange between themselves for any necessary interpretation on the initiative and under the authority of the convenor of the working group.

To enable the secretariat responsible for a meeting to make any necessary arrangements for interpretation, the secretariat should be informed, at the same time as it is notified of attendance at the meeting, of the languages in which the delegates are able to express themselves and of any aid which delegates might be able to provide in the matter of interpretation.

In those cases where a meeting is conducted mainly in one language, the following practice should be adopted as far as is practicable in order to assist delegates having the other language:

a) the decision taken on one subject should be announced in both languages before passing to the next subject;

b) whenever a change to an existing text is approved in one language, time should be allowed for delegates to consider the effect of this change on the other language version;

c) a summary of what has been said should be provided in the other language if a delegate so requests.

E.7.3 Interpretation into English and French of statements made in other languages

When at a meeting of a technical committee or a subcommittee a participant wishes, in view of exceptional circumstances, to speak in any language other than English or French, the chair of the session shall be entitled to authorize this, for the session in question, provided that a means of interpretation has been secured.

In order to give all experts an equal opportunity to express their views at meetings of technical committees and subcommittees, a very flexible application of this provision is recommended.

Annex F
(normative)

Options for development of a project

F.1 Simplified diagram of options

Project stage	Normal procedure	Draft submitted with proposal	"Fast-track procedure" [a]	Technical Specification [b]	Technical Report [c]	Publicly Available Specification [d]
Proposal stage (see 2.3)	Acceptance of proposal	Acceptance of proposal	Acceptance of proposal [a]	Acceptance of proposal		Acceptance of proposal [g]
Preparatory stage (see 2.4)	Preparation of working draft	*Study by working group* [e]		Preparation of draft		Approval of draft PAS
Committee stage (see 2.5)	Development and acceptance of committee draft	*Development and acceptance of committee draft* [e]		Acceptance of draft	Acceptance of draft	
Enquiry stage (see 2.6 and in IEC, see IEC Supplement E.3.1)	Development and acceptance of enquiry draft	Development and acceptance of enquiry draft	Acceptance of enquiry draft			
Approval stage (see 2.7)	*Approval of FDIS* [f]	*Approval of FDIS* [f]	*Approval of FDIS* [f]			
Publication stage (see 2.8)	Publication of International Standard	Publication of International Standard	Publication of International Standard	Publication of Technical Specification	Publication of Technical Report	Publication of PAS

Stages in *italics*, enclosed by dotted circles may be omitted.

[a] See F.2.
[b] See 3.1.
[c] See 3.3.
[d] See 3.2.
[e] According to the result of the vote on the new work item proposal, both the preparatory stage and the committee stage may be omitted.
[f] May be omitted if the approval criteria of 2.6.3 are met and no technical changes are to be included.
[g] See ISO and IEC Supplements for details on proposals for PAS.

F.2 "Fast-track procedure"

F.2.1 Proposals to apply the fast-track procedure may be made as follows.

F.2.1.1 Any P-member or category A liaison organization of a concerned technical committee or subcommittee may propose that an **existing standard from any source** be submitted for vote as an enquiry draft. The proposer shall obtain the agreement of the originating organization before making a proposal. The criteria for proposing an existing standard for the fast-track procedure are a matter for each proposer to decide.

F.2.1.2 An international standardizing body recognized by the ISO or IEC council board may propose that a **standard developed by that body** be submitted for vote as a final draft International Standard.

F.2.1.3 An organization having entered into a formal technical agreement with ISO or IEC may propose, in agreement with the appropriate technical committee or subcommittee, that a **draft standard developed by that organization** be submitted for vote as an enquiry draft within that technical committee or subcommittee.

F.2.2 The proposal shall be received by the Chief Executive Officer, who shall take the following actions:

a) settle the copyright and/or trademark situation with the organization having originated the proposed document, so that it can be freely copied and distributed to National Bodies without restriction, and advise the organization that the ISO/IEC intellectual property policies shall apply to the proposed document, see in particular 2.13 and 2.14;

b) for cases F.2.1.1 and F.2.1.3, assess in consultation with the relevant secretariats which technical committee/subcommittee is competent for the subject covered by the proposed document; where no technical committee exists competent to deal with the subject of the document in question, the Chief Executive Officer shall refer the proposal to the technical management board, which may request the Chief Executive Officer to submit the document to the enquiry stage and to establish an ad hoc group to deal with matters subsequently arising;

c) ascertain that there is no evident contradiction with other International Standards;

d) distribute the proposed document as an enquiry draft (F.2.1.1 and F.2.1.3) in accordance with 2.6.1, or as a final draft International Standard (case F.2.1.2) in accordance with 2.7.1, indicating (in cases F.2.1.1 and F.2.1.3) the technical committee/subcommittee to the domain of which the proposed document belongs.

F.2.3 The period for voting and the conditions for approval shall be as specified in 2.6 for an enquiry draft and 2.7 for a final draft International Standard. In the case where no technical committee is involved, the condition for approval of a final draft International Standard is that not more than one-quarter of the total number of votes cast are negative.

In ISO the voting period on a proposal submitted as a final draft International Standard shall be 20 weeks.

F.2.4 If, for an enquiry draft, the conditions of approval are met, the draft standard shall progress to the approval stage (2.7). If not, the proposal has failed and any further action shall be decided upon by the technical committee/subcommittee to which the document was attributed in accordance with F.2.2 b).

The committee leadership can decide whether or not to skip the FDIS vote, and go straight to publication — see 2.6.4.

If, for a final draft International Standard, the conditions of approval are met, the document shall progress to the publication stage (2.8). If not, the proposal has failed and any further action shall be decided upon by the technical committee/subcommittee to which the FDIS was attributed in accordance with F.2.2 b), or by discussion between the originating organization and the office of the CEO if no technical committee was involved.

If the standard is published, its maintenance shall be handled by the technical committee/subcommittee to which the document was attributed in accordance with F.2.2 b), or, if no technical committee was involved, the approval procedure set out above shall be repeated if the originating organization decides that changes to the standard are required.

Annex G
(normative)

Maintenance agencies

G.1 A technical committee or subcommittee developing an International Standard that will require a maintenance agency shall inform the Chief Executive Officer at an early stage in order that an ISO/TMB or IEC Council Board decision may be taken in advance of the publication of the International Standard.

G.2 The ISO/TMB or IEC Council Board designates maintenance agencies in connection with International Standards, including appointment of their members, on the proposal of the technical committee concerned.

G.3 The secretariat of a maintenance agency should be attributed wherever possible to the secretariat of the technical committee or subcommittee that has prepared the International Standard.

G.4 The Chief Executive Officer shall be responsible for contacts with external organizations associated with the work of a maintenance agency.

G.5 The rules of procedure of maintenance agencies shall be subject to ISO/TMB or IEC Council Board approval and any requested delegation of authority in connection with the updating of the International Standard or the issuing of amendments shall be specifically authorized by the ISO/TMB or IEC Council Board.

G.6 Any charges for services provided by a maintenance agency shall be authorized by the council board.

Annex H
(normative)

Registration authorities

H.1 A technical committee or subcommittee developing an International Standard that will require a registration authority shall inform the Chief Executive Officer at an early stage, in order to permit any necessary negotiations and to allow the technical management board to take a decision in advance of the publication of the International Standard.

H.2 The technical management board designates registration authorities in connection with International Standards on the proposal of the technical committee concerned.

H.3 Registration authorities should be qualified and internationally acceptable bodies; if there is no such organization available, such tasks may be conferred upon the office of the CEO by decision of the technical management board.

H.4 Registration authorities should be required to indicate clearly in their operations that they have been designated by ISO or IEC (for example, by including appropriate wording in the letterhead of the designated body).

H.5 Registration functions undertaken by the registration authority under the provisions of the relevant International Standard shall require no financial contribution from ISO or IEC or their members. This would not preclude, however, the charging for services provided by the registration authority if duly authorized by the council board.

Annex I
(normative)

Guideline for Implementation of the Common Patent Policy for ITU-T/ITU-R/ISO/IEC

The latest edition of the Guidelines for Implementation of the Common Patent Policy for ITU-T/ITU-R/ISO/IEC are available on the ISO website through the following link (including the forms in Word or Excel formats):

http://www.iso.org/iso/home/standards_development/governance_of_technical_work/patents.htm

They are also available on the IEC website through the following link:

https://www.iec.ch/members_experts/tools/patents/patent_policy.htm

Guidelines for Implementation of the Common Patent Policy for ITU-T/ITU-R/ISO/IEC

Revision 2, effective 26 June 2015

Revision 1, effective 23 April 2012

Revises initial edition of 1 March 2007

CONTENTS

Part I – Common guidelines

I.1 Purpose

I.2 Explanation of terms

I.3 Patent disclosure

I.4 Patent Statement and Licensing Declaration Form

I.4.1 The purpose of the Declaration Form

I.4.2 Contact information

I.5 Conduct of meetings

I.6 Patent Information database

I.7 Assignment or Transfer of Patent Rights

Part II – Organization-specific provisions

II.1 Specific provisions for ITU

II.2 Specific provisions for ISO and IEC

ANNEX 1: COMMON PATENT POLICY FOR ITU-T/ITU-R/ISO/IEC

ANNEX 2: PATENT STATEMENT AND LICENSING DECLARATION FORM FOR ITU-T OR ITU-R RECOMMENDATION | ISO OR IEC DELIVERABLE

ANNEX 3: GENERAL PATENT STATEMENT AND LICENSING DECLARATION FORM FOR ITU-T OR ITU-R RECOMMENDATION

Guidelines for Implementation of the Common Patent Policy for ITU-T/ITU-R/ISO/IEC

Revision 2, effective 26 June 2015

Part I – Common guidelines

I.1 Purpose

ITU, in its Telecommunication Standardization Sector (ITU-T) and its Radiocommunication Sector (ITU-R), ISO and IEC have had patent policies for many years, the purpose being to provide in simple words practical guidance to the participants in their Technical Bodies in case patent rights matters arise.

Considering that the technical experts are normally not familiar with the complex issue of patent law, the Common Patent Policy for ITU-T/ITU-R/ISO/IEC (hereafter referred to as the "Patent Policy") was drafted in its operative part as a checklist, covering the three different cases which may arise if a Recommendation | Deliverable requires licences for Patents to be practiced or implemented, fully or partly.

The Guidelines for Implementation of the Common Patent Policy for ITU-T/ITU-R/ISO/IEC (hereafter referred to as the "Guidelines") are intended to clarify and facilitate implementation of the Patent Policy, a copy of which can be found in Annex 1 and also on the web site of each Organization.

The Patent Policy encourages the early disclosure and identification of Patents that may relate to Recommendations | Deliverables under development. In doing so, greater efficiency in standards development is possible and potential patent rights problems can be avoided.

The Organizations should not be involved in evaluating patent relevance or essentiality with regards to Recommendations | Deliverables, interfere with licensing negotiations, or engage in settling disputes on Patents; this should be left - as in the past - to the parties concerned.

Organization-specific provisions are contained in Part II of this document. However, it is understood that those Organization-specific provisions shall contradict neither the Patent Policy nor the Guidelines.

I.2 Explanation of terms

Contribution: Any document submitted for consideration by a Technical Body.

Free of Charge: The words "Free of Charge" do not mean that the Patent Holder is waiving all of its rights with respect to the Patent. Rather, "Free of Charge" refers to the issue of monetary compensation; *i.e.*, that the Patent Holder will not seek any monetary compensation as part of the licensing arrangement (whether such compensation is called a royalty, a one-time licensing fee, etc.). However, while the Patent Holder in this situation is committing to not charging any monetary amount, the Patent Holder is still entitled to require that the implementer of the relevant Recommendation | Deliverable sign a license agreement that contains other reasonable terms and conditions such as those relating to governing law, field of use, warranties, etc.

Organizations: ITU, ISO and IEC.

Patent: The word "Patent" means those claims contained in and identified by patents, utility models and other similar statutory rights based on inventions (including applications for any of these) solely to the extent that any such claims are essential to the implementation of a Recommendation | Deliverable. Essential patents are patents that would be required to implement a specific Recommendation | Deliverable.

Patent Holder: Person or entity that owns, controls and/or has the ability to license Patents.

Reciprocity: The word "Reciprocity" means that the Patent Holder shall only be required to license any prospective licensee if such prospective licensee will commit to license its Patent(s) for implementation of the same relevant Recommendation | Deliverable Free of Charge or under reasonable terms and conditions.

Recommendations | Deliverables: ITU-T and ITU-R Recommendations are referred to as "Recommendations", ISO deliverables and IEC deliverables are referred to as "Deliverables". The various types of Recommendation(s) | Deliverable(s) are referred to as "Document types" in the Patent Statement and Licensing Declaration Form (hereafter referred to as "Declaration Form") attached as Annex 2.

Technical Bodies: Study Groups, any subordinate groups and other groups of ITU-T and ITU-R and technical committees, subcommittees and working groups in ISO and IEC.

I.3 Patent disclosure

As mandated by the Patent Policy in its paragraph 1, any party participating [2] in the work of the Organizations should, from the outset, draw their attention to any known Patent or to any known pending Patent application, either its own or that of other organizations.

In this context, the words "from the outset" imply that such information should be disclosed as early as possible during the development of the Recommendation | Deliverable. This might not be possible when the first draft text appears since at this time, the text might be still too vague or subject to subsequent major modifications. Moreover, that information should be provided in good faith and on a best effort basis, but there is no requirement for patent searches.

In addition to the above, any party not participating in Technical Bodies may draw the attention of the Organizations to any known Patent, either their own and/or of any third-party.

When disclosing their own Patents, Patent Holders have to use the Patent Statement and Licensing Declaration Form (referred to as the "Declaration Form") as stated in Section 4 of these Guidelines.

Any communication drawing the attention to any third-party Patent should be addressed to the concerned Organization(s) in writing. The potential Patent Holder will then be requested by the Director/CEO of the relevant Organization(s) to submit a Declaration Form, if applicable.

The Patent Policy and these Guidelines also apply to any Patent disclosed or drawn to the attention of the Organizations subsequent to the approval of a Recommendation | Deliverable.

Whether the identification of the Patent took place before or after the approval of the Recommendation | Deliverable, if the Patent Holder is unwilling to license under paragraph 2.1 or 2.2 of the Patent Policy, the Organizations will promptly advise the Technical Bodies responsible for the affected Recommendation | Deliverable so that appropriate action can be taken. Such action will include, but may not be limited to, a review of the Recommendation | Deliverable or its draft in order to remove the potential conflict or to further examine and clarify the technical considerations causing the conflict.

I.4 Patent Statement and Licensing Declaration Form

I.4.1 The purpose of the Declaration Form

To provide clear information in the Patent Information databases of each Organization, Patent Holders have to use the Declaration Form, which is available on the web site of each Organization (the Declaration Form is included in Annex 2 for information purposes). They must be sent to the Organizations for the attention, for ITU, of the Directors of the TSB or the BR or, for ISO or IEC, of the CEOs. The purpose of the Declaration Form is to ensure a standardized submission to the respective Organizations of the declarations being made by Patent Holders.

---

2) In the case of ISO and IEC, this includes any recipient of a draft standard at any stage in the standards development process.

The Declaration Form gives Patent Holders the means of making a licensing declaration relative to rights in Patents required for implementation of a specific Recommendation | Deliverable. Specifically, by submitting this Declaration Form the submitting party declares its willingness to license (by selecting option 1 or 2 on the Form) /or its unwillingness to license (by selecting option 3 on the Form), according to the Patent Policy, Patents held by it and whose licence would be required to practice or implement part(s) or all of a specific Recommendation | Deliverable.

If a Patent Holder has selected the licensing option 3 on the Declaration Form, then, for the referenced relevant ITU Recommendation, the ITU requires the Patent Holder to provide certain additional information permitting patent identification. In such a situation, for any relevant ISO or IEC Deliverable, the ISO and IEC strongly encourage (but do not require) the Patent Holder to provide certain additional information permitting patent identification.

Multiple Declaration Forms are appropriate if the Patent Holder wishes to identify several Patents and classifies them in different options of the Declaration Form for the same Recommendation | Deliverable or if the Patent Holder classifies different claims of a complex patent in different options of the Declaration Form.

Information contained in a Declaration Form may be corrected in case of obvious errors, such as a typographical mistake in a standard or patent reference number. The licensing declaration contained in the Declaration Form remains in force unless it is superseded by another Declaration Form containing more favourable licensing terms and conditions from a licensee's perspective reflecting (a) a change in commitment from option 3 to either option 1 or option 2, (b) a change in commitment from option 2 to option 1 or (c) un-checking one or more sub-options contained within option 1 or 2.

I.4.2   Contact information

In completing Declaration Forms, attention should be given to supplying contact information that will remain valid over time. Where possible, the "Name and Department" and e-mail address should be generic. Also it is preferable, where possible, that parties, particularly multinational organizations, indicate the same contact point on all Declaration Forms submitted.

With a view to maintaining up-to-date information in the Patent Information database of each Organization, it is requested that the Organizations be informed of any change or corrections to the Declaration Form submitted in the past, especially with regard to the contact person.

I.5   Conduct of meetings

Early disclosure of Patents contributes to the efficiency of the process by which Recommendations | Deliverables are established. Therefore, each Technical Body, in the course of the development of a proposed Recommendation | Deliverable, will request the disclosure of any known Patents essential to the proposed Recommendation | Deliverable.

Chairmen of Technical Bodies will, if appropriate, ask, at an appropriate time in each meeting, whether anyone has knowledge of patents, the use of which may be required to practice or implement the Recommendation | Deliverable being considered. The fact that the question was asked shall be recorded in the meeting report, along with any affirmative responses.

As long as the Organization concerned has received no indication of a Patent Holder selecting paragraph 2.3 of the Patent Policy, the Recommendation | Deliverable may be approved using the appropriate and respective rules of the Organization concerned. It is expected that discussions in Technical Bodies will include consideration of including patented material in a Recommendation | Deliverable, however the Technical Bodies may not take position regarding the essentiality, scope, validity or specific licensing terms of any claimed Patents.

I.6 Patent Information database

In order to facilitate both the standards-making process and the application of Recommendations | Deliverables, each Organization makes available to the public a Patent Information database composed of information that was communicated to the Organizations by the means of Declaration Forms. The Patent Information database may contain information on specific patents, or may contain no such information but rather a statement about compliance with the Patent Policy for a particular Recommendation | Deliverable.

The Patent Information databases are not certified to be either accurate or complete, but only reflect the information that has been communicated to the Organizations. As such, the Patent Information databases may be viewed as simply raising a flag to alert users that they may wish to contact the entities who have communicated Declaration Forms to the Organizations in order to determine if patent licenses must be obtained for use or implementation of a particular Recommendation | Deliverable.

I.7 Assignment or transfer of patent rights

The rules governing the assignment or transfer of Patent rights are contained in the patent statement and licensing declaration forms (see Annexes 2 and 3). By complying with these rules, the Patent Holder has discharged in full all of its obligations and liability with regards to the licensing commitments after the transfer or assignment. These rules are not intended to place any duty on the Patent Holder to compel compliance with the licensing commitment by the assignee or transferee after the transfer occurs.

Part II – Organization-specific provisions

II.1 Specific provisions for ITU

ITU-1 General Patent Statement and Licensing Declaration Form

Anyone may submit a General Patent Statement and Licensing Declaration Form which is available on the web sites of ITU-T and ITU-R (the form in Annex 3 is included for information purposes). The purpose of this form is to give Patent Holders the voluntary option of making a general licensing declaration relative to material protected by Patents contained in any of their Contributions. Specifically, by submitting its form, the Patent Holder declares its willingness to license its Patents owned by it in case part(s) or all of any proposals contained in its Contributions submitted to the Organization are included in Recommendation(s) and the included part(s) contain items for which Patents have been filed and whose licence would be required to practice or implement Recommendation(s).

The General Patent Statement and Licensing Declaration Form is not a replacement for the "individual" (see clause 4 of Part I) Declaration Form, which is made per Recommendation, but is expected to improve responsiveness and early disclosure of the Patent Holder's compliance with the Patent Policy. Therefore, in addition to its existing General Patent Statement and Licensing Declaration in respect of its Contributions, the Patent Holder should, when appropriate (e.g. if it becomes aware that it has a Patent for a specific Recommendation), also submit an "individual" Patent Statement and Licensing Declaration Form:

— for the Patents contained in any of its Contributions submitted to the Organization which are included in a Recommendation, any such "individual" Patent Statement and Licensing Declarations may contain either the same licensing terms and conditions as in the General Patent Statement and Licensing Declaration Form, or more favourable licensing terms and conditions from a licensee's perspective as defined in the "individual" (see clause 4.1 of Part I) Declaration Form; and

— for the Patents that the Patent Holder did not contribute to the Organization which are included in a Recommendation, any such "individual" Patent Statement and Licensing Declarations may contain any of the three options available on the Form (see clause 4.1 of Part I), regardless of the commitment in its existing General Patent Statement and Licensing Declaration.

The General Patent Statement and Licensing Declaration remains in force unless it is superseded by another General Patent Statement and Licensing Declaration form containing more favourable licensing terms and conditions from a licensee's perspective reflecting (a) a change in commitment from option 2 to option 1 or (b) un-checking one or more sub-options contained within option 1 or 2.

The ITU Patent Information database also contains a record of General Patent Statement and Licensing Declarations.

ITU-2 Notification

Text shall be added to the cover sheets of all new and revised ITU-T and ITU-R Recommendations, where appropriate, urging users to consult the ITU Patent Information database. The wording is:

"ITU draws attention to the possibility that the practice or implementation of this Recommendation may involve the use of a claimed Intellectual Property Right. ITU takes no position concerning the evidence, validity or applicability of claimed Intellectual Property Rights, whether asserted by ITU members or others outside of the Recommendation development process.

As of the date of approval of this Recommendation, ITU [had/had not] received notice of intellectual property, protected by patents, which may be required to implement this Recommendation. However, implementers are cautioned that this may not represent the latest information and are therefore strongly urged to consult the ITU Patent Information database."

II.2 Specific provisions for ISO and IEC

ISO/IEC-1 Consultations on draft Deliverables

All drafts submitted for comment shall include on the cover page the following text:

"Recipients of this draft are invited to submit, with their comments, notification of any relevant patent rights of which they are aware and to provide supporting documentation."

ISO/IEC-2 Notification

A published document, for which no patent rights are identified during the preparation thereof, shall contain the following notice in the foreword:

"Attention is drawn to the possibility that some of the elements of this document may be the subject of patent rights. ISO [and/or] IEC shall not be held responsible for identifying any or all such patent rights."

A published document, for which patent rights have been identified during the preparation thereof, shall include the following notice in the introduction:

"The International Organization for Standardization (ISO) [and/or] International

Electrotechnical Commission (IEC) draws attention to the fact that it is claimed that compliance with this document may involve the use of a patent concerning (... subject matter ...) given in (... subclause ...).

ISO [and/or] IEC take[s] no position concerning the evidence, validity and scope of this patent right.

The holder of this patent right has assured the ISO [and/or] IEC that he/she is willing to negotiate licences under reasonable and non-discriminatory terms and conditions with applicants throughout the world. In this respect, the statement of the holder of this patent right is registered with ISO [and/or] IEC. Information may be obtained from:

> name of holder of patent right ...
>
> address ...

Attention is drawn to the possibility that some of the elements of this document may be the subject of patent rights other than those identified above. ISO [and/or] IEC shall not be held responsible for identifying any or all such patent rights."

ISO/IEC-3 National Adoptions

Patent Declarations in ISO, IEC and ISO/IEC Deliverables apply only to the ISO and/or IEC documents indicated in the Declaration Forms. Declarations do not apply to documents that are altered (such as through national or regional adoption). However, implementations that conform to identical national and regional adoptions and the respective ISO and/or IEC Deliverables, may rely on Declarations submitted to ISO and/or IEC for such Deliverables.

ANNEX 1

COMMON PATENT POLICY FOR ITU-T/ITU-R/ISO/IEC

The following is a "code of practice" regarding patents covering, in varying degrees, the subject matters of ITU-T Recommendations, ITU-R Recommendations, ISO deliverables and IEC deliverables (for the purpose of this document, ITU-T and ITU-R Recommendations are referred to as "Recommendations", ISO deliverables and IEC deliverables are referred to as "Deliverables"). The rules of the "code of practice" are simple and straightforward. Recommendations | Deliverables are drawn up by technical and not patent experts; thus, they may not necessarily be very familiar with the complex international legal situation of intellectual property rights such as patents, etc.

Recommendations | Deliverables are non-binding; their objective is to ensure compatibility of technologies and systems on a worldwide basis. To meet this objective, which is in the common interests of all those participating, it must be ensured that Recommendations | Deliverables, their applications, use, etc. are accessible to everybody.

It follows, therefore, that a patent embodied fully or partly in a Recommendation | Deliverable must be accessible to everybody without undue constraints. To meet this requirement in general is the sole objective of the code of practice. The detailed arrangements arising from patents (licensing, royalties, etc.) are left to the parties concerned, as these arrangements might differ from case to case.

This code of practice may be summarized as follows:

1 The ITU Telecommunication Standardization Bureau (TSB), the ITU Radio-communication Bureau (BR) and the offices of the CEOs of ISO and IEC are not in a position to give authoritative or comprehensive information about evidence, validity or scope of patents or similar rights, but it is desirable that the fullest available information should be disclosed. Therefore, any party participating in the work of ITU, ISO or IEC should, from the outset, draw the attention of the Director of ITU-TSB, the Director of ITU-BR, or the offices of the CEOs of ISO or IEC, respectively, to any known patent or to any known pending patent application, either their own or of other organizations, although ITU, ISO or IEC are unable to verify the validity of any such information.

2 If a Recommendation | Deliverable is developed and such information as referred to in paragraph 1 has been disclosed, three different situations may arise:

   2.1 The patent holder is willing to negotiate licences free of charge with other parties on a non-discriminatory basis on reasonable terms and conditions. Such negotiations are left to the parties concerned and are performed outside ITU-T/ITU-R/ISO/IEC.

   2.2 The patent holder is willing to negotiate licences with other parties on a non-discriminatory basis on reasonable terms and conditions. Such negotiations are left to the parties concerned and are performed outside ITU-T/ITU-R/ISO/IEC.

   2.3 The patent holder is not willing to comply with the provisions of either paragraph 2.1 or paragraph 2.2; in such case, the Recommendation | Deliverable shall not include provisions depending on the patent.

3 Whatever case applies (2.1, 2.2 or 2.3), the patent holder has to provide a written statement to be filed at ITU-TSB, ITU-BR or the offices of the CEOs of ISO or IEC, respectively, using the appropriate "Patent Statement and Licensing Declaration" Form. This statement must not include additional provisions, conditions, or any other exclusion clauses in excess of what is provided for each case in the corresponding boxes of the form.

ANNEX 2

PATENT STATEMENT AND LICENSING DECLARATION FORM FOR ITU-T OR ITU-R RECOMMENDATION | ISO OR IEC DELIVERABLE

**Patent Statement and Licensing Declaration
for ITU-T or ITU-R Recommendation | ISO or IEC Deliverable**

This declaration does not represent an actual grant of a license

Please return to the relevant organization(s) as instructed below per document type:

Director Telecommunication Standardization Bureau International Telecommunication Union Place des Nations CH-1211 Geneva 20 Switzerland Fax: +41 22 730 5853 Email: tsbdir@itu.int	Director Radiocommunication Bureau International Telecommunication Union Place des Nations CH-1211 Geneva 20 Switzerland Fax: +41 22 730 5785 Email: brmail@itu.int	Secretary-General International Organization for Standardization 8 chemin de Blandonnet CH-1214 Vernier, Geneva Switzerland Fax: +41 22 733 3430 Email: patent.statements@iso.org	General Secretary International Electrotechnical Commission 3 rue de Varembé CH-1211 Geneva 20 Switzerland Fax: +41 22 919 0300 Email: inmail@iec.ch

---

Patent Holder:

Legal Name _____

Contact for license application:

Name & Department _____

Address _____

Tel. _____

Fax _____

E-mail _____

URL (optional) _____

Document type:

☐ **ITU-T Rec. (*)**   ☐ **ITU-R Rec. (*)**   ☐ **ISO Deliverable (*)**   ☐ **IEC Deliverable (*)**
(please return the form to the relevant Organization)

☐ **Common text or twin text (ITU-T Rec. | ISO/IEC Deliverable (*))** (for common text or twin text, please return the form to each of the three Organizations: ITU-T, ISO, IEC)

☐ **ISO/IEC Deliverable (*)** (for ISO/IEC Deliverables, please return the form to both ISO and IEC)

(*) Number _____

(*) Title _____

Licensing declaration:

The Patent Holder believes that it holds granted and/or pending applications for Patents, the use of which would be required to implement the above document and hereby declares, in accordance with the Common Patent Policy for ITU-T/ITU-R/ISO/IEC, that (check <u>one</u> box only):

☐ 1. The Patent Holder is prepared to grant a <u>Free of Charge</u> license to an unrestricted number of applicants on a worldwide, non-discriminatory basis and under other reasonable terms and conditions to make, use, and sell implementations of the above document.

Negotiations are left to the parties concerned and are performed outside the ITU-T, ITU-R, ISO or IEC.

Also mark here __ if the Patent Holder's willingness to license is conditioned on <u>Reciprocity</u> for the above document.

Also mark here __ if the Patent Holder reserves the right to license on reasonable terms and conditions (but not <u>Free of Charge</u>) to applicants who are only willing to license their Patent, whose use would be required to implement the above document, on reasonable terms and conditions (but not <u>Free of Charge</u>).

☐ 2. The Patent Holder is prepared to grant a license to an unrestricted number of applicants on a worldwide, non-discriminatory basis and on reasonable terms and conditions to make, use and sell implementations of the above document.

Negotiations are left to the parties concerned and are performed outside the ITU-T, ITU-R, ISO or IEC.

Also mark here __ if the Patent Holder's willingness to license is conditioned on <u>Reciprocity</u> for the above document.

☐ 3. The Patent Holder is unwilling to grant licenses in accordance with provisions of either 1 or 2 above.

In this case, the following information must be provided to ITU, and is strongly desired by ISO and IEC, as part of this declaration:

— granted patent number or patent application number (if pending);
— an indication of which portions of the above document are affected;
— a description of the Patents covering the above document.

<u>Free of Charge</u>: The words "Free of Charge" do not mean that the Patent Holder is waiving all of its rights with respect to the Patent. Rather, "Free of Charge" refers to the issue of monetary compensation; *i.e.*, that the Patent Holder will not seek any monetary compensation as part of the licensing arrangement (whether such compensation is called a royalty, a one-time licensing fee, etc.). However, while the Patent Holder in this situation is committing to not charging any monetary amount, the Patent Holder is still entitled to require that the implementer of the same above document sign a license agreement that contains other reasonable terms and conditions such as those relating to governing law, field of use, warranties, etc.

<u>Reciprocity</u>: The word "Reciprocity" means that the Patent Holder shall only be required to license any prospective licensee if such prospective licensee will commit to license its Patent(s) for implementation of the same above document Free of Charge or under reasonable terms and conditions.

<u>Patent</u>: The word "Patent" means those claims contained in and identified by patents, utility models and other similar statutory rights based on inventions (including applications for any of these) solely to the extent that any such claims are essential to the implementation of the same above document. Essential patents are patents that would be required to implement a specific Recommendation | Deliverable.

<u>Assignment/transfer of Patent rights</u>: Licensing declarations made pursuant to Clause 2.1 or 2.2 of the Common Patent Policy for ITU-T/ITU-R/ISO/IEC shall be interpreted as encumbrances that bind all successors-in-interest as to the transferred Patents. Recognizing that this interpretation may not apply in all jurisdictions, any Patent Holder who has submitted a licensing declaration according to the Common Patent Policy - be it selected as option 1 or 2 on the Patent Declaration form - who transfers ownership of a Patent that is subject to such licensing declaration shall include appropriate provisions in the relevant transfer documents to ensure that, as to such transferred Patent, the licensing declaration is binding on the transferee and that the transferee will similarly include appropriate provisions in the event of future transfers with the goal of binding all successors-in-interest.

Patent Information (desired but not required for options 1 and 2; required in ITU for option 3 (NOTE))

No.	Status [granted / pending]	Country	Granted Patent Number or Application Number (if pending)	Title
1				
2				
3				
4				
5				
6				
7				
8				
9				
10				

☐ Check here if additional patent information is provided on additional pages.

NOTE   For option 3, the additional minimum information that shall also be provided is listed in the option 3 box above.

Signature (include on final page only):	
Patent Holder	_____
Name of authorized person	_____
Title of authorized person	_____
Signature	_____
Place, Date	_____

FORM: 26 June 2015

ANNEX 3

GENERAL PATENT STATEMENT AND LICENSING DECLARATION FORM FOR
ITU-T OR ITU-R RECOMMENDATION

ITU
International Telecommunication Union

General Patent Statement and Licensing Declaration for ITU-T or ITU-R Recommendation

This declaration does not represent an actual grant of a license

Please return to the relevant bureau:

Director	Director
Telecommunication Standardization Bureau	Radiocommunication Bureau
International Telecommunication Union	International Telecommunication Union
Place des Nations	Place des Nations
CH-1211 Geneva 20	CH-1211 Geneva 20
Switzerland	Switzerland
Fax: +41 22 730 5853	Fax: +41 22 730 5785
Email: tsbdir@itu.int	Email: brmail@itu.int

Patent Holder:

Legal Name _____

Contact for license application:

Name & Department _____

Address _____

Tel. _____

Fax _____

E-mail _____

URL (optional) _____

Licensing declaration:

In case part(s) or all of any proposals contained in Contributions submitted by the Patent Holder above are included in ITU-T/ITU-R Recommendation(s) and the included part(s) contain items for which Patents have been filed and whose use would be required to implement ITU-T/ITU-R Recommendation(s), the above Patent Holder hereby declares, in accordance with the Common Patent Policy for ITU-T/ITU-R/ISO/IEC (check <u>one</u> box only):

☐ 1. The Patent Holder is prepared to grant a <u>Free of Charge</u> license to an unrestricted number of applicants on a worldwide, non-discriminatory basis and under other reasonable terms and conditions to make, use, and sell implementations of the relevant ITU-T/ITU-R Recommendation.

	Negotiations are left to the parties concerned and are performed outside the ITU-T/ITU-R. *Also mark here ___ if the Patent Holder's willingness to license is conditioned on Reciprocity for the above ITU-T/ITU-R Recommendation.* *Also mark here ___ if the Patent Holder reserves the right to license on reasonable terms and conditions (but not Free of Charge) to applicants who are only willing to license their patent claims, whose use would be required to implement the above ITU-T/ITU-R Recommendation, on reasonable terms and conditions (but not Free of Charge).*
☐	2. The Patent Holder is prepared to grant a license to an unrestricted number of applicants on a worldwide, non-discriminatory basis and on reasonable terms and conditions to make, use and sell implementations of the relevant ITU-T/ITU-R Recommendation. Negotiations are left to the parties concerned and are performed outside the ITU-T/ITU-R. *Also mark here ___ if the Patent Holder's willingness to license is conditioned on Reciprocity for the above ITU-T/ITU-R Recommendation.*

Free of Charge: The words "Free of Charge" do not mean that the Patent Holder is waiving all of its rights with respect to the Patent. Rather, "Free of Charge" refers to the issue of monetary compensation; *i.e.*, that the Patent Holder will not seek any monetary compensation as part of the licensing arrangement (whether such compensation is called a royalty, a one-time licensing fee, etc.). However, while the Patent Holder in this situation is committing to not charging any monetary amount, the Patent Holder is still entitled to require that the implementer of the relevant ITU-T/ITU-R Recommendation sign a license agreement that contains other reasonable terms and conditions such as those relating to governing law, field of use, Reciprocity, warranties, etc.

Reciprocity: The word "Reciprocity" means that the Patent Holder shall only be required to license any prospective licensee if such prospective licensee will commit to license its Patent(s) for implementation of the relevant ITU-T/ITU-R Recommendation Free of Charge or under reasonable terms and conditions.

Patent: The word "Patent" means those claims contained in and identified by patents, utility models and other similar statutory rights based on inventions (including applications for any of these) solely to the extent that any such claims are essential to the implementation of the relevant Recommendation | Deliverable. Essential patents are patents that would be required to implement the relevant Recommendation | Deliverable.

Assignment/transfer of Patent rights: Licensing declarations made pursuant to Clause 2.1 or 2.2 of the Common Patent Policy for ITU-T/ITU-R/ISO/IEC shall be interpreted as encumbrances that bind all successors-in-interest as to the transferred Patents. Recognizing that this interpretation may not apply in all jurisdictions, any Patent Holder who has submitted a licensing declaration according to the Common Patent Policy - be it selected as option 1 or 2 on the Patent Declaration form - who transfers ownership of a Patent that is subject to such licensing declaration shall include appropriate provisions in the relevant transfer documents to ensure that, as to such transferred Patent, the licensing declaration is binding on the transferee and that the transferee will similarly include appropriate provisions in the event of future transfers with the goal of binding all successors-in-interest.

Signature:	
Patent Holder	_____
Name of authorized person	_____
Title of authorized person	_____
Signature	_____
Place, Date	

FORM: 26 June 2015

Annex J
(normative)

Formulating scopes of technical committees and subcommittees

J.1 Introduction

The scope of a technical committee or subcommittee is a statement precisely defining the limits of the work of that committee. As such it has a number of functions:

— it assists those with queries and proposals relating to a field of work to locate the appropriate committee;

— it prevents overlapping the work programmes of two or more ISO and/or IEC committees;

— it can also help guard against moving outside the field of activities authorized by the parent committee.

J.2 Formulation of scopes

Basic rules for the formulation of scopes of technical committees and subcommittees are given in 1.5.10.

The order of the elements of a scope shall be:

— basic scope;

— in the ISO, horizontal functions, where applicable;

— in the IEC, horizontal and/or group safety functions where applicable;

— exclusions (if any);

— notes (if any).

J.3 Basic scope

Scopes of technical committees shall not refer to the general aims of international standardization or repeat the principles that govern the work of all technical committees.

In exceptional cases, explanatory material may be included if considered important to the understanding of the scope of the committee. Such material shall be in the form of "Notes".

J.4 Exclusions

Should it be necessary to specify that certain topics are outside the scope of the technical committee, these shall be listed and be introduced by the words "Excluded ..."

Exclusions shall be clearly specified.

Where the exclusions are within the scope of one or more other existing ISO or IEC technical committees, these committees shall also be identified.

EXAMPLE 1   "Excluded: Those … covered by ISO/TC …".

EXAMPLE 2   "Excluded: Standardization for specific items in the field of … (ISO/TC …), … (IEC/TC …), etc.".

It is *not* necessary to mention self-evident exclusions.

EXAMPLE 3   "Excluded: Products covered by other ISO or IEC technical committees".

EXAMPLE 4   "Excluded: … Specifications for electrical equipment and apparatus, which fall within the scope of IEC committees".

J.5 Scopes of committees related to products

Scopes of committees related to products shall clearly *indicate the field, application area or market sector* which they intend to cover, in order to easily ascertain whether a particular product is, or is not, within that field, application area or market sector.

EXAMPLE 1   "Standardization of … and … used in …".

EXAMPLE 2   "Standardization of materials, components and equipment for construction and operation of … and … as well as equipment used in the servicing and maintenance of …".

The limits of the scope can be defined by *indicating the purpose* of the products, or by *characterizing* the products.

The scope *should not enumerate the types* of product covered by the committee since to do so might suggest that other types can be, or are, standardized by other committees. However, if this is the intention, then it is preferable to list those items which are excluded from the scope.

The *enumeration of aspects* such as terminology, technical requirements, methods of sampling, test methods, designation, marking, packaging, dimensions, etc. suggests a restriction in the scope to those particular aspects, and that other aspects may be standardized by other committees. The aspects of the products to be standardized should therefore not be included in the scope unless it is intended that the scope is limited to those particular aspects.

If the scope makes no mention of any aspect, this means that the subject *in its entirety* is covered by the committee.

NOTE   The coverage does not necessarily mean the need for preparing a standard. It only means that standards on any aspect, if needed, will be prepared by that committee and no other.

An example of unnecessary enumeration of aspects is as follows:

EXAMPLE 3   "Standardization of classification, terminology, sampling, physical, chemical or other test methods, specifications, etc.".

Mention of priorities, whether referring to type of product or aspect, shall not appear in the scope since these will be indicated in the programme of work.

J.6 Scopes of committees not related to products

If the scope of a committee is intended to be limited to *certain aspects* which are unrelated, or only indirectly related to products, the scope shall only indicate the aspect to be covered (e.g. safety colours and signs, non-destructive testing, water quality).

The term *terminology* as a possible aspect of standardization should not be mentioned unless this aspect is the only task to be dealt with by the committee. If this is not the case, the mention of terminology is superfluous since this aspect is a logical part of any standardization activity.

Annex K
(normative)

Project committees

K.1 Proposal stage

A new work item proposal not falling within the scope of an existing technical committee shall be presented using the appropriate form and fully justified (see 2.3.4) by one of the bodies authorized to make new work item proposals (see 2.3.2).

The office of the CEO may decide to return the proposal to the proposer for further development before circulation for voting. In this case, the proposer shall make the changes suggested or provide justification for not making the changes. If the proposer does not make the changes and requests that its proposal be circulated for voting as originally presented, the technical management board will decide on appropriate action. This could include blocking the proposal until the changes are made or accepting that it be balloted as received.

In all cases, the office of the CEO may also include comments and recommendations to the proposal form.

For details relating to justification of the proposal, see Annex C.

In the case of a proposal to establish a project committee to prepare management system standards, see Annex L.

It shall be submitted to the secretariat of the technical management board which shall arrange for it to be submitted to all National Bodies for voting.

Proposers are also encouraged to indicate the date of the first meeting of the project committee (see K.3).

If the proposal was not submitted by a National Body, the submission to the National Bodies shall include a call for offers to assume the secretariat of a project committee.

Votes shall be returned within 12 weeks.

Acceptance requires:

— approval by a 2/3 majority of the National Bodies voting;

— a commitment to participate actively by at least five National Bodies that approved the new work item proposal and nominated technical experts.

K.2 Establishment of a project committee

The technical management board shall review the results of voting on the new work item proposal and if the approval criteria are met, shall establish a project committee (the reference number shall be the next available number in the technical committee/project committee sequence).

The secretariat of the project committee shall be allocated to the National Body that submitted the proposal, or the technical management board shall decide on the allocation amongst the offers received if the proposal did not originate from a National Body.

National Bodies that approved the new work item proposal and nominated (a) technical expert(s) shall be registered as P-members of the project committee. National Bodies that approved the new work item proposal but did not make a commitment to participate actively shall be registered as O-members.

National Bodies that voted negatively, but nevertheless indicated that they would participate actively if the new work item was approved, shall be registered as P-members. National Bodies voting negatively without indicating a wish to participate shall be registered as O-members.

The office of the CEO shall announce to the National Bodies the establishment of the project committee and its membership.

National Bodies will be invited to confirm/change their membership status by informing the office of the CEO.

The secretariat will contact any potential liaison organizations identified in the new work item proposal or in National Body comments thereon and will invite them to indicate whether they have an interest in the work and, if so, which category of liaison they would be interested in. Requests for liaison will be processed according to the existing procedures.

K.3 First meeting of a project committee

The procedure for calling a project committee meeting shall be carried out in accordance with Clause 4, with the exception that a six weeks' notice period may be used if the date of the first meeting was communicated at the time of submission of the proposal.

The chair of the project committee shall be the project leader nominated in the new work item proposal or shall be nominated by the secretariat if no project leader was nominated in the new work item proposal.

The first meeting shall confirm the scope of the new work item. In case revision is necessary (for purposes of clarification but not extension of the scope), the revised scope shall be submitted to the technical management board for approval. It shall also confirm the project plan and in ISO the development track and decide on any substructures needed to carry out the work.

If it is determined that the project needs to be subdivided to produce two or more publications, this is possible provided that the subdivisions of the work lie fully within the scope of the original new work item proposal. If not, a new work item will need to be prepared for consideration by the technical management board.

NOTE   Project committees are exempted from the requirement to establish a strategic business plan.

K.4 Preparatory stage

The preparatory stage shall be carried out in accordance with 2.4.

K.5 Committee, enquiry, approval and publication stages

The committee, enquiry, approval and publication stages shall be carried out in accordance with 2.5 to 2.8.

K.6 Disbanding of a project committee

Once the standard(s) is/are published, the project committee shall be disbanded.

K.7 Maintenance of standard(s) prepared by a project committee

The National Body which held the secretariat shall assume responsibility for the maintenance of the standard(s) according to the procedures given in 2.9 unless the project committee has been transformed into a technical committee (see 1.10) in which case the technical committee shall be given the responsibility for the maintenance of the standard.

Annex L
(normative)

Proposals for management system standards

L.1 General

Whenever a proposal is made to prepare a new management system standard (MSS), including sector-specific MSS, a justification study (JS) shall be carried out in accordance with Appendix 1 to this annex.

NOTE   No JS is needed for the revision of an existing MSS whose development has already been approved and provided the scope is confirmed (unless it was not provided during its first development).

To the extent possible, the proposer shall endeavour to identify the full range of deliverables which will constitute the new or revised MSS family, and a JS shall be prepared for each of the deliverables.

L.2 Terms and definitions

For the purposes of this annex, the following terms and definitions apply.

L.2.1
management system
See definition contained in Appendix 2 (clause 3.4) of this annex.

L.2.2
management system standard
MSS
standard for *management systems* (L.5.1)

Note 1 to entry: For the purposes of this document, this definition also applies to other ISO and IEC deliverables (e.g. TS, PAS).

L.2.3
generic MSS
MSS designed to be widely applicable across economic sectors, various types and sizes of organizations and diverse geographical, cultural and social conditions

L.2.4
sector-specific MSS
MSS that provides additional requirements or guidance for the application of a generic MSS to a specific economic or business sector

L.2.5
Type A MSS
MSS providing requirements

EXAMPLES

— Management system requirements standards (specifications).

— Management system sector-specific requirements standards.

L.2.6
Type B MSS
MSS providing guidelines

EXAMPLES

— Guidance on the use of management system requirements standards.

— Guidance on the establishment of a management system.

— Guidance on the improvement/enhancement of a management system.

L.2.7
high level structure
HLS
outcome of the work of the ISO/TMB/JTCG "Joint technical Coordination Group on MSS" which refers to high level structure (HLS), identical subclause titles, identical text and common terms and core definitions. See Appendix 2 to this annex.

L.3 Obligation to submit a JS

All MSS [including sector-specific MSS (L.2.4), see Annex M] proposals and their JS shall be identified by the relevant TC/SC/PC (or SyC, in IEC) leadership and the JS shall be sent to the TMB (or its MSS task force) for evaluation and approval before the NP ballot takes place. It is the responsibility of the relevant TC/SC/PC secretariat to identify all MSS proposals, without exception, so that there will be no MSS proposals which fail (with knowledge or without knowledge) to carry out the JS or which fail to be sent to the ISO/TMB for evaluation.

No JS is required for a Type B MSS providing guidance on a specific Type A MSS for which a JS has already been submitted and approved.

EXAMPLE   ISO/IEC 27003:2010 (*Information technology — Security techniques — Information security management system implementation guidance*) does not need to have JS submitted as ISO/IEC 27001:2013 (*Information technology — Security techniques — Information security management systems — Requirements*) has already had a JS submitted and approved.

L.4 Cases where no JS have been submitted

MSS proposals which have not been submitted for TMB evaluation before the NP ballot will be sent to the TMB for evaluation and no new ballot should take place before the TMB decision (project on hold). It is considered good practice that the TC/SC/PC (and/or SyC, in IEC) members endorse the JS prior it being sent to the TMB.

NOTE   Already published MSS which did not have a JS submitted will be treated as new MSS at the time of revision, i.e. a JS is to be presented and approved before any work can begin.

L.5 Applicability of this annex

The above procedures apply to all ISO and IEC deliverables, including IWAs.

L.6 General principles

All projects for new MSS (or for MSS which are already published but for which no JS was completed) shall undergo a JS (see L.1 and Note to L.3). The following general principles provide guidance to assess the market relevance of proposed MSS and for the preparation of a JS. The justification criteria questions in Appendix 1 to this annex are based on these principles. The answers to the questions will form part of the JS. An MSS should be initiated, developed and maintained only when all of the following principles are observed.

1)	**Market relevance**	— Any MSS should meet the needs of, and add value for, the primary users and other affected parties.
2)	**Compatibility**	— Compatibility between various MSS and within an MSS family should be maintained.
3)	**Topic coverage**	— A generic MSS (L.5.3) should have sufficient application coverage to eliminate or minimize the need for sector-specific variances.
4)	**Flexibility**	— An MSS should be applicable to organizations in all relevant sectors and cultures and of every size. An MSS should not prevent organizations from competitively adding to or differentiating from others, or enhancing their management systems beyond the standard.
5)	**Free trade**	— An MSS should permit the free trade of goods and services in line with the principles included in the WTO Agreement on Technical Barriers to Trade.
6)	**Applicability of conformity assessment**	— The market need for first-, second- or third-party conformity assessment, or any combination thereof, should be assessed. The resulting MSS should clearly address the suitability of use for conformity assessment in its scope. An MSS should facilitate joint audits.
7)	**Exclusions**	— An MSS should not include directly related product (including services) specifications, test methods, performance levels (i.e. setting of limits) or other forms of standardization for products produced by the implementing organization.
8)	**Ease of use**	— It should be ensured that the user can easily implement one or more MSS. An MSS should be easily understood, unambiguous, free from cultural bias, easily translatable, and applicable to businesses in general.

L.7 Justification study process and criteria

L.7.1 General

This clause describes the justification study (JS) process for justifying and evaluating the market relevance of proposals for an MSS. Appendix 1 to this annex provides a set of questions to be addressed in the justification study.

L.7.2 Justification study process

The JS process applies to any MSS project and consists of the following:

a) the development of the JS by (or on behalf of) the proposer of an MSS project;

b) an approval of the JS by the TMB (or in ISO, the ISO/TMB MSS task force).

The JS process is followed by the normal ISO or IEC balloting procedure for new work item approval as appropriate.

L.7.3 Justification study criteria

Based on Annex C and the general principles stated above, a set of questions (see Appendix 1 to this annex) shall be used as criteria for justifying and assessing a proposed MSS project and shall be answered by the proposer. This list of questions is not exhaustive and any additional information that is relevant to the case should be provided. The JS should demonstrate that all questions have been considered. If it is decided that they are not relevant or appropriate to a particular situation, then the reasons for this decision should be clearly stated. The unique aspect of a particular MSS may require consideration of additional questions in order to assess objectively its market relevance.

L.8 Guidance on the development process and structure of an MSS

L.8.1 General

The development of an MSS will have effects in relation to

— the far-reaching impact of these standards on business practice,

— the importance of worldwide support for the standards,

— the practical possibility for involvement by many, if not all, National Bodies, and

— the market need for compatible and aligned MSS.

This clause provides guidance in addition to the procedures laid down in other clauses of the ISO/IEC Directives, in order to take these effects into account.

All MSS (whether they are Type A or Type B MSS, generic or sector-specific) shall, in principle, use consistent structure, common text and terminology so that they are easy to use and compatible with each other. The guidance and structure given in Appendix 2 to this annex shall, in principle, also be followed.

A Type B MSS which provides guidance on another MSS of the same MSS family should follow the same structure (i.e. clause numbering). Where MSS providing guidance (Type B MSS) are involved, it is important that their functions be clearly defined together with their relationship with the MSS providing requirements (Type A MSS), for example:

— guidance on the use of the requirements standard;

— guidance on the establishment/implementation of the management system;

— guidance on improvement/enhancement of the management system.

Where the proposed MSS is sector-specific:

— it should be compatible and aligned with the generic MSS;

— rules and principles specified in Annex M shall be followed;

— the relevant committee responsible for the generic MSS may have additional requirements to be met or procedures to be followed (see Annex M);

— other committees may need to be consulted, as well as ISO CASCO and IEC CAB on conformity assessment issues.

In the case of sector specific documents, their function and relationship with the generic MSS should be clearly defined (e.g. additional sector-specific requirements; elucidation; or both as appropriate).

Sector-specific documents should always show clearly (e.g. by using different typographical styles) the kind of sector-specific information being provided.

NOTE 1   Where the identical text or any of the requirements cannot be applied in a specific MSS, due to special circumstances, this should be reported to the TMB through the TMB Secretary at tmb@iso.org (see L.9.3) or the IEC/SMB Secretary.

L.8.2 MSS development process

L.8.2.1 General

In addition to the JS, the development of an MSS should follow the same requirements as other ISO and IEC deliverables (see Clause 2).

L.8.2.2 Design specification

To ensure that the intention of the standard, as demonstrated by the justification study, will be maintained, a design specification may be developed before a working draft is prepared.

The responsible committee will decide whether the design specification is needed and in case it is felt necessary, it will decide upon its format and content that is appropriate for the MSS and should set up the necessary organization to carry out the task.

The design specification should typically address the following.

User needs	The identification of the users of the standard and their associated needs, together with the costs and benefits for these users.
Scope	The scope and purpose of the standard, the title and the field of application.
Compatibility	How compatibility within this and with other MSS families will be achieved, including identification of the common elements with similar standards, and how these will be included in the recommended structure (see Appendix 2 to this annex).
Consistency	Consistency with other documents (to be) developed within the MSS family.

NOTE   Most, if not all of the information on user needs and scope will be available from the justification study.

The design specification should ensure that

a) the outputs of the justification study are translated correctly into requirements for the MSS,

b) the issues of compatibility and alignment with other MSS are identified and addressed,

c) a basis for verification of the final MSS exists at appropriate stages during the development process,

d) the approval of the design specification provides a basis for ownership throughout the project by the members of the TC/SC(s), and/or SyC in IEC,

e) account is taken of comments received through the NP ballot phase, and

f) any constraints are taken into account.

The Committee developing the MSS should monitor the development of the MSS against the design specification in order to ensure that no deviations happen in the course of the project.

L.8.2.3 Producing the deliverables

L.8.2.3.1 Monitoring output

In the drafting process, the output should be monitored for compatibility and ease of use with other MSS, by covering issues such as

— the high level structure (HLS), identical subclause titles, identical text and common terms and core definitions,

— the need for clarity (both in language and presentation), and

— avoiding overlap and contradiction.

L.8.2.4 Transparency of the MSS development process

MSS have a broader scope than most other types of standard. They cover a large field of human endeavour and have an impact on a wide range of user interests.

Committees preparing MSS should accordingly adopt a highly transparent approach to the development of the standards, ensuring that

— possibilities for participation in the process of developing standards are clearly identified, and

— the development processes being used are understood by all parties.

Committees should provide information on progress throughout the life cycle of the project, including

— the status of the project to date (including items under discussion),

— contact points for further information,

— communiqués and press releases on plenary meetings, and

— regular listings of frequently asked questions and answers.

In doing this, account needs to be taken of the distribution facilities available in the participating countries.

Where it may be expected that users of a Type A MSS are likely to demonstrate conformity to it, the MSS shall be so written that conformity can be assessed by a manufacturer or supplier (first party, or self-declaration), a user or purchaser (second party) or an independent body (third party, also known as certification or registration).

Maximum use should be made of the resources of the ISO Central Secretariat or IEC Central Office to facilitate the transparency of the project and the committee should, in addition, consider the establishment of a dedicated open-access website.

Committees should involve the national member bodies to build up a national awareness of the MSS project, providing drafts as appropriate for different interested and affected parties, including accreditation bodies, certification bodies, enterprises and the user community, together with additional specific information as needed.

The committee should ensure that technical information on the content of the MSS under development is readily available to participating members, especially those in developing countries.

L.8.2.5 Process for interpretation of a standard

The committee may establish a process to handle interpretation questions related to its standards from the users, and may make the resulting interpretations available to others in an expedient manner. Such a mechanism can effectively address possible misconceptions at an early stage and identify issues that may require improved wording of the standard during the next revision cycle. Such processes are considered, in ISO, to be "committee specific procedures" [see Foreword f)]. In IEC, the committee shall use the Interpretation Sheet process (see IEC Supplement).

L.9 High level structure, identical core text and common terms and core definitions for use in management systems standards

L.9.1 Introduction

The aim of this document is to enhance the consistency and alignment of MSS by providing a unifying and agreed upon high level structure, identical core text and common terms and core definitions. The aim is that all Type A MSS (and B where appropriate) are aligned and the compatibility of these standards is enhanced. It is envisaged that individual MSS will add additional "discipline-specific" requirements as required.

NOTE   In L.9.1 and L.9.4, "discipline-specific" is used to indicate specific subject(s) to which a management system standard refers, e.g. energy, quality, records, environment etc.

The intended audience for this document is Technical Committees (TC), Subcommittees (SC) and Project Committees (PC) (and in IEC, SyCs) and others that are involved in the development of MSS.

This common approach to new MSS and future revisions of existing standards will increase the value of such standards to users. It will be particularly useful for those organizations that choose to operate a single (sometimes called "integrated") management system that can meet the requirements of two or more MSS simultaneously.

Appendix 2 to this annex sets out the high level structure, identical core text and common terms and core definitions that form the nucleus of future and revised Type A MSS and Type B MSS when possible.

Appendix 3 to this annex sets out guidance to the use of Appendix 2 to this annex.

L.9.2   Use

MSS include the high level structure and identical core text as presented in Appendix 2 to this annex. The common terms and core definitions are either included or normatively referenced an international standard where they are included.

NOTE   The high level structure includes the main clauses (1 to 10) and their titles, in a fixed sequence. The identical core text includes numbered subclauses (and their titles) as well as text within the subclauses.

L.9.3   Non applicability

If due to exceptional circumstances the high level structure or any of the identical core text, common terms and core definitions cannot be applied in the management system standard, then the TC/PC/SC needs to explain its rationale for review by:

a)   providing an initial deviation report to ISO/CS or IEC/CO with the DIS submission;

b)   providing a final deviation report to TMB (through the ISO/TMB Secretary at tmb@iso.org) or IEC/SMB Secretary upon submission of the final text of the standard for publication.

The TC/PC/SC (or in IEC, SyC) shall use the ISO or IEC commenting template to provide its deviation reports.

NOTE 1   The final deviation report can be an updated version of the initial deviation report.

NOTE 2   The TC/PC/SC (or in IEC, SyC) strives to avoid any non-applicability of the high level structure or any of the identical core text, common terms and core definitions.

L.9.4   Using Appendix 2 to this annex

Discipline-specific text additions to Appendix 2 to this annex are managed as follows.

1. Discipline-specific additions are made by the individual TC, PC, SC (or in IEC, SyC) or other group that is developing the specific management system standard.

2. Discipline-specific text does not affect harmonization or contradict or undermine the intent of the high level structure, identical core text, common terms and core definitions.

3. Insert additional subclauses, or sub-subclauses (etc.) either ahead of an identical text subclause (or sub-subclause etc.), or after such a subclause (etc.) and renumbered accordingly.

NOTE 1   Hanging paragraphs are not permitted (see ISO/IEC Directives, Part 2).

NOTE 2   Attention is drawn to the need to check cross referencing.

4. Add or insert discipline-specific text within Appendix 2 to this annex. Examples of additions include:

   a) new bullet points;

   b) discipline-specific explanatory text (e.g. Notes or Examples), in order to clarify requirements;

   c) discipline-specific new paragraphs to subclauses (etc.) within the identical text;

   d) adding text that enhances the existing requirements in Appendix 2 to this annex.

5. Avoid repeating requirements between identical core text and discipline-specific text by adding text to the identical core text, taking account of point 2 above.

6. Distinguish between discipline-specific text and identical core text from the start of the drafting process. This aids identification of the different types of text during the development and balloting stages.

NOTE 1 Distinguishing options include by colour, font, font size, italics, or by being boxed separately, etc.

NOTE 2 Identification of distinguishing text is not necessarily carried into the published version.

7. Understanding of the concept of "risk" may be more specific than that given in the definition under 3.9 of Appendix 2 to this annex. In this case, a discipline-specific definition may be needed. The discipline-specific terms and definitions are differentiated from the core definitions, e.g. (XXX) risk.

NOTE The above can also apply to a number of other definitions.

8. Common terms and core definitions will be integrated into the listing of terms and definitions in the discipline-specific management system standard, consistent with the concept system of that standard.

L.9.5 Implementation

Follow the sequence, high level structure, identical core text, common terms and core definitions for any new management system standard and for any revisions to existing management system standards.

L.9.6 Guidance

Find supporting guidance in Appendix 3 to this annex.

Appendix 1

(normative)

Justification criteria questions

1. General

The list of questions to be addressed in the justification study are in line with the principles listed in L.6. This list is not exhaustive. Additional information not covered by the questions should be provided if it is relevant to the case.

Each general principle should be given due consideration and, ideally, when preparing the JS, the proposer should provide a general rationale for each principle, prior to answering the questions associated with the principle.

The principles to which the proposer of the MSS should pay due attention when preparing the justification study are:

1. Market relevance
2. Compatibility
3. Topic coverage
4. Flexibility
5. Free trade
6. Applicability of conformity assessment
7. Exclusions

NOTE   No questions directly refer to the principle 8 ("Ease of use"), but it should guide the development of the deliverable.

Basic information on the MSS proposal

1	What is the proposed purpose and scope of the MSS? Is the document supposed to be a guidance document or a document with requirements?
2	Does the proposed purpose or scope include product (including service) specifications, product test methods, product performance levels, or other forms of guidance or requirements directly related to products produced or provided by the implementing organization?
3	Is there one or more existing committee or non-ISO and non-IEC organization that could logically have responsibility for the proposed MSS? If so, identify.
4	Have relevant reference materials been identified, such as existing guidelines or established practices?
5	Are there technical experts available to support the standardization work? Are the technical experts direct representatives of the affected parties from the different geographical regions?
6	What efforts are anticipated as being necessary to develop the document in terms of experts needed and number/duration of meetings?
7	Is the MSS intended to be a guidance document, contractual specification or regulatory specification for an organization?

Principle 1: Market relevance

8	Have all the affected parties been identified? For example:
	a) organizations (of various types and sizes): the decision-makers within an organization who approve work to implement and achieve conformance to the MSS;
	b) customers/end-users, i.e. individuals or parties that pay for or use a product (including service) from an organization;
	c) supplier organizations, e.g. producer, distributor, retailer or vendor of a product, or a provider of a service or information;
	d) MSS service provider, e.g. MSS certification bodies, accreditation bodies or consultants;
	e) regulatory bodies;
	f) non-governmental organizations.
9	What is the need for this MSS? Does the need exist at a local, national, regional or global level? Does the need apply to developing countries? Does it apply to developed countries? What is the added value of having an ISO or IEC document (e.g. facilitating communication between organizations in different countries)?
10	Does the need exist for a number of sectors and is thus generic? If so, which ones? Does the need exist for small, medium or large organizations?
11	Is the need important? Will the need continue? If yes, will the target date of completion for the proposed MSS satisfy this need? Are viable alternatives identified?
12	Describe how the need and importance were determined. List the affected parties consulted and the major geographical or economical regions in which they are located.
13	Is there known or expected support for the proposed MSS? List those bodies that have indicated support. Is there known or expected opposition to the proposed MSS? List those bodies that have indicated opposition.
14	What are the expected benefits and costs to organizations, differentiated for small, medium and large organizations if applicable?
	Describe how the benefits and the costs were determined. Provide available information on geographic or economic focus, industry sector and size of the organization. Provide information on the sources consulted and their basis (e.g. proven practices), premises, assumptions and conditions (e.g. speculative or theoretical), and other pertinent information.
15	What are the expected benefits and costs to other affected parties (including developing countries)?
	Describe how the benefits and the costs were determined. Provide any information regarding the affected parties indicated.
16	What will be the expected value to society?
17	Have any other risks been identified (e.g. timeliness or unintended consequences to a specific business)?

Principle 2: Compatibility

18	Is there potential overlap or conflict with (or what is the added value in relation to) other existing or planned ISO, IEC, non-ISO or non-IEC international standards, or those at the national or regional level? Are there other public or private actions, guidance, requirements and regulations that seek to address the identified need, such as technical papers, proven practices, academic or professional studies, or any other body of knowledge?

19	Is the MSS or the related conformity assessment activities (e.g. audits, certifications) likely to add to, replace all or parts of, harmonize and simplify, duplicate or repeat, conflict with, or detract from the existing activities identified above? What steps are being considered to ensure compatibility, resolve conflict or avoid duplication?
20	Is the proposed MSS likely to promote or stem proliferation of MSS at the national or regional level, or by industry sectors?

Principle 3: Topic coverage

21	Is the MSS for a single specific sector?
22	Will the MSS reference or incorporate an existing, non-industry-specific MSS (e.g. from the ISO 9000 series of quality management standards)? If yes, will the development of the MSS conform to the ISO/IEC Sector Policy (see ISO/IEC Directives, Part 2), and any other relevant policy and guidance procedures (e.g. those that may be made available by a relevant committee)?
23	What steps have been taken to remove or minimize the need for particular sector-specific deviations from a generic MSS?

Principle 4: Flexibility

24	Will the MSS allow an organization competitively to add to, differentiate or encourage innovation of its management system beyond the standard?

Principle 5: Free trade

25	How would the MSS facilitate or impact global trade? Could the MSS create or prevent a technical barrier to trade?
26	Could the MSS create or prevent a technical barrier to trade for small, medium or large organizations?
27	Could the MSS create or prevent a technical barrier to trade for developing or developed countries?
28	If the proposed MSS is intended to be used in government regulations, is it likely to add to, duplicate, replace, enhance or support existing governmental regulations?

Principle 6: Applicability of conformity

29	If the intended use is for contractual or regulatory purposes, what are the potential methods to demonstrate conformance (e.g. first party, second party or third party)? Does the MSS enable organizations to be flexible in choosing the method of demonstrating conformance, and to accommodate for changes in its operations, management, physical locations and equipment?
30	If third-party registration/certification is a potential option, what are the anticipated benefits and costs to the organization? Will the MSS facilitate joint audits with other MSS or promote parallel assessments?

Principle 7: Exclusions

31	Does the proposed purpose or scope include product (including service) specifications, product test methods, product performance levels, or other forms of guidance or requirements directly related to products produced or provided by the implementing organization?

Appendix 2

(normative)

High level structure, identical core text, common terms and core definitions

NOTE   In the identical text proposals, XXX = an MSS discipline specific qualifier (e.g. energy, road traffic safety, IT security, food safety, societal security, environment, quality) that needs to be inserted. Blue italicized text is given as advisory notes to standards drafters.

Introduction

DRAFTING INSTRUCTION   *Specific to the discipline.*

This text has been prepared using the "high-level structure" (i.e. clause sequence, identical core text and common terms and core definitions) provided in Annex L, Appendix 2 of the ISO/IEC Directives, Part 1. This is intended to enhance alignment among ISO and IEC management system standards, and to facilitate their implementation for organizations that need to meet the requirements of two or more such standards.

HLS is highlighted in the text (clauses 1 to 10) by the use of blue font. Black represents the ISO or IEC specific discipline text. Strikeout is used to show agreed deletions within the HLS text. The use of blue text and strikeout is only to facilitate analysis and will not be incorporated after the Draft International Standard stage of development for this document.

1. Scope

DRAFTING INSTRUCTION   *Specific to the discipline.*

2. Normative references

DRAFTING INSTRUCTION   *Clause Title shall be used. Specific to the discipline.*

3. Terms and definitions

DRAFTING INSTRUCTION 1   *Clause Title shall be used. Terms and definitions may either be within the standard or in a separate document.*

Common terms and core definitions shall be stated as well as others that are discipline specific.

The arrangement of terms and definitions should preferably be listed according to the hierarchy of the concepts (i.e. systematic order). Alphabetical order is the least preferred order.

For the purposes of this document, the following terms and definitions apply.

DRAFTING INSTRUCTION 2   *The following terms and definitions constitute an integral part of the "common text" for management systems standards. Additional terms and definitions may be added as needed. Notes may be added or modified to serve the purpose of each standard.*

DRAFTING INSTRUCTION 3   *Italics type in a definition indicates a cross-reference to another term defined in this clause, and the number reference for the term is given in parentheses.*

DRAFTING INSTRUCTION 4   *Where the text "XXX" appears throughout this clause, the appropriate reference should be inserted depending on the context in which these terms and definitions are being applied. For example: "an XXX objective" could be substituted as "an information security objective".*

3.1
organization
person or group of people that has its own functions with responsibilities, authorities and relationships to achieve its *objectives* (3.8)

Note 1 to entry: The concept of organization includes, but is not limited to, sole-trader, company, corporation, firm, enterprise, authority, partnership, charity or institution, or part or combination thereof, whether incorporated or not, public or private.

3.2
interested party (preferred term)
stakeholder (admitted term)
person or *organization* (3.1) that can affect, be affected by, or perceive itself to be affected by a decision or activity

3.3
requirement
need or expectation that is stated, generally implied or obligatory

Note 1 to entry: "Generally implied" means that it is custom or common practice for the organization and interested parties that the need or expectation under consideration is implied.

Note 2 to entry: A specified requirement is one that is stated, e.g. in documented information.

3.4
management system
set of interrelated or interacting elements of an *organization* (3.1) to establish *policies* (3.7) and *objectives* (3.8) and *processes* (3.12) to achieve those objectives

Note 1 to entry: A management system can address a single discipline or several disciplines.

Note 2 to entry: The system elements include the organization's structure, roles and responsibilities, planning and operation.

Note 3 to entry: The scope of a management system can include the whole of the organization, specific and identified functions of the organization, specific and identified sections of the organization, or one or more functions across a group of organizations.

3.5
top management
person or group of people who directs and controls an *organization* (3.1) at the highest level

Note 1 to entry: Top management has the power to delegate authority and provide resources within the organization.

Note 2 to entry: If the scope of the *management system* (3.4) covers only part of an organization, then top management refers to those who direct and control that part of the organization.

3.6
effectiveness
extent to which planned activities are realized and planned results achieved

3.7
policy
intentions and direction of an *organization* (3.1), as formally expressed by its *top management* (3.5)

3.8
objective
result to be achieved

Note 1 to entry: An objective can be strategic, tactical, or operational.

Note 2 to entry: Objectives can relate to different disciplines (such as financial, health and safety, and environmental goals) and can apply at different levels (such as strategic, organization-wide, project, product and *process* (3.12)).

Note 3 to entry: An objective can be expressed in other ways, e.g. as an intended outcome, a purpose, an operational criterion, as an XXX objective, or by the use of other words with similar meaning (e.g. aim, goal, or target).

Note 4 to entry: In the context of XXX management systems, XXX objectives are set by the organization, consistent with the XXX policy, to achieve specific results.

3.9
risk
effect of uncertainty

Note 1 to entry: An effect is a deviation from the expected — positive or negative.

Note 2 to entry: Uncertainty is the state, even partial, of deficiency of information related to, understanding or knowledge of, an event, its consequence, or likelihood.

Note 3 to entry: Risk is often characterized by reference to potential "events" (as defined in ISO Guide 73) and "consequences" (as defined in ISO Guide 73), or a combination of these.

Note 4 to entry: Risk is often expressed in terms of a combination of the consequences of an event (including changes in circumstances) and the associated "likelihood" (as defined in ISO Guide 73) of occurrence.

3.10
competence
ability to apply knowledge and skills to achieve intended results

3.11
documented information
information required to be controlled and maintained by an *organization* (3.1) and the medium on which it is contained

Note 1 to entry: Documented information can be in any format and media, and from any source.

Note 2 to entry: Documented information can refer to:

— the *management system* (3.4), including related *processes* (3.12);

— information created in order for the organization to operate (documentation);

— evidence of results achieved (records).

3.12
process
set of interrelated or interacting activities which transforms inputs into outputs

3.13
performance
measurable result

Note 1 to entry: Performance can relate either to quantitative or qualitative findings.

Note 2 to entry: Performance can relate to managing activities, *processes* (3.12), products (including services), systems or *organizations* (3.1).

3.14
outsource (verb)
make an arrangement where an external *organization* (3.1) performs part of an organization's function or *process* (3.12)

Note 1 to entry: An external organization is outside the scope of the *management system* (3.4), although the outsourced function or process is within the scope.

3.15
monitoring
determining the status of a system, a *process* (3.12) or an activity

Note 1 to entry: To determine the status, there can be a need to check, supervise or critically observe.

3.16
measurement
process (3.12) to determine a value

3.17
audit
systematic, independent and documented *process* (3.12) for obtaining audit evidence and evaluating it objectively to determine the extent to which the audit criteria are fulfilled

Note 1 to entry: An audit can be an internal audit (first party) or an external audit (second party or third party), and it can be a combined audit (combining two or more disciplines).

Note 2 to entry: An internal audit is conducted by the organization itself, or by an external party on its behalf.

Note 3 to entry: "Audit evidence" and "audit criteria" are defined in ISO 19011.

3.18
conformity
fulfilment of a *requirement* (3.3)

3.19
nonconformity
non-fulfilment of a *requirement* (3.3)

3.20
corrective action
action to eliminate the cause(s) of a *nonconformity* (3.19) and to prevent recurrence

3.21
continual improvement
recurring activity to enhance *performance* (3.13)

4. Context of the organization

4.1 Understanding the organization and its context

The organization shall determine external and internal issues that are relevant to its purpose and that affect its ability to achieve the intended outcome(s) of its XXX management system.

4.2 Understanding the needs and expectations of interested parties

The organization shall determine:

— the interested parties that are relevant to the XXX management system;

— the relevant requirements of these interested parties.

4.3 Determining the scope of the XXX management system

The organization shall determine the boundaries and applicability of the XXX management system to establish its scope.

When determining this scope, the organization shall consider:

— the external and internal issues referred to in 4.1;

— the requirements referred to in 4.2.

The scope shall be available as documented information.

4.4 XXX management system

The organization shall establish, implement, maintain and continually improve an XXX management system, including the processes needed and their interactions, in accordance with the requirements of this document.

5. Leadership

5.1 Leadership and commitment

Top management shall demonstrate leadership and commitment with respect to the XXX management system by:

— ensuring that the XXX policy and XXX objectives are established and are compatible with the strategic direction of the organization;

— ensuring the integration of the XXX management system requirements into the organization's business processes;

— ensuring that the resources needed for the XXX management system are available;

— communicating the importance of effective XXX management and of conforming to the XXX management system requirements;

— ensuring that the XXX management system achieves its intended outcome(s);

— directing and supporting persons to contribute to the effectiveness of the XXX management system;

— promoting continual improvement;

— supporting other relevant managerial roles to demonstrate their leadership as it applies to their areas of responsibility.

NOTE　Reference to "business" in this document can be interpreted broadly to mean those activities that are core to the purposes of the organization's existence.

5.2　Policy

Top management shall establish a XXX policy that:

a)　is appropriate to the purpose of the organization;

b)　provides a framework for setting XXX objectives;

c)　includes a commitment to satisfy applicable requirements;

d)　includes a commitment to continual improvement of the XXX management system.

The XXX policy shall:

— be available as documented information;

— be communicated within the organization;

— be available to interested parties, as appropriate.

5.3　Roles, responsibilities and authorities

Top management shall ensure that the responsibilities and authorities for relevant roles are assigned and communicated within the organization.

Top management shall assign the responsibility and authority for:

a)　ensuring that the XXX management system conforms to the requirements of this document;

b)　reporting on the performance of the XXX management system to top management.

6.　Planning

6.1　Actions to address risks and opportunities

When planning for the XXX management system, the organization shall consider the issues referred to in 4.1 and the requirements referred to in 4.2 and determine the risks and opportunities that need to be addressed to:

— give assurance that the XXX management system can achieve its intended outcome(s);

— prevent, or reduce, undesired effects;

— achieve continual improvement.

The organization shall plan:

a)　actions to address these risks and opportunities;

b)　how to:

　　— integrate and implement the actions into its XXX management system processes;

　　— evaluate the effectiveness of these actions.

6.2　XXX objectives and planning to achieve them

The organization shall establish XXX objectives at relevant functions and levels.

The XXX objectives shall:

a) be consistent with the XXX policy;

b) be measurable (if practicable);

c) take into account applicable requirements;

d) be monitored;

e) be communicated;

f) be updated as appropriate.

The organization shall retain documented information on the XXX objectives.

When planning how to achieve its XXX objectives, the organization shall determine:

— what will be done;

— what resources will be required;

— who will be responsible;

— when it will be completed;

— how the results will be evaluated.

7. Support

7.1 Resources

The organization shall determine and provide the resources needed for the establishment, implementation, maintenance and continual improvement of the XXX management system.

7.2 Competence

The organization shall:

— determine the necessary competence of person(s) doing work under its control that affects its XXX performance;

— ensure that these persons are competent on the basis of appropriate education, training, or experience;

— where applicable, take actions to acquire the necessary competence, and evaluate the effectiveness of the actions taken;

— retain appropriate documented information as evidence of competence.

NOTE   Applicable actions can include, for example, the provision of training to, the mentoring of, or the re-assignment of currently employed persons; or the hiring or contracting of competent persons.

7.3 Awareness

Persons doing work under the organization's control shall be aware of:

— the XXX policy;

— their contribution to the effectiveness of the XXX management system, including the benefits of improved XXX performance;

— the implications of not conforming with the XXX management system requirements.

7.4 Communication

The organization shall determine the internal and external communications relevant to the XXX management system, including:

— on what it will communicate;

— when to communicate;

— with whom to communicate;

— how to communicate.

7.5 Documented information

7.5.1 General

The organization's XXX management system shall include:

a) documented information required by this document;

b) documented information determined by the organization as being necessary for the effectiveness of the XXX management system.

NOTE   The extent of documented information for a XXX management system can differ from one organization to another due to:

— the size of organization and its type of activities, processes, products and services;

— the complexity of processes and their interactions;

— the competence of persons.

7.5.2 Creating and updating

When creating and updating documented information the organization shall ensure appropriate:

— identification and description (e.g. a title, date, author, or reference number);

— format (e.g. language, software version, graphics) and media (e.g. paper, electronic);

— review and approval for suitability and adequacy.

7.5.3 Control of documented information

Documented information required by the XXX management system and by this document shall be controlled to ensure:

a) it is available and suitable for use, where and when it is needed;

b) it is adequately protected (e.g. from loss of confidentiality, improper use, or loss of integrity).

For the control of documented information, the organization shall address the following activities, as applicable:

— distribution, access, retrieval and use;

— storage and preservation, including preservation of legibility;

— control of changes (e.g. version control);

— retention and disposition.

Documented information of external origin determined by the organization to be necessary for the planning and operation of the XXX management system shall be identified, as appropriate, and controlled.

NOTE   Access can imply a decision regarding the permission to view the documented information only, or the permission and authority to view and change the documented information.

8. Operation

8.1 Operational planning and control

DRAFTING INSTRUCTION   This subclause heading will be deleted if no additional subclauses are added to Clause 8.

The organization shall plan, implement and control the processes needed to meet requirements, and to implement the actions determined in 6.1, by:

— establishing criteria for the processes;

— implementing control of the processes in accordance with the criteria;

— keeping documented information to the extent necessary to have confidence that the processes have been carried out as planned.

The organization shall control planned changes and review the consequences of unintended changes, taking action to mitigate any adverse effects, as necessary.

The organization shall ensure that outsourced processes are controlled.

9. Performance evaluation

9.1 Monitoring, measurement, analysis and evaluation

The organization shall determine:

— what needs to be monitored and measured;

— the methods for monitoring, measurement, analysis and evaluation, as applicable, to ensure valid results;

— when the monitoring and measuring shall be performed;

— when the results from monitoring and measurement shall be analysed and evaluated.

The organization shall retain appropriate documented information as evidence of the results.

The organization shall evaluate the XXX performance and the effectiveness of the XXX management system.

9.2 Internal audit

9.2.1   The organization shall conduct internal audits at planned intervals to provide information on whether the XXX management system:

a) conforms to:

— the organization's own requirements for its XXX management system;

— the requirements of this document;

b) is effectively implemented and maintained.

9.2.2 The organization shall:

a) plan, establish, implement and maintain an audit programme(s) including the frequency, methods, responsibilities, planning requirements and reporting, which shall take into consideration the importance of the processes concerned and the results of previous audits;

b) define the audit criteria and scope for each audit;

c) select auditors and conduct audits to ensure objectivity and the impartiality of the audit process;

d) ensure that the results of the audits are reported to relevant managers;

e) retain documented information as evidence of the implementation of the audit programme(s) and the audit results.

9.3 Management review

Top management shall review the organization's XXX management system, at planned intervals, to ensure its continuing suitability, adequacy and effectiveness.

The management review shall include consideration of:

a) the status of actions from previous management reviews;

b) changes in external and internal issues that are relevant to the XXX management system;

c) information on the XXX performance, including trends in:

— nonconformities and corrective actions;

— monitoring and measurement results;

— audit results;

d) opportunities for continual improvement.

The outputs of the management review shall include decisions related to continual improvement opportunities and any need for changes to the XXX management system.

The organization shall retain documented information as evidence of the results of management reviews.

10. Improvement

10.1 Nonconformity and corrective action

When a nonconformity occurs, the organization shall:

a) react to the nonconformity and, as applicable:

— take action to control and correct it;

— deal with the consequences;

b) evaluate the need for action to eliminate the cause(s) of the nonconformity, in order that it does not recur or occur elsewhere, by:

— reviewing the nonconformity;

— determining the causes of the nonconformity;

— determining if similar nonconformities exist, or can potentially occur;

c) implement any action needed;

d) review the effectiveness of any corrective action taken;

e) make changes to the XXX management system, if necessary.

Corrective actions shall be appropriate to the effects of the nonconformities encountered.

The organization shall retain documented information as evidence of:

— the nature of the nonconformities and any subsequent actions taken;

— the results of any corrective action.

10.2 Continual improvement

The organization shall continually improve the suitability, adequacy and effectiveness of the XXX management system.

Appendix 3

(informative)

Guidance on high level structure, identical core text, common terms and core definitions

Guidance on the high level structure, identical core text, common terms and core definitions is provided at the following URL:

Annex L Guidance documents (http://isotc.iso.org/livelink/livelink?func=ll&&objId=16347818&&objAction=browse&&viewType=1).

Annex M
(normative)

Policy for the development of sector-specific management standards and sector-specific management system standards (MSS)

M.1 General

Any technical committee or subcommittee, project committee (SyC in IEC) or International Workshop that proposes development of a sector-specific management standard (M.2.2) or a sector-specific management system standard (MSS) (M.2.4) shall follow the directions specified in this annex. It includes, as applicable, committee specific policies (M.5) which may not be limited to sector-specific management standards or sector-specific management system standards.

M.2 Terms and definitions

M.2.1
generic management standard
management standard designed to be widely applicable across economic sectors, various types and sizes of organizations and diverse geographical, cultural and social conditions

M.2.2
sector-specific management standard
management standard that provides additional requirements or guidance for the application of a *generic management standard* (M.2.1) to a specific economic or business sector

M.2.3
generic management system standard
generic MSS
MSS designed to be widely applicable across economic sectors, various types and sizes of organizations and diverse geographical, cultural and social conditions

M.2.4
sector-specific management system standard (MSS)
sector-specific MSS
MSS that provides additional requirements or guidance for the application of a *generic MSS* (M.2.3) to a specific economic or business sector

M.3 Sector-specific management standards and sector-specific management system standards

Any new proposal for a sector-specific management standard (M.2.2) or sector-specific MSS (M.2.4) shall:

— clearly demonstrate its market relevance and alignment through the completion of appropriate ISO or IEC project approval procedures by means of ISO Form 4, *New Work Item Proposal*, or IEC Form NP,

— [in the case of the development of a sector-specific MSS (M.2.4)] clearly demonstrate that all the rules and principles in Annex L have been followed, including the approval of the justification study (see Annex L), and

— clearly demonstrate that the liaison with the committee responsible for the generic management standard or generic MSS concerned is effective,

— if applicable, conform with the committee specific policies set out below.

M.4 Drafting rules

Sector-specific management standards (M.2.2) and sector-specific MSS (M.2.4) shall respect the following rules:

a) Normative reference shall be made to the generic management standard (M.2.1) or generic MSS (M.2.3). Alternatively, the clauses and subclauses may be reproduced verbatim.

b) If text from the generic management standard (M.2.1) or generic MSS (M.2.3) is reproduced in the sector-specific standard, it shall be distinguished from the other elements of the sector-specific standard.

c) Terms and definitions specified in the generic management standard (M.2.1) or generic MSS (M.2.3) shall be referred to in a normative manner or reproduced verbatim.

M.5 Committee specific policies

M.5.1 General

Sector-specific management standards (M.2.2) and sector-specific MSS (M.2.4) shall not interpret, change, or subtract from the requirements of the generic management standard or generic MSS.

M.5.2 Environment

M.5.2.1 Terms and definitions

The following terms and definitions are applicable to environmental policy:

M.5.2.1.1
sector-specific environmental management standard
standard that provides additional requirements or guidance for the application of a generic environmental management standard to a specific economic or business sector

EXAMPLE   The application of an environmental management system (ISO 14001) or life-cycle assessment (ISO 14044) to agri-food or energy sectors.

M.5.2.1.2
aspect-specific environmental management standard
standard that provides additional requirements or guidance for the application of a generic environmental management standard for a specific environmental aspect or aspects within its scope

EXAMPLE   The application of an environmental management system (ISO 14001) for greenhouse gas (aspect) management or life-cycle assessment (ISO 14044) for the water (aspect) footprint of products.

M.5.2.1.3
element-specific environmental management standard
standard that provides additional requirements or guidance for the application of a generic environmental management standard for a specific element or elements within its scope

EXAMPLE   Communications or emergency management (elements) within an environmental management system (ISO 14001) or data collection or critical review (elements) within a life-cycle assessment (ISO 14044).

M.5.2.2 General

Any technical committee, subcommittee, project committee (SyC in IEC) or International Workshop that proposes development of a sector-, aspect- or element-specific environmental management standard shall clearly demonstrate its market relevance and alignment through the completion of appropriate project approval procedures, including:

— ISO Form 4, *New Work Item Proposal* for sector-, aspect- or element-specific specific application of generic environmental management system standards, environmental labeling, life-cycle assessment and greenhouse gas management standards, and

— Annex L *Proposals for management system standards (MSS)* for sector-, aspect- or element-specific specific application of generic environmental MSS.

Approval documentation should include specific justification as to why the relevant generic ISO 14000 series standard(s) insufficiently address sector-, aspect- or element-specific needs and how the proposed new standard would effectively resolve identified issues. Proposers should critically assess whether additional sector-, aspect- or element-specific requirements are needed as opposed to the provision of additional guidance to the generic environmental management standard(s).

M.5.2.3 Any technical committee, subcommittee, project committee (SyC in IEC) or International Workshop that proposes development of a sector-, aspect- or element-specific environmental management standard should consider and reflect the needs of developing countries, economies in transition, small- and medium- enterprises and organizations operating across a variety of sectors.

M.5.2.4 ISO/TC 207 will cooperate in or, where appropriate and as decided by the Technical Management Board, lead joint projects with technical committee, subcommittee, project committee (SyC in IEC) or International Workshop developing sector-, aspect- or element-specific environmental management standards to avoid duplication of effort and promote consistency and alignment. There is no intention to restrict the development of market relevant standards in committees outside of ISO/TC 207.

M.5.2.5 Technical committee, subcommittee, project committee (SyC in IEC) or International Workshop developing sector-, aspect- or element-specific environmental management standards shall:

— include the normative reference of the appropriate generic ISO 14000 series environmental management systems, environmental auditing, environmental labeling, life-cycle assessment and greenhouse gas management standards;

— include the normative reference of the appropriate generic ISO 14050 terms and definitions;

— distinguish ISO 14000 series text if it is reproduced; and

— not interpret, change, or subtract from the requirements of the generic ISO 14000 series environmental management systems, environmental auditing, environmental labeling, life-cycle assessment and greenhouse gas management standards.

M.5.2.6 Any requests for guidance on this sector-, aspect- or element-specific policy or for interpretation of generic ISO 14000 series standards or ISO 14050 terms and definitions or for guidance on a sector-, aspect- or element-specific document shall be submitted to the ISO Central Secretariat as well as the relevant TC 207 subcommittee.

M.5.3 Quality

When an technical committee, subcommittee, project committee (SyC in IEC) or International Workshop wishes to develop quality management system requirements or guidance for a particular product or industry/economic sector it shall respect the following rules.

a) Normative reference shall be made to ISO 9001 in its entirety. Alternatively, the clauses and subclauses may be reproduced verbatim.

b) If text from ISO 9001 is reproduced in the sector document, it shall be distinguished from the other elements of the sector document [see d)].

c) Terms and definitions specified in ISO 9000 shall be referred to in a normative manner or reproduced verbatim.

d) The guidance and criteria provided in Quality management systems – Guidance and criteria for the development of documents to meet needs of specific product and industry/economic sectors, approved by ISO/TC 176, shall be considered not only when determining the need for a sector-specific requirements or guidance document but also in the document development process.

Any requests for guidance on this sector policy or for interpretation of ISO 9000 terms and definitions, ISO 9001 or ISO 9004 shall be submitted to the secretariat of ISO/TC 176.

M.5.4 Asset management

When an technical committee, subcommittee, project committee (SyC in IEC) or International Workshop wishes to develop asset management system requirements or guidance for a particular product or industry/economic sector it shall respect the following rules:

a) Normative reference shall be made to ISO 55001 in its entirety. Alternatively, the clauses and subclauses may be reproduced verbatim.

b) If text from ISO 55001 is reproduced in the sector document, it shall be distinguished from the other elements of the sector document.

c) Terms and definitions specified in ISO 55000 shall be referred to in a normative manner or reproduced verbatim.

Any requests for guidance on a sector-specific document or for interpretation of ISO 55000 terms and definitions or ISO 55001 shall be submitted to the secretariat of ISO/TC 251.

M.5.5 Risk

When a technical committee, subcommittee, project committee (SyC in IEC) or International Workshop wishes to develop risk management requirements or guidance for a particular product or industry/economic sector it shall respect the following rules:

a) Reference shall be made to ISO 31000 in its entirety. Alternatively, the clauses and subclauses may be reproduced verbatim.

b) If text from ISO 31000 is reproduced in the sector document, it shall be distinguished from the other elements of the sector document.

c) Terms and definitions specified in ISO 31000 shall be referred to in a normative manner or reproduced verbatim.

Any requests for guidance on a sector-specific document or for interpretation of ISO 31000 terms and definitions shall be submitted to the secretariat of ISO/TC 262.

M.5.6 Social responsibility

When a technical committee, subcommittee, project committee (SyC in IEC) or International Workshop wishes to develop social responsibility requirements or guidance for a particular product or industry/economic sector it shall respect the following rules:

a) Reference shall be made to ISO 26000 in its entirety. Alternatively, the clauses and subclauses may be reproduced verbatim.

b) If text from ISO 26000 is reproduced in the sector document, it shall be distinguished from the other elements of the sector document.

c) Terms and definitions specified in ISO 26000 shall be referred to in a normative manner or reproduced verbatim.

ANNEXES APPLICABLE TO ISO ONLY

Annex SA
(normative)

ISO Code of Conduct

The ISO Code of Conduct is available on the Governance of technical work page (http://www.iso.org/iso/home/standards_development/governance_of_technical_work.htm) of www.iso.org at:

https://www.iso.org/publication/PUB100397.html

The Code is to facilitate ISO's work which is carried out in an international, multi-stakeholder, multi-sector environment. It applies to people who choose to participate in an ISO committee, working group or other consensus group. The Code is an obligation for participation in the above groups that work in the framework of the ISO/IEC Directives.

Tips for implementing the Code are available at:

http://www.iso.org/iso/suggestions_for_implementation_of_the_iso_code_of_conduct.pdf

Annex SB
(normative)

Document distribution

SB.1 Document distribution

Details showing document distribution are shown in the table.

SB.2 Electronic notifications of document delivery

The information included in electronic notifications of document availability shall comprise

— **Subject line**: committee reference

— **Content** (the precise format may vary):

　1) N-number (automatically generated)

　2) Official ISO reference number (e.g. CD 1234 for a committee draft, NP 2345 for a new work item proposal)

　3) Title of the document

　4) Expected action

　5) Due date for expected action

　6) Type of document (information, ballot, comment, notice of meeting, etc.)

　7) Date and location, if it is a meeting document

　8) Note field, for special information related to a particular document

　9) Hyperlink to the document

ISO/IEC 导则　第 1 部分　ISO 技术工作程序合集

DOCUMENTS \ PARTY(IES) CONCERNED	Proposer	TC or SC secretariat and chairs	TC or SC P-members	TC or SC O-members	Category A & B and internal liaisons	ISO Central Secr.	WG convener	WG experts	National Bodies
Proposal stage									
New work item proposal	★	●				○			
NP ballot		★¹⁾	●	○	○	○			
Completed ballot		●	★			○			
Result of voting	○	★¹⁾	○	○	○	■			
Preparatory stage									
Working draft(s) (WD)		○				○	★	●	
Final working draft		●	○			■	★	○	
Committee stage									
Committee draft(s) (CD)		★	●	○	○	○			
Comments/Vote		●	★						
Compilation of comments + proposal		★	●	○	○	○	○		
Final committee draft		★¹⁾	○	○	○	■	○		
Enquiry stage									
Draft International Standard		○			○	★			●
Completed ballot					★	●			★
Result of voting + comments		●¹⁾	○	○	○	★			○
Report of voting		★	○	○	○	●	○		○
Text for final draft International Standard		★¹⁾	○	○	○	■	○		
Approval stage									
Final draft International Standard + ballot		○			○	★			●
Completed ballot					★	●			★
Result of voting		○	○	○	○	★			○
Final proof		●				★			
Proof corrections		★				●			
Publication stage									
International Standard		○				★			○
Systematic review									
SR ballot		○				★			●
Ballot						●			★
Completed ballot		●	●	○	○			○	
Report of voting + proposal		★	○		○	■			

★ Sender of document
● Recipient for action
○ Recipient for information
■ Recipient for registration action
☆ Optional action
¹⁾ In the case of an SC, a copy is also sent to the TC secretariat for information

Annex SC
(normative)

Strategic business plans

SC.1 Objectives of a strategic business plan (SBP)

1) To demonstrate in an objective manner the specific benefits which result from, or are expected from, the work of this technical committee.

NOTE 1   These benefits can vary significantly between different fields in which committees are involved: they can be economic (cost savings, reduced time to market, easier access to certain regional markets, lower sales prices), they can be social (improvement of safety for workers, measured in the reduction of accidents) or they can be the improvement of the environmental impact, for example.

2) To support prioritization and to improve the management of the technical work in a committee.

NOTE 2   This includes the definition of target dates and the planning of resources for the work of the technical committee (and its SCs) in relation to the development of new and the maintenance of existing documents.

3) To increase the transparency in relation to the market forces and the percentage of market share represented within a committee.

All current SBP may be viewed online at http://www.iso.org/bp.

SC.2 Development tools and additional guidance

To facilitate the preparation of SBP, ISO Central Secretariat has developed a document template. This template and additional guidance may be downloaded from the ISOTC server at www.iso.org/forms.

SC.3 Procedure for the development of a strategic business plan

SC.3.1   The committee secretary, in cooperation with committee members and/or the committee chair, shall prepare a draft SBP. The draft SBP shall be submitted to an internal consultation within the technical committee. It shall be formally agreed upon by the technical committee by way of a resolution, before being submitted to the technical management board for review.

The technical committee shall submit the draft SBP (tmb@iso.org) in revisable (i.e. Word) and PDF-format to the ISO technical management board who will review the draft SBP against the set of minimum requirements for SBPs (see Table SC.1).

Table SC.1 — Requirements for strategic business plans of ISO technical committees

Item	Description of Requirements	Score
1	Descriptions of relevant dynamics in the business environment related to the work of the ISO committee, and quantitative indictors of trends in this business environment and the acceptance and implementation of the ISO committee's standards.	1 to 5
2	Descriptions of tangible benefits that the standardization programme is expected to achieve for the business environment.	1 to 5
3	Descriptions of identified objectives of the ISO committee and strategies to achieve those objectives. This should include descriptions of specific actions that will be taken or that will be proposed to the ISO committee to better respond to the needs and trends of the business environment.	1 to 5
4	Descriptions of factors that may negatively impact the ISO committee's ability to achieve its objectives and implement its strategies, including information on the representation of the major market forces in the committee (geographically as well as by type, e.g. manufacturer, government, etc.).	1 to 5
5	Objective information regarding the ISO committee and its work programme that is required: — link added in section 7 to the TC's page on ISO's website About (Secretariat, Secretary, Chair, Date of creation, Scope, etc.), Contact details, Structure (Subcommittees and working groups), Liaisons, Meetings, Tools, Work programme (published standards and standards under development); — priorities assigned to projects in the work programme (if the committee assigns priorities) with an explanation of the reasons/process for prioritization; — relationships of projects to European regional standardization (CEN); — time allocated to each project by working group convenors, project leaders/editors and for translation; and — the range of stakeholders that should be engaged based on the subject area being standardized.	No score (Yes/No) Mandatory information

SC.3.2 An average score of at least 10 is necessary for approval. If the SBP is not approved by the TMB, the technical management board scores and comments will be sent to the committee who will be asked to resubmit the SBP. If approved, the comments from the technical management board shall be taken into consideration during the SBP review at the next plenary meeting.

SC.3.3 Once approved by the technical management board, the SBP is made available to the general public for review and comment on a publicly accessible server (http://www.iso.org/bp) by the ISO Central Secretariat. Member bodies are encouraged to inform the interested public in their countries of the availability of the SBP for public review and, if appropriate, to set links to the SBPs.

SC.3.4 The committee shall conduct a review of the SBP, considering all comments received, preferably once per year, but at least once every 3 years. Updated SBP do not need to be submitted for TMB approval — they shall be approved by consensus within the committee (or, if consensus is not possible, by a simple majority of P-members casting a vote).

Annex SD
(normative)

Matrix presentation of project stages

SD.1 Introduction to the Harmonized Stage Code

The standardization process has a number of definite steps or stages which can be used both to describe the process and to indicate where in the process any one item has reached. In general terms the methods used to develop and publish standards via the formal standardization process operated by international, regional and national standards bodies are very similar no matter which body is overseeing the process. Thus, at a high level, it is possible to have a common view of the standardization process and with it a common set of stages. There are differences between the processes of individual bodies, however, and this has led to the development of different stage systems for each body.

This Harmonized Stage Code (HSC) system is used in ISO's databases for tracking standards development projects. Its purpose is to provide a common framework for the transfer of core data. The system allows tracking of the development of a given project in the same way in databases being used at international, regional and national levels and the matrix is so constructed that it can easily be adapted to new requirements.

SD.2 Design of the stage code matrix

A series of "stages" representing procedural sequences common to different organizations has been established. These represent the main stages of standards development.

A series of "sub-stages" has been established within each stage, using a consistent logical system of concepts. The terms "stage" and "sub-stage" are hence used to designate the respective axes of the resulting matrix.

Principal stages and sub-stages are each coded by a two-digit number from 00 to 90, in increments of 10. Individual cells within the generic matrix are coded by a four-digit number made up of its stage and sub-stage coordinates. For visual presentation (although not necessarily for the purposes of database operations), the pair of coordinates are separated by a point (e.g. 10.20 for stage 10, sub-stage 20).

All unused stage codes are reserved for future use, to allow for interpolation of additional phases that might be identified, e.g. stage codes 10, 30, 40, 50 and 80.

SD.3 Basic guidelines for using the system

— Other information concerning, for example, document source or document type, should be recorded in separate database fields and should not be reflected in stage codes.

— There is no sub-code to indicate that a project is dormant at any particular stage. It is recommended to use another database field to address this issue.

— The HSC system allows for the cyclical nature of the standards process and for the repeating of either the current phase or an earlier phase. Events that may be repeated in the life of a project are recordable by repetition of the same stage codes.

— Freezing a project at any point is possible by using the code the project has reached. Projects that have been suspended should have this information recorded in a separate database field.

— The HSC system is not concerned with recording either target or actual dates for achieving stages.

ISO/IEC 导则 第 1 部分 ISO 技术工作程序合集

STAGE	SUB-STAGE						
	00 Registration	**20** Start of main action	**60** Completion of main action	**90** Decision			
				92 Repeat an earlier phase	**93** Repeat current phase	**98** Abandon	**99** Proceed
00 Preliminary stage	00.00 Proposal for new project received	00.20 Proposal for new project under review	00.60 Close of review			00.98 Proposal for new project abandoned	00.99 Approval to ballot proposal for new project
10 Proposal stage	10.00 Proposal for new project registered	10.20 New project ballot initiated	10.60 Close of voting	10.92 Proposal returned to submitter for further definition		10.98 New project rejected	10.99 New project approved
20 Preparatory stage	20.00 New project registered in TC/SC work programme	20.20 Working draft (WD) study initiated	20.60 Close of comment period			20.98 Project cancelled	20.99 WD approved for registration as CD
30 Committee stage	30.00 Committee draft (CD) registered	30.20 CD study/ballot initiated	30.60 Close of voting/ comment period	30.92 CD referred back to Working Group		30.98 Project cancelled	30.99 CD approved for registration as DIS
40 Enquiry stage	40.00 DIS registered	40.20 DIS ballot initiated: *12 weeks*	40.60 Close of voting	40.92 Full report circulated: DIS referred back to TC or SC	40.93 Full report circulated: decision for new DIS ballot	40.98 Project cancelled	40.99 Full report circulated: DIS approved for registration as FDIS
50 Approval stage	50.00 Final text received or FDIS registered for formal approval	50.20 Proof sent to secretariat or FDIS ballot initiated: *8 weeks*	50.60 Close of voting. Proof returned by secretariat	50.92 FDIS or proof referred back to TC or SC		50.98 Project cancelled	50.99 FDIS or proof approved for publication
60 Publication stage	60.00 International Standard under publication		60.60 International Standard published				
90 Review stage		90.20 International Standard under systematic review	90.60 Close of review	90.92 International Standard to be revised	90.93 International Standard confirmed		90.99 Withdrawal of International Standard proposed by TC or SC
95 Withdrawal stage		95.20 Withdrawal ballot initiated	95.60 Close of voting	95.92 Decision not to withdraw International Standard			95.99 Withdrawal of International Standard

Annex SE
(normative)

Numbering of documents

SE.1 Working documents (including committee drafts)

Each document (including committee drafts — see also SE.2) relating to the work of an ISO technical committee or subcommittee circulated to all or some of the member bodies shall have a "N-number" generated by the system that appears at the top right-hand corner of recto pages and at the top left-hand corner of verso pages. The first page of the document shall also bear, immediately under the reference number, the date, written in accordance with ISO 8601 (*Data elements and interchange formats — Information interchange — Representation of dates and times*), on which the document was compiled.

Once a reference number is used for a working document, it cannot be used again for a document with differing wording and/or different contents. The same number may be used for a proposal and a voting form (optional). If a document replaces an earlier one, the system generates a new N-number on the first page of the new document.

The N-number is made up of the following two parts separated by the letter N:

1) the number of the technical committee (TC) and, when applicable, the number of the subcommittee (SC) to which the working document belongs;

2) an overall serial number.

As an example, the reference number will look as follows: **ISO/TC** *a*/**SC** *b* **N***n*

SE.2 Working drafts (WD), committee drafts (CD), draft International Standards (DIS), final draft International Standards (FDIS) and International Standards

When a new project is registered by ISO Central Secretariat (see 2.3.5), the latter will allocate an ISO reference number to the project. The number allocated will remain the same for the ensuing WD, CD, DIS and FDIS and for the published International Standard. The number allocated is purely a registration and reference number and has no meaning whatsoever in the sense of classification or chronological order. The number allocated to a withdrawn project or International Standard will not be used again.

If the project represents a revision or amendment of an existing International Standard, the registered project will be allocated the same number as the existing International Standard (with, in the case of an amendment, a suffix indicating the nature of the document).

If, however, the scope is substantially changed, the project shall be given a different number.

Successive WDs or CDs on the same subject shall be marked "first working draft", "second working draft", etc., or "first committee draft", "second committee draft", etc., as well as with the working document number in accordance with SE.1.

Successive DIS on the same subject will carry the same number but will be distinguished by a numerical suffix (.2, .3, etc.).

Annex SF
(normative)

Hosting meetings

SF.1 Who may host an ISO meeting?

Meetings may be hosted by:

— any ISO member (member body, correspondent member, subscriber member);

— for working group meetings only, any liaison member of the committee concerned.

For technical committee or subcommittee meetings, the ISO member body in the country shall be the host and its pre-approval of the meeting is required.

For working group meetings, the ISO member body in the country shall be informed of the meeting before the meeting is confirmed.

It is not necessary for a host to be a direct participant in the work of the committee concerned, although that will typically be the case.

A potential host shall first ascertain that there are no restrictions imposed by its country or the hosting venue to the entry of representatives of any P-member of the technical committee or subcommittee for the purpose of attending the meeting. In some cases it may be necessary for special permission to be obtained for attendance — wherever possible the host should assist in determining if this is the case, but it is the joint responsibility of the secretariat or group leader, and the participant to determine any restrictions.

Given that some hosts may not have sufficient meeting facilities and/or other resources of their own, the main host — say an ISO member — may accept an invitation from another organization to assist by, for instance, providing a meeting location or organizing a welcoming event.

Irrespective of the actual host of a meeting, responsibility for the meeting concerned rests with the secretary (in the case of a technical committee or subcommittee, or similar groups) or group leader (in the case of working groups, ad-hoc groups, etc.). Thus, potential hosts should liaise directly with these individuals.

SF.2 Sponsorship of meetings

Large ISO meetings can be expensive and/or complicated to stage and support, and the resource requirements may exceed those that an ISO member can itself provide. It is therefore acceptable for a meeting to have one or more sponsors contributing to its organization and cost. Nevertheless, whilst it is acceptable for sponsors to be identified, and for their support to be recognized in the meeting, an ISO meeting shall not be used as an opportunity for the promotion for commercial or other reasons of the products or services of any sponsor (see also "Fees for delegates at ISO meetings").

SF.3 Proposing or withdrawing support as a meeting host

The *ISO/IEC Directives, Part 1*, request that the following advance notice be given:

Meeting of ...	Advance notice of meeting date	Advance notice of meeting location
Technical committee or subcommittee	2 years	4 months
Working group	—	6 weeks

In general, the more notice the better in order that participants may make appropriate travel and business arrangements, and also obtain briefing, where appropriate.

If a host finds that they need to withdraw an offer, these notice periods should also be respected.

SF.4 Meeting locations

In planning meetings, account should be taken of the possible advantage of grouping meetings of technical committees and subcommittees dealing with related subjects, in order to improve communication and to limit the burden of attendance at meetings by delegates who participate in several technical committees or subcommittees.

SF.5 Facilities to provide at meetings

Any member body offering to host an ISO technical committee or subcommittee meeting undertakes to provide appropriate meeting facilities, including any provisions for interpretation (see SF.6), for the TC/SC, taking into account the normal attendance at such meetings and any requirements foreseen by the committee secretariat for facilities for parallel meetings (for example, of working groups, ad hoc groups or the drafting committee). The working facilities required during a meeting (i.e. additional to rooms) will vary from group to group. This may include printing, and internet access (teleconference facilities). Administrative support may also be needed. The host may be requested, but shall not be obliged, to provide separate facilities serving as the "chair's" or "secretariat" room.

The exact requirements should be determined by the committee secretariat or group leader. Some committee secretariats develop and maintain a standing document to advise member bodies of the facilities that will be required of any potential host of a meeting of the committee, which ensures clear understanding of what will be required. The host of a working group is also expected to provide all basic working facilities.

SF.6 Interpretation facilities at meetings

The chair and secretariat are responsible for dealing with the question of language at a meeting in a manner acceptable to the participants, following the general rules of ISO, as appropriate. The official languages at meetings are English, French and Russian, and meetings are conducted in any one or more of these. The typical language combinations are English and French, or English only. In the latter case, French may only be omitted with the explicit agreement of the francophone member(s). Thus, there may be an obligation on the host to provide interpretation between English and French. This needs to be determined in advance. Potential hosts may need to seek assistance from ISO Central Secretariat or from other P-members. The National Body for the Russian Federation provides all interpretation and translation into or from the Russian language.

So far as the capabilities of translators (when required) are concerned, the guidance is:

— for meetings where final decisions on draft standards are expected to be taken, the services of an adequately qualified interpreter are generally required;

— meetings where "intermediary" or procedural decisions are expected to be taken, brief interpretation may be provided by a member of the secretariat or a volunteer delegate;

— at working group meetings, the members should, whenever possible, arrange between themselves for any necessary interpretation on the initiative and under the authority of the convenor of the working group.

SF.7 Welcoming activities and social events

ISO is widely recognized and respected. It is therefore not unusual for formal ceremonies to be associated with, for example, the opening of a technical committee meeting. Such events are acceptable, but their incorporation within the meeting period is entirely at the option of the host, and typically subject to the approval of the secretary and chair of the group/s concerned.

A host may also offer social events of one kind or another, perhaps with the aid of co-sponsors. As for welcoming events, social events are entirely optional (see also SF.8).

It is recognized that the inclusion of welcoming and social events significantly increases the time and cost, etc. of organizing a meeting. It may also increase the cost of participation. Given the pressures on the resources of all participants there is merit in the simplification of meetings. Above all, meetings shall not become "competitive" events so far as hosts are concerned, in the sense that no host shall feel obliged, in any respect, to equal or exceed the welcoming and/or social facilities offered at any previous meeting.

SF.8 Fees for delegates at ISO meetings

Accredited delegates shall not be obliged to pay a fee as a condition of their participation in ISO meetings. However, in very exceptional cases (e.g. large and complex meetings), some charging mechanism may be necessary, but such mechanisms shall be approved on a case-by-case basis by the Technical Management Board.

There is no obligation on a host to provide social functions during ISO meetings but if a committee requests a host to organize social functions, the host shall have the right to require that the costs of holding such functions be borne by the delegates participating therein.

SF.9 Media participation policy

SF.9.1 General

This clause describes a policy for engaging with the media to raise the effectiveness of communication activities of ISO. It is vital to ensure transparency and openness by proactively and consistently delivering accurate information to the public, and to support the use of proper channels to deliver information to the public as effectively and efficiently as possible.

This policy describes how to raise awareness about ISO activities and ensure transparency. It is important that all involved in ISO activities provide accurate and appropriate information in a spirit of good will and responsibility. The media contributes to the shaping of attitudes and behaviours of the general public. Building good media relations is important for building awareness through accurate information and preventing isolated and unrepresentative views and opinions from misleading the public.

Engagement with the media should be sought through multiple vehicles at multiple organizational levels.

All engagement with the media relating to ISO activities will be consistent with the following:

— Interest by the press or other media in ISO work is to be welcomed and the ISO Central Secretariat and many ISO member bodies have public relations and promotion services able to provide information to the press concerning ISO, ISO standards, and work in progress.

— In recent years, there has been a growing interest on the part of some of the press to be present during meetings of particular ISO committees. Whilst this interest is welcome, the presence of the press during an ISO meeting may inhibit the free and open discussion of issues. For this reason, members of the press shall not be permitted to be present during working sessions of ISO meetings. However participation by members of the press may be permitted during opening and closing ceremonies of ISO plenary meetings.

SF.9.2 ISO Central Secretariat roles

Within ISO, the ISO Central Secretariat prepares and releases any and all official press releases. The ISO Central Secretariat reserves the right of final decision on any ISO press release.

SF.9.3 ISO committee and working group leadership roles

Leaders shall only express the consensus reached by the delegates/experts at press conferences held in conjunction with the ISO committee meetings. Between meetings, the secretariat will endeavour to provide prompt written responses to media requests.

SF.9.4 Member body and mirror committee roles

At a local or national level, mirror committees are the proper channel for media inquiries. Mirror committees are encouraged to arrange meetings with local media. This will improve local participation and awareness. At the national level, the relevant member body's media policy should apply.

SF.9.5 Delegate, expert and observer roles

These parties shall not comment to the media on the consensus achieved in the ISO committee, nor shall they comment on the contributions and comments made by other delegates and experts in the working sessions of the ISO committee. This could have a detrimental effect on the productive dialogue and trust necessary among delegates and experts for effective ISO standards development.

SF.9.6 Meetings and media participation

Media representatives present at the site of a meeting shall register with the ISO committee secretariat and the meeting host (name, organization, and contact number).

Media representatives will be given a media badge to wear or keep handy when present at the meeting site.

Registered media representatives with badges are allowed to attend the following events:

— Opening and closing ceremonies;

— Press conferences;

— Social events if invited by the host organization.

People from the media are not allowed to attend:

— Technical committee or subcommittee plenary meetings;

— Working group meetings;

— Chair's Advisory Group (CAG) meetings.

Press conferences may be scheduled for appropriate times (pre- and/or post-plenary) in a designated location by the ISO committee leadership in consultation with the public relations service of the host ISO member body. At the press conferences, the ISO committee leaders and ISO Central Secretariat representatives will make statements to the press. Delegates or experts may be invited to speak by the ISO committee leaders. Otherwise, delegates, experts and observers may attend, but are not permitted to speak at the press conferences.

SF.10 Recordings at ISO meetings

The recording of ISO meetings by the committee secretary is acceptable provided that at the outset of the meeting all participants are made aware that the meeting will be recorded and there are no objections. Individual participants shall have the right to require that the means of recording are switched off during a particular intervention if s/he so desires.

Such recordings are intended to aid the meeting secretary in preparing the minutes/report of the meeting. They may also be used to resolve disputes concerning what occurred at a meeting, which, in some instances, may require the preparation of transcripts of the recording.

The recordings and any transcripts are the property of the meeting secretary, who is expected to respect the confidentiality of ISO meetings. The recordings and transcripts shall accordingly not be divulged to third parties and should preferably be destroyed once the minutes have been approved.

Annex SG
(normative)

Second (and subsequent) language texts for ISO standards

SG.1 Introduction

Annex E of the *ISO/IEC Directives, Part 1*, specifies the general policy on the use of languages as applied in ISO. This annex describes additional requirements concerning obtaining second and subsequent language texts. (English, French and Russian are the official ISO languages.)

SG.2 English, French, Russian and other languages

SG.2.1 Responsibilities

It is the responsibility of the ISO committee secretariat to obtain (but not necessarily themselves prepare) texts in English in general. Translation of documents into French is typically (but not exclusively) carried out by the translation department of the member body of France (AFNOR), or by agents nominated to act on their behalf.

It is the responsibility of the Russian member body to determine those cases for which a Russian version is required, and to provide all translation into Russian.

When a non-official language has been approved by the committee for inclusion in the final deliverable (this is permitted only in the case of a vocabulary), the National Body supporting the related non-official language is responsible for the text, and its timely delivery (i.e. at the latest before the FDIS ballot).

SG.2.2 Equivalence of texts

The National Body supporting a language is responsible for ensuring the equivalence of the related language version to the official language text during the development of an ISO deliverable, if necessary with the assistance of other experts. A recommended way of ensuring equivalence of texts is to establish a multilingual editing committee.

SG.2.3 Timing of preparation of texts

The ISO/IEC Directives note that it is beneficial to start preparation of language versions as early as possible in the development cycle. It is preferable, but not required, that Committee Drafts are available in all envisaged language versions. Apart from aiding comprehension during development, this also serves to help avoid preparation-related delays at later stages.

Where a second or subsequent official language version is required, the ISO Central Secretariat will provide the member body responsible for the second or subsequent language with a copy of the first language version, English in general, when the text is submitted to ISO Central Secretariat for processing as a DIS.

SG.2.4 Timely delivery of language versions

In the case where a second or subsequent official language version is required, the National Body supporting the related language shall provide the ISO Central Secretariat with the related language version within 8 weeks of the submission of the first language version, English in general.

SG.2.5 Updating language versions after ballot

In the case of preparation of a second or subsequent DIS, or the FDIS text, in order to facilitate the preparation of the second or subsequent official language text (where required), TC and SC secretariats should preferably provide the ISO Central Secretariat with a marked-up copy of the revised first-language DIS showing the changes to be made, for use by the member body responsible for the second or subsequent language version.

SG.3 Other non-official languages

Languages other than the official languages of ISO may only be included in the body of vocabularies. This is dependent on the committee approving the business case presented by a National Body, which is then responsible for the related language version.

In those cases where an ISO project is a joint project with CEN under the *Agreement on technical cooperation between ISO and CEN (Vienna Agreement)*, the preparation of a German version is a requirement. For more guidance reference should be made to the separate Guidelines for Chairmen and Secretariats on the implementation of the Vienna Agreement. In all cases of multilingual documents, the cover page, copyright and foreword pages of the document will only mention the official ISO languages and, only in case of a joint project with CEN under the Vienna Agreement, German might be added.

SG.4 Single-language versions

It may be proposed that an international document need only be developed in a single official language. Further official language versions can always be prepared as a translation of the published text.

SG.5 Translations

Under the terms of the ISO Statutes, as supplemented by the agreement ISO policies and procedures for copyright, copyright exploitation rights and sales of ISO publications (ISO POCOSA), the member bodies of ISO may, under their sole responsibility, prepare translations into other languages of certain ISO publications and documents.

Annex SH
(normative)

Procedures for the standardization of graphical symbols

SH.1 Introduction

This annex describes the procedures to be adopted in the submission and subsequent approval and registration, when appropriate, of all graphical symbols appearing in ISO documents.

Within ISO the responsibility for the coordination of the development of graphical symbols has been subdivided into two principal areas, allocated to two ISO technical committees:

— ISO/TC 145 — all graphical symbols (except those for use in technical product documentation) (see ISO/TC 145 website);

— ISO/TC 10 — graphical symbols for technical product documentation (tpd) (see ISO/TC 10 website).

In addition there is coordination with IEC/TC 3 (Information structures, documentation and graphical symbols) and with IEC/TC 3/SC 3C (Graphical symbols for use on equipment).

The basic objectives of the standardization of graphical symbols are to:

— meet the needs of users;

— ensure that the interests of all concerned ISO committees are taken into account;

— ensure that graphical symbols are unambiguous and conform to consistent sets of design criteria;

— ensure that there is no duplication or unnecessary proliferation of graphical symbols.

The basic steps in the standardization of a new graphical symbol are:

— identification of need;

— elaboration;

— evaluation;

— approval, when appropriate;

— registration;

— publication.

All steps should be carried out by electronic means.

— Proposals for new or revised graphical symbols may be submitted by an ISO committee, a liaison member of an ISO committee or any ISO member organization (hereafter jointly called the "proposer").

— Each approved graphical symbol will be allocated a unique number to facilitate its management and identification through a register that provides information that can be retrieved in an electronic format.

— Conflicts with the relevant requirements and guidelines for graphical symbols shall be resolved by liaison and dialogue between ISO/TC 145 or ISO/TC 10 and the product committee concerned at the earliest possible stage.

SH.2 All graphical symbols except those for use in technical product documentation

SH.2.1 General

ISO/TC 145 is responsible within ISO for the overall coordination of standardization in the field of graphical symbols (except for tpd). This responsibility includes:

— standardization in the field of graphical symbols as well as of colours and shapes, whenever these elements form part of the message that a symbol is intended to convey, e.g. a safety sign;

— establishing principles for preparation, coordination and application of graphical symbols: general responsibility for the review and the coordination of those already existing, those under study, and those to be established.

The standardization of letters, numerals, punctuation marks, mathematical signs and symbols, and symbols for quantities and units is excluded. However, such elements may be used as components of a graphical symbol.

The review and co-ordination role of ISO/TC 145 applies to all committees that undertake the responsibility for creation and standardization of graphical symbols within their own particular fields.

ISO/TC 145 has allocated these responsibilities as follows:

— ISO/TC 145/SC 1: Public information symbols;

— ISO/TC 145/SC 2: Safety identification, signs, shapes, symbols and colours;

— ISO/TC 145/SC 3: Graphical symbols for use on equipment.

There is also liaison with ISO/TC 10 and with IEC, in particular with IEC/SC 3C, Graphical symbols for use on equipment.

Table SH.1 shows the categories of graphical symbols covered by each coordinating committee.

Table SH.1 — Categories of graphical symbols

	Basic message	Location	Target audience	Design principles	Overview	Responsible committee
Public information symbols	Location of service or facility	In public areas	General public	ISO 22727	ISO 7001	ISO/TC 145/SC 1
Safety signs (symbols)	Related to safety and health of persons	In workplaces and public areas and on products	a) General public or b) authorized and trained persons	ISO 3864-1 ISO 3864-2 ISO 3864-3	ISO 7010	ISO/TC 145/SC 2
Product safety labels	Related to safety and health of persons	On products	c) General public or d) authorized and trained persons	ISO 3864-2 ISO 3864-3	—	ISO/TC 145/SC 2
Graphical symbols for use on equipment	Related to equipment (function, control, identification)	On equipment	e) General public or f) authorized and trained persons	IEC 80416-1 ISO 80416-2 IEC 80416-3	ISO 7000 IEC 60417	ISO/TC 145/SC 3 IEC/TC 3/SC 3C
tpd symbols	(Product representation)	Technical product documentation (drawings, diagrams, etc.)	Trained persons	ISO 81714-1	ISO 14617 IEC 60617	ISO/TC 10/SC 10 IEC/TC 3

Table SH.2 — Examples of different types of graphical symbols shown in their context of use

Public information symbols	Telephone ISO 7001 – PI PF 017	Aircraft ISO 7001 – PI TF 001	Sporting activities ISO 7001 – PI SA 001	Gasoline station ISO 7001 – PI CF 009	Direction ISO 7001 – PI PF 030
Safety signs (symbols)	Means of escape and emergency equipment sign: Emergency exit (left hand) ISO 7010 – E001	Fire equipment sign: Fire extinguisher ISO 7010 – F001	Mandatory action sign: General mandatory action sign ISO 7010 – M001	Prohibition sign: No smoking ISO 7010 – P002	Warning sign: Warning; Explosive material ISO 7010 – W002
Graphical symbols for use on equipment	Ventilating fan: Air-circulating fan ISO 7000 – 0089	General symbol for recovery/recyclable ISO 7000 - 1135	Weight ISO 7000 – 0430	Lamp; lighting; illumination IEC 60417 – 5012	Brightness/Contrast IEC 60417 – 5435
tpd symbols	Two-way valve ISO 14617-8 – 2101	Liquid ISO 14617-15 – 321	Hydraulic motor ISO 14617-10 – 2405	Amplifier ISO 14617-5– 892	Boiler/steam generator ISO 14617-11 – 2531

SH.2.2 Submission of proposals

Proposers shall submit their proposals on the relevant application form as soon as possible to the secretariat of the appropriate ISO/TC 145 subcommittee in order to allow for timely review and comment. It is strongly recommended that this submission be made by proposers at the CD stage, but it shall be no later than the first enquiry stage (i.e. DIS or DAM) in the case of an International Standard.

Prior to submitting a graphical symbol proposal, the proposer should:

— be able to demonstrate the need for the proposed graphical symbol;

— have reviewed the relevant ISO and/or IEC standards of graphical symbols, in order to avoid ambiguity and/or overlap with existing standardized graphical symbols, and to check for consistency with any related graphical symbol or family of graphical symbols already standardized;

— create the proposed graphical symbol in accordance with the relevant standards and instructions, these include design principles and criteria of acceptance.

SH.2.3 Standardization procedure for proposed graphical symbols

Upon receipt of a proposal, the ISO/TC 145 sub-committee concerned shall review the application form, within two months, to check whether it has been correctly completed and the relevant graphics file(s) has been correctly provided. If necessary, the proposer will be invited to modify the application, and to re-submit it.

Upon receipt of a correctly completed application form, a formal review process shall be commenced to review the proposal for consistency with standardized graphical symbols, the relevant design principles and criteria of acceptance.

When this formal review process has been completed, the results shall be transmitted to the proposer, together with any recommendations. The proposer will, where appropriate, be invited to modify the proposal, and to re-submit it for a further review.

The procedures outlined on the relevant ISO/TC 145 sub-committee website shall be followed:

— ISO/TC 145/SC 1: Public information symbols (www.iso.org/tc145/sc1);

— ISO/TC 145/SC 2: Safety identification, signs, shapes, symbols and colours (www.iso.org/tc145/sc2);

— ISO/TC 145/SC 3: Graphical symbols for use on equipment (www.iso.org/tc145/sc3).

These websites also provide application forms for the submission of proposals.

Graphical symbols approved by ISO/TC 145 shall be assigned a definitive registration number and included in the relevant ISO/TC 145 standard.

NOTE　　In exceptional cases, unregistered symbols may be included in ISO standards subject to TMB approval.

SH.3 Graphical symbols for use in technical product documentation (tpd) (ISO/TC 10)

ISO/TC 10 is responsible for the overall responsibility for standardization in the field of graphical symbols for technical product documentation (tpd). This responsibility includes

— maintenance of ISO 81714-1: Design of graphical symbols for use in the technical documentation of products — Part 1: Basic rules, in co-operation with IEC;

— standardization of graphical symbols to be used in technical product documentation, co-ordinated with IEC;

— establishing and maintaining a database for graphical symbols including management of registration numbers.

Included is the standardization of symbols for use in diagrams and pictorial drawings.

ISO/TC 10 has allocated these responsibilities to ISO/TC 10/SC 10. The Secretariat of ISO/TC 10/SC 10 is supported by a maintenance group.

Any committee identifying the need for new or revised graphical symbols for tpd shall as soon as possible submit their proposal to the secretariat of ISO/TC 10/SC 10 for review and — once approved — allocation of a registration number.

Annex SI
(normative)

Procedure for the development of International Workshop Agreements (IWA)

SI.1 Proposals to develop IWAs

A proposal to hold an ISO workshop for the purpose of developing one or more IWAs on a particular subject may come from any source, including ISO member bodies, liaison organizations, corporate bodies etc. An organization that is not an ISO member body or liaison organization, or is not international in scope, shall inform the ISO member body in its country of its intent to submit such a proposal.

Whenever practicable, proposers shall provide details concerning:

— Purpose and justification of the proposal;

— Relevant documents; and

— Cooperation and liaison,

in accordance with the ISO/IEC Directives, Part 1, Annex C.

Additionally, wherever possible, proposals shall include indication of an ISO Member Body willing to provide secretariat support to the IWA Workshop. If it is considered likely that participation in the workshop will need to be limited (see SJ.5.2), this shall also be indicated.

In some circumstances, it may be considered that several meetings may be needed in order to reach a consensus. In such cases, the proposer is encouraged, or may be required by the ISO Technical Management Board, to develop a business plan which would give details concerning meeting schedules, expected dates of availability of draft documents, the possibility that a workshop may establish project teams to progress work between meetings of the workshop, the expected date of availability of any IWA, etc.

SI.2 Review of proposals

Proposals will be referred to the ISO Technical Management Board for approval. Proposals should contain an analysis of the global interest in the subject being proposed for the IWA in different countries and by stakeholders.

If the proposal is accepted, the ISO/TMB will initiate consultations with member bodies to identify a candidate willing to act as the organizer and to provide administrative and logistics support to the proposer. Preference will normally be given to:

— The member body from the country of the proposer, if the proposer is not a member body; or

— Member bodies holding secretariats in fields related to that covered by the proposal.

If there is more than one offer, the ISO/TMB will formally designate the member body assigned to act as the workshop secretariat. The assigned ISO member body may establish financial arrangements with the proposer to cover administrative and logistics support costs for the workshop. If a member body is not willing to act as workshop secretariat, the ISO/TMB may authorize the ISO Central Secretariat to fulfil this role, provided all associated costs are recovered by workshop registration fees.

An informative checklist for estimating IWA workshop costs is provided in SI.10.

The workshop secretariat and the proposer shall designate the chair of the workshop.

SI.3 Announcement

Once the workshop secretariat and the proposer have agreed on a date and venue for the first meeting of the workshop, these shall be communicated to the ISO member bodies. These details shall be further announced by the workshop secretariat, the ISO Central Secretariat and by any other interested member bodies in the most appropriate way(s) to achieve the widest possible circulation (e.g. a publicly accessible website). This may include a number of different announcement options and media, but the intent is to ensure that the broadest range of relevant interested parties worldwide are informed of the workshop and have the opportunity to attend.

The proposer and workshop secretariat will ensure that any ISO committees with projects relevant to the subject will be invited to be represented at the workshop.

A registration fee may be applied to help support preparation and hosting of the workshop. Any registration fees shall be stated in the workshop announcement.

NOTE   When the subject matter of a workshop is likely to be of interest to developing countries, it is recommended either that a funding mechanism other than a registration fee be applied to facilitate participation from such countries, or that a number of "free" registrations be permitted.

The announcement shall be made at least 90 days in advance of the agreed date to allow potential attendees adequate time to plan on attending the workshop. The announcement shall be accompanied by a registration form to allow potential participants to register for the workshop. Registration forms shall be returned to the workshop Secretariat.

SI.4 Workshop information

A workshop programme detailing workshop objectives, deliverables, agenda, draft documents and any other relevant details for the workshop shall be available, and circulated to registered participants, no later than six weeks prior to the workshop date. Registered participants may submit their own contributions to the workshop secretariat for further distribution to other participants.

SI.5 Participation

SI.5.1 Workshop chairs

The proposer and workshop secretary shall designate the chair of any particular workshop. The chair shall act in a purely international and neutral capacity and in particular shall

— Ensure that all points of view expressed during a workshop are adequately summed up so that they are understood by all present,

— Conduct the workshop with a view to reaching consensus,

— Ensure that all decisions are clearly formulated and, if needed, made available to the participants before closure of the meeting of the workshop.

SI.5.2 Registered participants

Any organization may register as a participant in a workshop and participation will be open to the registered participants only. Participants are not required to be appointed by the ISO member body in their country. The workshop secretariat, chair and proposer shall endeavour to ensure that the broadest range of interests is represented in any workshop and that there is an appropriate balance of representation. If needed, this may require that some limitation be placed on participation (for example no more than two registered participants from the same corporate body or organization). If the need to limit participation is expected at the outset, this shall be indicated in the proposal submitted to the ISO Technical Management Board. If a need for limitation becomes apparent after announcement of the workshop, this shall be authorized by the TMB secretariat following consultation with the ISO/TMB chair and, if needed, other ISO/TMB members.

Workshop secretariats are responsible for ensuring that the outcome of the workshop is truly international, i.e. it is attended by more than one country.

SI.5.3 Project teams

In cases in which more than one meeting will be required to reach consensus, a workshop may establish one or more project teams to progress work between meetings of the workshop. The workshop shall designate the membership of such project teams, ensuring that their working methods will allow all interests to participate fully.

SI.6 Workshop procedures and management oversight

Workshops will be permitted to work in a practically autonomous manner using very flexible procedures.

However, there are a number of general ISO policies which need to be respected, in particular those concerning intellectual property rights and the use of SI units. It shall be the responsibility of the workshop secretariat to ensure that the appropriate policies are known to registered participants and are respected.

Management oversight will be kept to the minimum required to ensure coordination with existing standardization activities if relevant and to ensure that appropriate resource is provided by the ISO system. It will be the responsibility of the workshop chair to determine when consensus of the workshop participants has been reached on a particular item or deliverable. For the purposes of determining consensus, the workshop chair shall apply the definition of consensus in clause 2.5.6.

It should be noted that an IWA workshop may arrive at the consensus that an IWA deliverable is not necessary.

The workshop deliverables shall contain a description of the workshop consensus achieved including any recommendations for possible future actions or revisions to the workshop deliverables. The deliverable resulting from the workshop will proceed to publication based on the consensus of the workshop without additional reviews or approvals by any other body, except in the case of an appeal on such a deliverable (see immediately below).

SI.7 Appeals

Any parties affected by the deliverable resulting from the workshop shall have the right of appeal for the following reasons:

— The workshop and the process to arrive at its deliverable have not complied with these procedures;

— The deliverable resulting from the workshop is not in the best interests of international trade and commerce, or such public factors as safety, health or the environment; or

— The contents of the deliverable resulting from the workshop conflict with existing or draft ISO standard(s) or may be detrimental to the reputation of ISO.

Such appeals shall be submitted within two months of the date of the workshop and shall be considered by the ISO Technical Management Board which in such circumstances will take the final decision concerning publication of an IWA.

SI.8 Workshop deliverables and publication

Workshops will decide on the content of their own deliverables. Deliverables shall be sent to the ISO Central Secretariat for publication as International Workshop Agreements. They will be numbered in a special IWA series. Each IWA shall include the list of participants in the workshop that developed it.

IWAs may be published in one of the official ISO languages only and competing IWAs on the same subject are permitted. The technical content of an IWA may compete with the technical content of an existing ISO or IEC standard, or the proposed content of an ISO or IEC standard under development, but conflict is not normally permitted unless expressly authorized by the ISO/TMB.

SI.9 Review of IWAs

If there is an existing ISO committee whose scope covers the topic, the published IWA shall automatically be allocated to this committee for maintenance.

An IWA shall be reviewed three years after its publication. This review shall be organized by the member body which provided the workshop secretariat, or the committee which has been allocated the responsibility for the maintenance of the IWA. The review shall include the consultation of interested market players. The result of the review may be to confirm the IWA for a further three year period, to withdraw the IWA or to submit it for further processing as another ISO deliverable in accordance with Part 1 of the ISO/IEC Directives.

An IWA may be further processed to become a Publicly Available Specification, a Technical Specification or an International Standard, according to the market requirement.

An IWA may exist for a maximum of six years following which it shall either be withdrawn or be converted into another ISO deliverable.

SI.10 Checklist to estimate costs associated with hosting an ISO IWA Workshop (WS)

Completion of this checklist is NOT mandatory for the proposed hosting of an IWA. It is intended simply as a tool to assist proposers and assigned ISO member bodies in determining major costs associated with hosting an ISO IWA Workshop. Some of the sample costs provided may not be applicable to particular IWA Workshops, or may be covered in a different manner than is described below.

SI.10.1 Overall measures to consider

Expected number of deliverables	
Expected number of Project Groups	
Expected number of participants	
Expected number of WS plenary meetings	
Expected number of Project Group meetings	

SI.10.2 Planned Resources — Human Resource Requirements & Functions

Function	Who	Estimated Cost	Covered by ... (Who)	Commitment (Y/N)
WS Secretariat				
— organizing WS plenary meetings				
— producing WS and project meeting reports and action lists				
— administrative contact point for WS projects				
— managing WS (and project group) membership lists				
— managing WS (and project group) document registers				
— follow-up of action lists				
— if the Workshop works mainly by electronic means, assist Chair in monitoring and follow-up of electronic discussion				
WS Chair				
— chairing WS plenary meetings				
— responsible for overall WS/project management				
— producing project management progress reports				
— acting as formal liaison with related WS/projects and liaison body				
— if the Workshop works mainly by electronic means, follow and steer electronic discussions that take place between meetings, take necessary decisions				
WS Vice-Chair				
If your WS will have "Vice-Chair(s)", you should list them here, along with their particular functions and fill-in the right side of this table				
Project Leader(s) (where applicable)				
— chairing project group meetings				
— producing project group meeting				
— reports and action lists				
— preparing project group progress reports				
— if the Workshop works mainly by				
— electronic means, follow and steer				
— electronic discussions that take place				
— regarding his/her particular project				
IWA Editor				
— editing of the IWA texts				
— attend meetings				
— follow discussions relating to the IWA (s) that s/he is editing				

SI.10.3 Planned Resources — Material Resource Requirements

Material Resource costs	Numbers	Estimated Cost per meeting	Covered by … (Who)	Commitment (Y/N)
WS Meetings				
Meeting rooms				
Logistics (IT support, photocopies, etc.)				
Food and beverage				
Interpretation during meetings				

Annex SJ
(normative)

Forms

These are available in electronic format (typically MS Word) for download from www.iso.org/forms.

Annex SK
(normative)

Deadlines for posting committee and working group meeting documents

SK.1 Why it is important to respect these deadlines

National mirror committees need time to prepare their national positions on the various issues discussed at the meeting.

SK.2 Ensuring that a document will be ready for circulation by the deadlines stipulated

This is the responsibility of the secretariat. They shall carefully consider the steps needed and leave enough time in advance of the meeting deadlines to make sure that the document is ready. For example, to discuss a DIS, the secretariat will have to start the process 7 months before the committee meeting (i.e. 2 weeks document preparation by the ISO Central Secretariat + 20 weeks for vote and translation + 6 weeks deadline before the meeting).

SK.3 How to deal with issues that come up at the meeting for which no document was issued 6 weeks in advance

It is possible that committees make decisions on points for which no information was provided in advance of the meeting within the deadlines indicated below. In such cases, the committee leadership should ask the P-members present if they are comfortable and prepared to take such decisions in principle or if they would prefer time to consult further with other national stakeholders. The committee leaders should ensure that they are seeking to engage all P members in decisions, even if those P members that were not able to attend the meeting. A ballot can then be launched after the meeting via the ISO Committee Internal Ballot (CIB) mechanism to confirm any decisions.

Item	Time (− before meeting, + after meeting)
Agenda & meeting logistics (e.g. location, accommodation, visa information)	−16 weeks
Documents for decision in meetings such as:	
Working group reports	−6 weeks
Liaison reports	−6 weeks
Leadership elections/renewals	−6 weeks
Creation/disbandment of subcommittees and working group	−6 weeks
Strategic Business Plans	−6 weeks
Voting results and compilation of comments for CD, DIS, DTR and DTS	−6 weeks
Voting results for SR and any CIB ad hoc enquiries requiring discussion or follow-up	−6 weeks
Resolutions	+48 hours
Minutes (including list of attendees)	+4 weeks

SK.4 Deadlines for sending working group documents

The ISO/IEC Directives, Part 1 state only that the agenda shall be sent out 6-weeks in advance of working group meetings. The timeframe for the circulation of other documents is to be decided by the working group members. WG convenors should ensure that their working groups take decisions on the acceptable time frame for circulating documents before meetings to ensure working group experts can be well-prepared.

Table SK.1 — Deadlines for WGs

Type of document, announcement or activity	Deadline (– before meeting, + after meeting)
First WG meeting after a TC decision	+12 weeks
Meeting announcement	(To be decided by the WG)
Agenda	–6 weeks
Documents for the WG meeting	(To be decided by the WG) Recommended no later than –4 to –6 weeks
Minutes (including list of attendees)	+4 weeks

SK.5 Time frames for circulating documents in advance of a web meeting or teleconference

Again, the principle should be that agendas and documents are circulated with enough time before any meeting in order that participants can be well-prepared for informed discussions and decisions. Web or teleconference meetings of TCs, PCs, SCs and WGs for limited aspects of their work should follow the same time frames detailed above. P-members involved should decide on appropriate timeframes for circulation of documents before Chairman's Advisory Groups, Advisory Groups, Task Forces, etc.

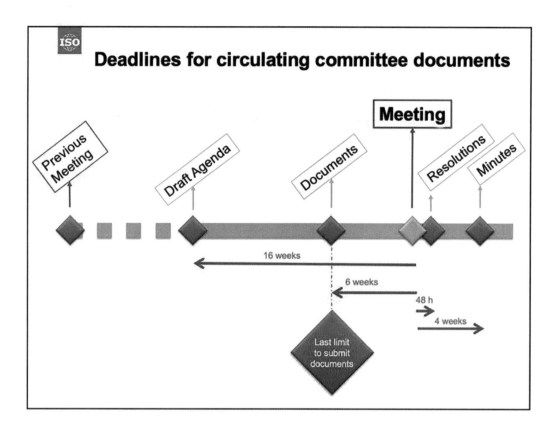

Annex SL

(Currently blank — Placeholder for new annexes)

Annex SM
(normative)

Global relevance of ISO technical work and publications

SM.1 Introduction

The formation of the WTO and the subsequent adoption of the WTO Technical Barriers to Trade Agreement (WTO/TBT), placed an obligation on ISO to ensure that the International Standards it develops, adopts and publishes are globally relevant. In Annex 4, paragraph 10 of the Second Triennial Review of the Agreement, the following criteria state that a globally relevant standard should:

— Effectively respond to regulatory and market needs (in the global marketplace)

— Respond to scientific and technical developments in various countries

— Not distort the market

— Have no adverse effects on fair competition

— Not stifle innovation and technological development

— Not give preference to characteristics or requirements of specific countries or regions when different needs or interests exist in other countries or regions

— Be performance based as opposed to design prescriptive

Hence the development and adoption of an International Standard that fails to meet these requirements is open to being challenged as creating a barrier to free trade.

Noting the need to provide fuller advice to committees on global relevance, and following a request from the ISO Council, the ISO/TMB established a Global Relevance Task Force. This task force and, subsequently, the ISO/TMB have agreed on the set of principles that follows.

SM.2 Terms and definitions

Standard

Document, established by consensus and approved by a recognized body, that provides, for common and repeated use, rules, guidelines or characteristics for activities or their results, aimed at the achievement of the optimum degree of order in a given context.

NOTE Standards should be based on the consolidated results of science, technology and experience, and aimed at the promotion of optimum community benefits.

(ISO/IEC Guide 2, ISO/IEC Directives, Part 2)

international standard

Standard that is adopted by an international standardizing/standards organization and made available to the public.

(ISO/IEC Guide 2, ISO/IEC Directives, Part 2)

International Standard

International standard where the international standards organization is ISO or IEC.

(ISO/IEC Directives, Part 2)

Global relevance

Required characteristic of an International Standard that it can be used/implemented as broadly as possible by affected industries and other stakeholders in markets around the world.

SM.3 Principles

SM.3.1 The status and meaning of an International Standard shall be respected.

Any International Standard shall respect the above definitions and shall to the extent possible represent a unique international solution. In cases where unique international solutions are not possible for specific provisions of an International Standard at the current time due to legitimate market and essential differences, International Standards may present options to accommodate these differences where justified.

SM.3.2 The commitment to participate in the development of and the feasibility of preparing International Standards shall be demonstrated at the outset of a standards development project.

It is recognized that in some instances various solutions exist to meet unique aspects of the local markets in different regions and countries. With globalization and the unification of markets, these market differences should be minimized over time and evolve into one global market. Simply projecting one solution that accommodates one market (but not others) as the International Standard will not force markets to evolve and coalesce. In such cases, the markets and their related industries will look elsewhere for standards that better accommodate their needs, and ISO will lose its relevance for those markets and industries. Rather than force such a situation, ISO committees should ascertain at the outset of a project whether:

— a globally relevant International Standard presenting one unique international solution in all of its provisions is feasible;

— an International Standard is feasible that presents options in specific provisions to accommodate existing and legitimate market differences where justified; or

— the preparation of a globally relevant International Standard is not feasible and work should not be undertaken in such circumstances.

Additional practical guidance for committee leaders and delegates/experts may be found in the ISO/TMB's Global Relevance Implementation Guidance document.

SM.3.3 Preference shall be given to preparing performance rather than prescriptive standards.

Please note the following: Annex 3 of the WTO/TBT Agreement.

"Wherever appropriate, the standardizing body shall specify standards based on product requirements in terms of performance rather than design or descriptive characteristics."

ISO/IEC Directives, Part 2, clause 5.4 Performance principle (Excerpt) "Whenever possible, requirements shall be expressed in terms of performance rather than design or descriptive characteristics. This principle allows maximum freedom for technical development and reduces the risk of undesirable market impacts (e.g. restriction of innovative solutions)."

Given these quotations, the use of the performance-based approach is widely recognized as supporting the development of globally relevant ISO standards. In the case of design-based standards, the freedom for further technical innovation is most limited, while performance-based standards provide for maximum freedom for further innovation. However, in practice, there may be cases where inclusion of design requirements for some provisions within a performance-based standard is appropriate. There may also be other cases where development of a completely design-based standard may be appropriate and will result in a globally relevant ISO standard.

Thus, which approach is most appropriate depends on the technical matter in question.

Additional practical guidance for committee leaders and delegates/experts may be found in the ISO/TMB's Global Relevance Implementation Guidance document.

SM.3.4 Given existing and legitimate market differences, an International Standard may pass through an evolutionary process, with the ultimate objective being to publish, at a later point, an International Standard that presents one unique international solution in all of its provisions.

Under this principle, a committee may wish to consider how it addresses current and potentially changeable differences in markets (based on factors such as legislation, economies, social conditions, trade patterns, market needs, scientific theories, design philosophies, etc.) in the ISO deliverables it produces.

Additional practical guidance for committee leaders and delegates/experts may be found in the ISO/TMB's Global Relevance Implementation Guidance document.

SM.3.5 Essential differences consistent with Annex 3 to the WTO Agreement on Technical Barriers to Trade can be included in International Standards, but specific rules shall be applied if a committee wishes to introduce such differences and special authorization needs to be given by the ISO/TMB in instances not covered by these rules.

Under this principle, a committee may wish to consider how it addresses essential differences in markets around the world, that is, factors that are not expected to change over time, such as imbedded technological infrastructures, climatic, geographical or anthropological differences.

Additional practical guidance for committee leaders and delegates/experts may be found in the ISO/TMB's Global Relevance Implementation Guidance document.

SM.3.6 Committees can only ensure the global relevance of the International Standards they produce if they are aware of all the factors that may affect a particular standard's global relevance.

Additional practical guidance for committee leaders and delegates/experts may be found in the ISO/TMB's Global Relevance Implementation Guidance document.

Annex SN
(normative)

Registration Authority ("RA") Policy

SN.1 Scope

A number of International Standards developed by ISO technical committees require the assignment of unique registration elements, and describe the methodology for the assignment of these elements. The elements themselves are not part of the standard but are assigned by an appointed RA, who also maintains an accurate register of the elements that have been assigned. The RA is a competent body with the requisite infrastructure for ensuring the effective allocation of these registration elements. These bodies are designated by ISO to serve as the unique RA for particular standards, which creates a de facto monopoly situation.

This Policy is mandatory and shall be read in conjunction with Annex H of Part 1 of the ISO/IEC Directives.

Where ISO/CS becomes aware of a RA Standard that has not followed this Policy, it shall stop the publication process to allow time to implement this Policy before the RA Standard is published. For this reason committees are encouraged to make the ISO Technical Program Manager (TPM) aware of a project requiring an RA as early in the development process as possible to avoid delays in publication.

SN.2 Definitions

SN.2.1 RA Standard: The standard for which the RA is providing the Registration Services.

SN.2.2 Registration Services: Services provided by the RA in the implementation of the RA Standard and which shall be described in the RA Standard.

SN.2.3 Registration Authority ("RA"): Entity appointed by ISO to fulfill the Registration Services in a RA Standard.

SN.2.4 Registration Agencies: Third parties (e.g. national or regional sub-entities) to which the RA delegates part of the Registration Services. Even when delegated to Registration Agencies, the Registration Services remain under the responsibility of the RA.

SN.2.5 Registration Authority Agreement ("RAA"): Agreement based on the RAA template signed by the RA and the ISO Secretary-General on behalf of ISO, which details the functions, roles and legal obligations of the parties involved. A RAA shall be signed before a RA Standard is published (including revisions).

SN.2.6 Registration Elements: Unique elements, the methodology for which is described in the standard but which themselves are not part of the standard.

SN.2.7 Technical Programme Managers (TPM): Individual within ISO/CS assigned to work with a given committee.

SN.3 Procedure

SN.3.1 Chronology

This Policy addresses the various aspects of a RA in the order of the life cycle of a typical RA noting that some stages may be done in parallel. Each stage is addressed as follows:

— Identifying the need for a RA (SN.3.2)

— Drafting a RA Standard (SN.3.3)

— Selecting a RA (SN.3.4)

— Appointing a RA (SN.3.5)

— Signing the RAA (SN.3.6)

— Terminating a RA (SN.3.7)

— Implementing a RA Standard (SN.3.8)

　— The Role of the RA (SN.3.8.1)

　— The Role of the Committee (SN.3.8.2)

　— The Role of the ISO Central Secretariat (SN.3.8.2)

SN.3.2 Identifying the need for RA

The committee confirms its decision that a standard needs a RA for its implementation by way of a resolution.

The committee secretariat completes the RA Confirmation ("RAC") Form (See Annex SJ) and submits it to the TPM as soon as possible, but no later than the beginning of the preparatory (WD) stage.

SN.3.3 Drafting a RA Standard

The following shall be included in all RA Standards:

— A description of the identification scheme or the mechanism for generating unique registration elements.

— A description of the Registration Services, and the responsibility of the RA for providing the Registration Services.

— The link to the page on iso.org where users can find out the name and contact information of the RA for a given standard. This should also provide links to the RA's website which contains more information on the Registration Services.

The following shall not be included in RA Standards:

— In accordance with clause 4 of the ISO/IEC Directives, Part 2, contractual or other legal elements shall not be included in the RA Standard.

— Procedures concerning the provision of the Registration Services (e.g. a Handbook made available by the RA) shall not be included in the RA Standard. These are to be publicly provided by the RA (e.g. via a website).

— The name of the RA shall not be mentioned in the RA Standard. Instead, the link to the ISO website shall be provided (see above).

— References to the selection or confirmation process for the RA shall not be included in the RA Standard.

— In case of delegation of Registration Services by the RA to third parties (e.g. national or regional sub-entities referred to as "Registration Agencies") as agreed under the RAA, the RA Standard may mention the fact that some Registration Services have been delegated. However, any further details about Registration Agencies shall not be included in the RA Standard.

The TPM is responsible for coordinating with the committee to ensure that the right elements are included in the RA Standard. Any questions about what should be included in the RA Standard are to be addressed to the TPM.

SN.3.4 Selecting a RA

The selection process of the RA applies to new RA Standards and existing RA Standards in cases where the existing RA ceases to be the RA for any reason.

In the case of revisions, the committee shall decide whether the existing RA should continue or if a selection process should be launched to select additional RA candidates. The decision to launch a selection process should be supported by a rationale. The committee shall confirm its decision by resolution.

The selection process shall be completed before the project reaches the DIS stage. It is important for the DIS to contain details about the nature of the Registration Services needed.

The committee establishes the criteria for the selection of the RA and confirms these by resolution. The minimum elements of the selection process shall be:

— **Selection criteria** – these shall be clearly explained and with sufficient details for possible RA candidates to assess their ability to meet the criteria and apply on this basis. Included in the selection criteria shall be the requirement that the RA provide the following information in writing:

  — Proof (e.g. Statutes) that it is a legal entity which means that is an organization formed under the laws of a jurisdiction and that it is therefore subject to governance related rules.

  — Expression of willingness to take responsibility for the Registration Services.

  — Confirmation that the RA is technically and financially able to carry out the RA Services described in the RA Standard and the RAA on an international level.

  — Expression of willingness to sign a RAA, the ISO/CS RAA template for which shall be shared with RA candidates.

  — Confirmation of whether it intends to delegate part of the Registration Services to Registration Agencies.

  — Whether it will charge fees for the services and, if it charges fees, confirmation that any such fees will be on a cost recovery basis.

— **Public call for RA candidates** – committees shall take the appropriate steps needed to post the call for competent RA applicants to as broad a market as possible, also targeting possible organizations by inviting them to apply.

— **Evaluation** – RA candidates shall provide their responses in writing. The committee (or a subset thereof) shall decide on objective evaluation criteria based on selection criteria, including the relative weight given to each selection criteria.

— **Record-keeping** – the committee secretariat shall keep records of all documents in the selection process, including the call for candidates, applications, evaluation, decision, etc.

The committee shall then confirm its recommendation for appointment of the organization selected to be RA via a resolution.

SN.3.5 Appointing a RA

The information that is provided by the committee in the RAC (see SN.3.2 above) is needed to launch the TMB ballot appointing the RA, as well as the ISO Council ballot if the RA intends to charge fees. In the case of revisions, approval from the TMB or ISO Council is not needed if the committee decides that the same RA should continue (see SN.3.4).

The ISO/IEC Directives state that RA may charge fees for the Registration Services subject to authorization by the ISO Council, and as long as the basis of charging fees is strictly on a cost recovery basis.

The information shall therefore be fully complete. In the case of JTC 1 RA Standards, a copy of the RAC Form shall also be provided to the IEC since RA appointments shall all be confirmed by the IEC/SMB (and Council Board) where fees are charged.

SN.3.6 Signing a RAA

Only after the TMB (and Council if fees are charged) has appointed the RA (and in the case of JTC 1 RA Standards, involving the IEC) can the RAA be signed. Signing a RAA based on the ISO/CS template is mandatory for all RAs. The RAA shall be signed before publication of a new or revised RA Standard. If a RAA is not signed, the new or revised RA Standard shall not be published.

Because a RA Standard can change throughout the phases of development, the RA shall be provided with the version of the RA Standard that is ready for publication. The list of RA tasks is referenced in the RAA.

To ensure consistency and equality of treatment between the different RAs, any requested deviations from the RAA template which ISO/CS considers to be significant in nature shall be submitted to the TMB for approval.

SN.3.7 Terminating a RA

Termination of RAs could occur when the RA resigns or because the committee and ISO/CS have come to the conclusion (due to withdrawal of the RA Standard, customer complaints, etc.) that the RA should no longer continue to provide the Registration Services. The committee shall notify the TPM of any issue that it becomes aware of that could lead to the end of the Registration Services by the RA.

The process detailed in SN.3.4 above should be followed in the selection of a replacement RA unless the committee has identified an alternative RA candidate that meets the selection criteria in 3.4 and going through the selection process for additional RA candidates would cause disruption in the RA Services.

The RAA defines a notice period in case of the termination of an RA. This notice period is to be used for the selection process of a new RA so that the transition from one RA to the other is seamless.

SN.3.8 Implementing a RA Standard

SN.3.8.1 Role of the RA

The RA provides the Registration Services by:

— providing the Registration Services described in the RA Standard, and

— respecting the provisions of the RAA.

SN.3.8.2 Role of the committee

Although RAA are signed by the RA and by the ISO Secretary-General, the signature of a RAA by the Secretary-General binds all components in the ISO system, including ISO members and ISO committees. The central role is played by committees. In addition to identifying the need for a RA standard (3.2), drafting the RA Standard (3.3) and selecting a RA (3.4) for both new and revised RA Standards, the committee has the main responsibility for the oversight of the RA as follows:

— **Answering questions:** The committee shall be available to the RA to answer questions about the RA Standard and clarify any expectations regarding their role in implementing the RA Standard.

— **Assessing RA's annual reports:** The RAA requires RA to provide annual reports with a specified list of elements. The committee shall ensure that these annual reports are provided and shall read through them.

— **Monitoring:** In addition to the annual RA report, the committee shall also collect any feedback from industry and users of the RA Standard. Based on all of these elements (RA report and other feedback), the committee shall report to ISO/CS (see below).

— **Reporting to ISO/CS:** At least once per year and based on the information collected under **Monitoring** above, the committee shall provide a report to the responsible TPM using the Annual Committee Report to TPM ("ACR") Form (See Annex SJ). The purpose of such reports is to confirm that the RA operates in accordance with the RAA or to raise any concerns (concerns can include: RA Standard not meeting industry or user needs, complaints about the quality of the Registration Services, etc.). Such reports shall be provided at least annually to the responsible TPM or more frequently if the committee deems it necessary. The TPM may also ask for ad hoc reports. If the report identifies concerns, it shall include the planned **Corrective measures** (see below) needed to address these concerns.

— **Corrective measures:**

   — By the committee: the committee is responsible for implementing the corrective measures that are within its area of responsibility. These could include: revising the RA Standard, providing advice and guidance to the RA, carrying out audits or recommending the termination of the RAA to ISO/CS in severe cases.

   — By ISO/CS: the corrective measures that fall within the responsibility of ISO/CS (e.g. updating or overseeing the RAA) will be coordinated by the TPM. The TPM may also recommend corrective measures.

   — By the RA: the RA is responsible for any corrective measure that are within its area of responsibility, which would include the Registration Services and the provisions described in the RAA.

— **Dispute resolution:** In case of a dispute with customers of RA Services, the RA shall act diligently to resolve it. The RA shall bring disputes that it cannot resolve to the attention of the committee (as stated in the RAA) who shall advise the responsible TPM. The role of the committee (and ISO/CS) is limited to supporting the RA address the dispute. The committee shall not take over responsibility for the dispute or become the appellate body for disputes against the RA because this may inadvertently give the impression that ISO is responsible for the Registration Services.

— **Maintenance of records:** All key communications and documentation from and about the RA shall be archived. The committee secretariat is responsible for ensuring that these are maintained in a separate folder on e-committees.

The committee may create a subgroup [often referred to as a Registration Management Group ("RMG")] in order to help them with the above. Committees (either directly or through the RMG) shall not participate or get involved in providing the Registration Services.

SN.3.8.3 Role of the ISO/CS

The committee's interface with ISO/CS is through the responsible TPM. The role of the TPM includes:

— Identification of RA Standards during the development process if not done by the committee.

— Providing guidance and advice for the drafting of RA Standards.

— Training committees on the RA Policy.

— Coordination with committees to ensure compliance with the RA Policies, quality of Registration Services, appropriate handling of complaints, addressing industry and users' needs, including addressing the concerns raised in the annual reports provided by committees (using the ACR Form) and recommending and assisting in the implementation of any corrective measures (see SN.3.8.2).

— Maintenance of records in relation to his or her involvement.

Annex SO
(informative)

Principles for developing ISO and IEC Standards related to or supporting public policy initiatives

SO.1 Background Context

The 2007 ISO General Assembly Open Session on International Standards and Public Policy addressed important dynamics confronting the ISO community — the relationship between ISO standards and public policy as well as the unique needs and concerns of one major class of ISO standards users, namely, governments. ISO's sister organization IEC, responsible for electrotechnical standardization, applauds the initiative and wishes to co-operate with ISO to develop common ISO/IEC principles and guidance to technical committees of both organizations. It is a reasonable goal that ISO and IEC wish to make their portfolio of standards more visible to public authorities and, equally important, ensure that its standards address the relevant needs and concerns of the public authorities. This will support their global relevance and applicability worldwide, as it has been shown that ISO and IEC standards are capable of providing valuable support to the implementation of public policy. ISO and IEC have been and will continue to be effective providers of voluntary standards that support the programs of government authorities, who need standards that meet the WTO TBT criteria and that support technical regulations and/or procurement actions.

It should be noted that, as private, voluntary organizations, ISO and IEC themselves are not directly representative of government interests. ISO and IEC consensus on ISO and IEC standards reflects agreement across a range of stakeholders at the standard drafting level, AND it reflects a consensus across national standards bodies at the approval level. National positions on ISO or IEC standards are not necessarily government positions, although government experts may participate in developing these positions with their counterparts from the private sector.

The following set of principles has been established to guide ISO and IEC committees developing standards related to or supporting public policy initiatives. These principles will ensure that ISO and IEC standards can properly support and be used by public authorities.

SO.2 Principles

a) ISO and IEC are committed to creating market-driven International Standards, based on objective information and knowledge on which there is global consensus, and not on subjective judgments, in order to provide credible technical tools that can support the implementation of regulation and public policy initiatives.

b) ISO and IEC are committed to developing International Standards that are market relevant, meeting the needs and concerns of all relevant stakeholders including public authorities where appropriate, without seeking to establish, drive or motivate public policy, regulations, or social and political agendas.

When ISO or IEC standards are anticipated to support a public policy initiative, the relationship between the standard(s) and the public policy initiative(s) should be clearly understood among all concerned parties. The interaction of standardization and public policy or regulation should be anticipated and the intervention of the public authorities in the standards development process should occur as early as possible. It should be noted that in many cases experts representing regulatory authorities are actively participating in the development of ISO and IEC International Standards both as members of the relevant international committees and/or at the pertinent national mirror committees to ISO and IEC technical bodies.

c) ISO and IEC recognize that the development of regulation, public policy and/or the development and interpretation of international treaties are the role of governments or treaty organizations.

d) ISO and IEC standards supporting regulation, regulatory cooperation and public policy are best developed within ISO and IEC structures and under operational approaches and participation models that have been proven successful and that are detailed in the ISO/IEC Directives.

The use of special committee structures, procedures or participation models may compromise the credibility and suitability of the resulting ISO/IEC standards supporting regulation and public policy.

SO.3 Implementation

— In addition to promoting these principles to ISO and IEC committees, their leaders and participants, ISO and IEC shall actively support and promote the principles for international standardization established in the WTO TBT Agreement and subsequent decisions of the TBT Committee regarding the development of International Standards.

— ISO member bodies and IEC National Committees should ensure that national governments, including their trade representatives, are aware of ISO's and IEC's portfolios, are informed about ISO and IEC as venues for standards development, and are engaged in ISO and IEC standards development whenever appropriate to reduce misunderstanding and inadvertent contradictions.

— The ISO Technical Management Board and the IEC Standardization Management Board should develop additional implementation guidance and case studies of successful ISO and IEC standards efforts that support public policy initiatives (for example, ISO standards related to medical devices and greenhouse gas emissions, and IEC standards related to radio interference, safety of household appliances, ships and marine technology).

Annex SP

(Currently blank — Placeholder for new annexes)

Annex SQ
(normative)

Selection criteria for people leading the technical work

SQ.1 Obligations of member bodies

Member bodies are responsible for ensuring that candidates for leadership positions (i.e. chairs, convenors, and secretaries) meet the requirements contained in SQ.3. Member bodies are also responsible for ensuring that any gaps in skills or knowledge of the selected leaders, as well as experts, are identified and filed through ongoing training.

SQ.2 Resources available to fill gaps in skills or knowledge

A number of resources are available to help member bodies fill any identified gaps in skills or knowledge:

— Training and other materials are available from the ISO Central Secretariat to ensure the effectiveness of the various roles in conducting/attending meetings, including understanding key concepts.

— To help train experts, it may be useful to schedule committee pre-sessions and information on existing resources before or in conjunction with committee plenaries.

— The support of the Technical Programme Manager, particularly in the case of new committees, is available to train leaders and experts. In the case of new committees, the Technical Programme Manager should attend the first meeting to provide an overall introduction of ISO and its processes.

— Exchange programmes could also be organized between member bodies or with the ISO Central Secretariat. Member bodies may also wish to consider twinnings.

— The services of an external facilitator for coaching and training could be considered as one of the options to assist the committee leadership to develop non-ISO specific skills, such as the skills needed to run effective meetings, general leadership skills, etc.

SQ.3 Selection Criteria for people leading the technical work

The success of any committee or working group is dependent on its leadership. This selection criteria applies to committee [3] chairs, working group convenors and committee secretaries. Members bodies are therefore required to apply this criteria when nominating people to these roles in order to ensure that the new ISO Code of Conduct for the technical work is upheld and that the ISO/IEC Directives are correctly applied.

SQ.3.1 Chairs and convenors

SQ.3.1.1 Competencies and attributes of good chairs and convenors:

— existing role and good reputation in the sector

— relevant professional experience with previous experience of chairmanship

— lead and inspire delegates and experts from the sector towards consensus

— understand the international nature of ISO's work and its resulting benefits

---

3) Technical Committees, Subcommittees, Project Committees.

— commit time and resources to their role

— develop solutions through innovative and creative thinking in a consensus environment

— act proactively and communicate diplomatically

— foster and value cooperation with other ISO and IEC Committees and partners including those from regulatory bodies

— act in a purely international capacity

SQ.3.1.2 Job specification for chairs and convenors:

— lead meetings effectively with a view to reaching agreement and to ensure that positions and decisions are clearly understood

— ensure that all positions and views (at meetings and by correspondence) are given equal treatment

— manage projects according to agreed target dates in accordance with the project plan from preparation to completion

— work to ensure that a full range of technical competence is available to the group

— be fully knowledgeable of the subject and market needs

— propose decisions to progress or to stop work on the basis of its market or global relevance

— have basic knowledge of ISO and its procedures

SQ.3.1.3 Additionally — chairs of committees are required to:

— take responsibility for the overall management of the committee, including any subcommittees and working groups

— advise the ISO Technical Management Board on important matters relating to the Committee

— ensure that the policy and strategic decisions of the ISO Technical Management Board are implemented by the Committee

— think strategically to promote ISO's work in the sector

SQ.3.1.4 Additionally — convenors of working groups are required to:

— have appropriate knowledge and capabilities in using MS Word based drafting tools and the ISO web-based applications, in order to support the working group's work

SQ.3.1.5 Supporting information

The following information should be considered by the member bodies in the nominations for chairs and convenors:

— current role in the sector

— education

— professional career

— leadership experience

— similar activities

— language skills

SQ.3.2 Secretaries and secretariats

SQ.3.2.1 Selection of secretaries and secretariats

The success of an ISO committee or working group is dependent on its secretariat and secretary (or convenor of a WG in case there is no secretary). The following list is based on the ISO/IEC Directives and shows the tasks expected to be performed in these roles. ISO Members should use this list when appointing organizations and professionals as secretariats and secretaries.

SQ.3.2.2 Good documents

Preparing drafts for the committee or working group, arranging for their distribution and the treatment of the comments received. In the case of committee secretaries, preparing of drafts, text and figures for circulation by ISO Central Secretariat (ISO/CS) for enquiry and final draft International Standards or for publication. Fulfilling the ISO/CS submission requirements of such documents when sending them.

SQ.3.2.3 Excellent project management

Assisting in the establishment of priorities and target dates for each project. Notifying the names of all working group convenors and project leaders to the ISO Central Secretariat. Initiating ballots. Proposing proactive solutions for projects that are running significantly overtime, and/or which appear to lack sufficient support.

SQ.3.2.4 Well prepared meetings

Establishing the agenda and arranging for its distribution as well as the distribution of all documents on the agenda, including reports of working groups, and indicating all other documents which are necessary for discussion during the meeting. Recording the decisions taken in a meeting and making these decisions available in writing for confirmation in the meeting. Preparing the minutes of meetings to be circulated in a timely manner after the meeting.

SQ.3.2.5 Good advice on ISO processes

Providing advice to the chair, project leaders, and convenors on the ISO/IEC Directives and in particular the procedures associated with the progression of projects. Contacting any subcommittees and working group regarding their activities.

SQ.3.2.6 Connecting and networking

Working in close liaison with the chair of the committee or convenor of the working group. Maintaining close contact with the ISO Central Secretariat and with the members of the committee or working group regarding its activities. Maintaining close contact with the secretary of any parent committee.

SQ.3.2.7 Proactive follow up of actions

Ensuring that all actions agreed at meetings or by correspondence are completed on time and in a transparent manner.

SQ.3.2.8 Good with IT

Appropriate knowledge and capabilities in using MS Word based drafting tools and the ISO web-based applications including Livelink-e-committees, the committee balloting application and the submission interface.

SQ.3.2.9 Supporting Information

The following information should be considered by the member body when appointing secretaries:

— education

— professional career

— experience in standardization work

— participation in training programmes on standardization

— experience with ISO's IT tools and IT infrastructure

— language skills

Annex SR
(normative)

Statements intended to limit the purpose or use of deliverables

SR.1 Principles

The ISO Technical Management Board adopted the following ISO/TMB Resolution 8/2012 regarding statements intended to limit the purpose or use of deliverables:

> Noting that ISO Council Resolution 9/2001 confirms ISO's compliance with the basic principles of the operation and implementation of the WTO Agreement on Technical Barriers to Trade (TBT),
>
> Agrees that statements intended to limit the purpose or use of deliverables in relation to barriers or obstacles to trade are not permitted,
>
> Further agrees that statements relating to contractual obligations or government regulation are also not permitted,
>
> Requests that any such statements be removed during the development of a deliverable (i.e. before the close of the DIS) and that any such statements in existing deliverables be removed when the deliverable is revised,
>
> Requires that any exceptions to the above in exceptional cases be approved by the ISO/TMB (...)

The ISO Technical Management Board also adopted the following ISO/TMB Resolution 69/2012 regarding statements intended to limit the purpose or use of deliverables in relation to conformity assessment ("conformity assessment disclaimers"):

> Recalling TMB Resolution 8/2012 *Statements intended to limit the purpose or use of deliverables*,
>
> (...)
>
> Decides that statements related to conformity assessment (for example certification) are not permitted in ISO deliverables,
>
> (...)
>
> Agrees that any conformity assessment statements must be removed during the development of a deliverable (i.e. before the close of the DIS) and that any such statements in existing deliverables must be removed when the deliverable is revised, to be replaced by the link in the Foreword, and
>
> Requires that any exceptions to the above be approved by the TMB.

SR.2 Further information regarding statements in relation to barriers or obstacles to trade

Various statements intended to limited purpose or use of deliverables (sometimes referred to as "disclaimers") in relation to barriers or obstacles to trade are included in a number of ISO deliverables. Though the wording differs slightly from case to case, they make the point that the deliverable in question is not intended to create barriers or obstacles to trade.

As explained on the ISO/IEC Information Centre, the Agreement on Technical Barriers to Trade (TBT) is one of the legal texts of the WTO Agreement. It obliges WTO Members to ensure that, *inter alia*, voluntary standards do not create unnecessary obstacles to trade. Specifically, Annex 3 of the TBT entitled *Code of Good Practice for the Preparation, Adoption and Application of Standards* (para. E) states: "The standardizing body shall ensure that standards are not prepared, adopted or applied with a view to or with the effect of, creating unnecessary obstacles to international trade."

Standardizing bodies within the territory of a WTO Member can notify their acceptance of the Code of Good Practice for the Preparation, Adoption and Application of Standards (Annex 3 to the TBT). A list is available of the standardizing bodies from the countries that have notified their acceptance. In accordance with the TBT, the onus is then on WTO Members to ensure that the TBT is respected within their respective territories.

The obligation to avoid obstacles to trade is therefore amply covered in the TBT. To add a TBT declaimer in some standards but not in others creates two classes of standards with no clear difference between the two. In fact, if the practice were allowed to continue, this could inadvertently create the impression that there are two different levels of compliance with the TBT.

In addition, the issue of ISO's general compliance with the basic principles of the operation and implementation of the TBT was considered by the ISO Council. By way of ISO Council Resolution 9/2001, the ISO Council confirmed that ISO was indeed compliant in this regard.

SR.3 Further information regarding statements in relation to contractual obligations or government regulation

ISO/TMB Resolution 8/2012 reinforces the application of clause 4 of the ISO/IEC Directives, Part 2 which states that "A document shall not include contractual requirements (e.g. concerning claims, guarantees, covering of expenses) and legal or statutory requirements."

The ISO Technical Management Board also adopted the following ISO/TMB Resolution 70/2018 regarding the interpretation of "statements relating to contractual obligations or government regulation":

> Noting the issues of interpretation related to Technical Management Board Resolution 8/2012 concerning the phrase "Further agrees that statements relating to contractual obligations or government regulation are also not permitted";
>
> Further noting that
>
> - text relating to compliance with contractual obligations, legal requirements and government regulations exists in many ISO standards; and
>
> - ISO deliverables can be used to complement such requirements and serve as useful tools for all related stakeholders (which can include government authorities and industry players);
>
> Further noting the responses received from the DMT consultation on this question;
>
> Clarifies that, for all ISO deliverables:
>
> a) Statements that include an explicit requirement or recommendation to comply with any specific law, regulation or contract (such as a normative reference to such requirements), or portion thereof, are not permitted;
>
> b) Statements related to legal and regulatory requirements that do not violate point a) are permitted;
>
> c) Factual examples of the content of specific laws or regulations for informative purposes are permitted; and
>
> d) No exceptions shall be granted to point a).

In addition, a document shall not include statements containing an explicit requirement or recommendation to comply with national law or statements that an ISO standard is not intended to conflict with national law. The application of ISO standards is voluntary and national law takes precedence in case of conflict. Furthermore, to add such statements in some standards but not in others creates two classes of standards. This could inadvertently create the impression that an ISO standard takes precedence over national law unless primacy of national law is expressly stated.

SR.4 Further information on do's and don'ts related to conformity assessment

Further information regarding conformity assessment for standard writers is available in the publication entitled "Conformity assessment for standards writers — Do's and Don'ts": http://www.iso.org/iso/PUB100303.pdf and in clause 33 of ISO/IEC Directives, Part 2. Additional guidance is also contained in the generic text, a link to which is included in the Foreword of ISO deliverables (www.iso.org/foreword-info).

SR.5 Requests for exceptions

Requests for exceptions to include a statement intended to limit the purpose or use of deliverables shall be approved by the ISO Technical Management Board who will carefully consider such requests on a case by case basis.

Annex SS
(normative)

Optional use of the Committee Draft (CD) stage — Guidance for committees

The following is guidance for committees regarding the optional use of the Committee Draft (CD) stage.

SS.1 Implementation guidance

— Decisions on whether to skip the CD should be made on a case by case basis.

— Decisions should be made on the basis of what is expected to be gained by a CD circulation (e.g. further understanding of an issue) compared to possible increases in costs and time (additional drafting and meetings).

— The proposal to skip the CD stage should be made by the Working Group Convenor/Project Leader following a consultation with the WG experts to prove consensus.

— The final decision should then be taken by the parent committee by consensus through a 4 week Committee Internal Ballot or at a meeting (for a definition of consensus, see clause 2.5.6).

— In cases where there are concerns that skipping of the CD stage may seriously compromise consensus, then skipping the CD stage should be avoided.

SS.2 Tips for Convenors and Project Leaders when proposing to skip the CD stage

1) Use stage 0 to build a draft document which reflects already known stakeholder views.

2) Ensure a high quality draft is circulated with the New Work Item Proposal.

3) Inform the parent committee secretary as soon as it is known that the Working Group (WG) would like the project to skip the CD stage.

4) Ensure P-members with a major interest in the subject have nominated experts to the WG and that consensus has been achieved on the document among the WG.

5) Ensure appropriate number of re-iterations of Working Drafts circulations to the experts.

6) Give regular updates about the status of the work to the secretary of the committee who in turn should provide updates to the P-members and circulate working drafts for information if necessary.

7) Ask WG Experts to regularly inform their nominating P-members of the status of the work and encourage the WG experts to conduct informal consultations at the national level.

8) Ensure the document submitted to the committee secretary is of high technical quality which is good enough for a DIS vote.

9) Always ensure a full review of the DIS comments after the DIS vote.

SS.3 Tips for committee Chairs and Secretaries

— Inform the members of the committee as soon as it is known that the Working Group would like to skip the CD stage.

— When submitting the DIS draft to ISO CS for preparation of the vote send the draft to the committee for information.

SS.4 Other information

— This does not alter the current standards development timeframes which remain at 2, 3 or 4 years. The optional use of the CD will help you meet or beat your target dates that are set under these timeframes.

— For Vienna Agreement projects the CD can also be used on an optional basis.

Annex ST
(normative)

Twinning Policy

ST.1 Scope

The objective of twinning is to build capacity and improve participation and performance in the twinned partner. The twinning objectives should support those priorities identified by the twinned partner and be integrated into their national development plans/strategies.

This policy is mandatory.

ST.2 Definitions

ST.2.1 Twinning: A partnership agreement between two ISO members for the purposes of capacity building, where capacity building is defined as the "development of competencies (at National Body-NSB level) to be successful in ISO standardization work.

ST.2.2 Lead partner: The ISO member body providing the guidance, expertise and training.

ST.2.3 Twinned partner: The ISO member body seeking to build capacity.

ST.2.4 Leadership twinning: Twinning arrangements that involve learning the competencies relevant to leadership roles in the development of International Standards. There are three types of leadership twinning arrangements; chair, convenor and secretary. Through leadership twinning, the twinned partner has the opportunity to learn the skills and responsibilities needed to assume the chairmanship/convenorship/the role of a committee secretary.

ST.2.5 Chair twinning: An arrangement between a Chair and a Twinned-Chair.

ST.2.6 Convenor twinning: An arrangement between a Convenor and a Twinned Convenor.

ST.2.7 Secretariat twinning: An arrangement between a Secretariat and a Twinned Secretariat.

ST.2.8 P-member twinning: The lead partner and the twinned partner both need to be P-members of the committee in question. P-member Twinning allows the twinned partner, that may not yet have the level of competency required to cooperate on a leadership position, to participate actively in the technical work of standards development. P-member twinning allows members to gain specific targeted experience related to the work of a committee at the international and national levels (including the role of national mirror committees) and insight into the obligations of a P-member.

ST.3 Requirements

To be eligible to participate in a twinning arrangement, the following requirements need to be met:

— Full membership of ISO (or participation in the ISO Council's New member rights pilot, in the case of P-member Twinning).

— Requirement to be a lead partner: demonstration of sufficient knowledge and experience (at NSB level) of the role in question.

— Requirement to be a twinned partner: demonstration of a need for capacity building, as evidenced by a lack of experience holding secretariats /appointing chairs or convenors/ participating actively in technical committees.

Selection criteria

The suggested experience for a lead partner is:

— Secretariat twinning: holding 5 or more committee secretariats.

— Chair or convenor twinning: having 5 or more appointed chairs or convenors.

— P-member twinning: being a P-member in 100 or more committees.

The suggested experience for a twinned partner is:

— Secretariat twinning: holding fewer than 5 committee secretariats.

— Chair or convenor twinning: having fewer than 5 appointed chairs or convenors.

— P-member twinning: being a P-member in fewer than 100 committees.

ST.4 General principles

1) **Voluntary in nature** — Twinning arrangements are to be entered into voluntarily.

2) **Objectives** — Both partners to the arrangement shall be committed to the primary objective of twinning: to build capacity and improve participation and performance in the twinned partner. Twinning objectives should support those priorities identified by the twinned partner and integrated into their national development plans/strategies.

3) **No expectation of financial support** — There is no expectation that the lead partner will provide financial support to the twinned partner.

4) **One country, one vote** — Twinning arrangements shall have no influence in terms of national position. There should be no expectation that one partner will support the positions and views of the other partner. Twinning arrangements shall not jeopardize the principle of "one country, one vote" in ISO (and proxy voting is not allowed).

5) **Cooperation** — The cooperation should be viewed as broadly as possible, with a flexible approach, and include technical, strategic and policy-level activities between the partners.

6) **Commitment** — The commitment of both partners to the twinning arrangement should be sustained for the duration of the arrangement. By mutual agreement of the two partners, the lead partner may also assume other specific administrative tasks beyond providing training and guidance.

7) **Continual improvement** — Twinning arrangements should embrace the continual improvement concept. There should be mechanisms in place to ensure that, for each action, there is a possibility for follow-up and, if necessary, corrective action. Projects should include indicators to measure/monitor the progress of a twinned partner's participation and leadership in ISO, with progress reviewed on an annual basis.

8) **Sustainability** — The knowledge and experience gained through the twinning arrangement should be used to train other staff in the twinned partner to build capabilities in a sustainable manner. Participation in twinning should reduce, or remove altogether, the need for twinning of the twinned partner so that, once the twinning arrangement is finished, the twinned country can participate independently.

ST.5 Procedure

ST.5.1 Chronology

This Policy addresses the various aspects of a Twinning arrangement in the order of the life cycle of a typical Twinning arrangement noting that some stages may be done in parallel. Each stage is addressed as follows:

— Identifying the need for a Twinning arrangement (ST.5.2)

— Selecting a Twinning partner (ST.5.3)

— Drafting/Submitting a Twinning Request (ST.5.4)

— Approving the Twinning Request (ST.5.5)

— Implementing the Twinning arrangement (ST.5.6)

— Terminating a twinning arrangement (ST.5.7)

— Role of the ISO CS (ST.5.8)

ST.5.2 Identifying the need for a Twinning arrangement

Twinning arrangements are to be entered voluntarily. The partners shall fill in the Request Form (See Annex SJ) once they have agreed on the type of twinning (P-member, Secretariat, Chair or Convenor). In the case of a Secretariat or Chair twinning, the request form should be submitted to the TMB Secretariat. In the case of a WG Convenor twinning arrangement, it should be submitted directly to the parent committee and the ISO Technical Program Manager (TPM) for processing. For P-member twinning arrangements, the request form should be submitted to the TPM for review and the committee secretariat for information.

ST.5.3 Selecting a Twinning partner

If a member body has identified the need for a twinning arrangement, but not a twinning partner, it can contact ISO/CS or the TPM of the committee in question, if a committee has been identified, and ask for help in identifying a partner to twin with.

ST.5.4 Drafting/Submitting a Twinning Request

The following shall be included in the request form:

— Contact details of the ISO member bodies making the request

— Committee in which a twinning relationship is requested

— Type of twinning requested

— Statement from the ISO member body (twinned partner) demonstrating the need for capacity building and explaining what kind of capacity-building this arrangement will target

— Statement from the ISO member body (lead partner) demonstrating competencies and resources to support capacity building

— Goals and expected outcome of the twinning arrangement

— Proposed time frame of the twinning arrangement

— In the case of a leadership twinning arrangement, the contact details and CV of the Secretary/Twinned Secretary, Chair/Twinned-Chair, Convenor/Twinned Convenor

Secretariat or Chair twinning arrangements at TC, PC or SC level shall be approved by the TMB.

For WG Convenor Twinning arrangements, the requesting partners submit the Twinning request form directly to the secretariat of the parent committee for approval.

For P-member Twinning arrangements, the requesting partners submit the Twinning request form directly to the ISO/CS TPM (copied to the committee secretariat for information).

ST.5.5 Approving the Twinning request:

- Secretariat and Chair twinning arrangements shall be submitted to the TMB Secretariat who will send the Twinning request form and supporting material to the TMB members for approval by correspondence (4-week CIB ballot). Supporting material for the ballot shall include an assessment from the TPM/ISO CS, plus any relevant statistics on NSB participation in ISO and performance in other committee leadership roles. The request is evaluated by the TMB according to the quality of the information, and against the above guidelines (suggested experience) for being a "lead" and "twinned" partner.

  If approved by the TMB, the Secretariat or Chair twinning arrangement will be registered in the Global Directory (GD).

- For WG Convenor Twinning arrangements, the requesting NSBs submit the Twinning request form directly to the secretariat of the parent committee for approval. The committee then submits its decision to the TPM for processing.

- For P-member Twinning arrangements, the requesting NSBs submit the Twinning request form directly to the ISO/CS TPM (copied to the committee secretariat for information). The TPM reviews the form and registers the Twinning arrangement in the GD.

Once approved, the twinned partners shall submit a completed Twinning Agreement (with end date, maximum time limit of 5 years) and Collaboration Plan to their ISO/CS TPM within 3 months of the registration of the Twinning arrangement. See Annex SJ for the Forms. The TPM reviews the Agreement and Collaboration Plan and works with the twinned partners to improve it, if necessary. If the partners do not submit a satisfactory Twinning Agreement and Collaboration plan within 3 months, the twinning arrangement shall not be started.

ST.5.6 Implementing the Twinning arrangement

Twinning arrangements shall be subject to an annual review and performance assessment, consisting of a self-assessment by the twinned partners, an assessment by the P-members of the committee, and an assessment by the TPM.

Annual performance assessment form: The twinned partners shall submit a completed form once per year. This form includes a section that will be completed by the TPM.

P-member assessments will be done as part of the regular post-meeting feedback survey process. The TPM reviews the results of these surveys and notes the P-member feedback on the Annual performance assessment form.

The TMB secretariat reviews the performance assessment forms and sends the completed forms back to the twinned partners for information. If there are no problems raised, the form is filed and the arrangement continues.

ST.5.7 Terminating a Twinning arrangement

Failure to submit annual performance assessments or negative assessments/problems raised shall be brought to the attention of the TMB. The TMB evaluates the information and decides on any action to be taken. This may include the cancellation of the twinning arrangement.

Once a twinning arrangement reaches the 5-year maximum time limit, it is automatically terminated. Extensions/renewals of Twinning arrangements are not possible.

There shall be no automatic transfer of leadership roles (secretariat, chair, convenor) from the lead to the twinned partner at the end of the twinning arrangement. The twinned partner leadership roles shall be removed from the GD.

In the case of secretariat twinning, if the lead partner wishes to relinquish the secretariat, the committee may take a resolution to support the assignment of the secretariat to the twinned partner. In the case of an SC secretariat, the TC will approve this assignment based on the SC resolution. In the case of a TC secretariat, the ISO/TMB will approve this assignment based on the TC's resolution.

Any other changes to leadership roles shall follow the approval process outlined in the ISO/IEC Directives Part 1.

ST.5.8 Role of ISO/CS

The Twinned partners' interface with ISO/CS is through the responsible TPM. The role of the TPM includes:

— Helping to identify a potential twinning partner in cases where a twinning partner/lead partner has not been identified

— Providing guidance and advice on how to implement the twinning policy

— Raising potential problems to the attention of the TMB

Reference documents

The following are links to reference documents for the ISO technical community on a number of important subjects.

— ISO's web page (www.iso.org)

— ISO/IEC Directives, Parts 1 & 2, Consolidated ISO Supplement to ISO/IEC Directives, Part 1, JTC 1 Supplement (www.iso.org/directives)

— Guidance for ISO national standards bodies — engaging stakeholders and building consensus (http://www.iso.org/iso/guidance_nsb.pdf). See also additional guidance from the TMB on stakeholder engagement: (http://www.iso.org/iso/additional_guidance_on_stakeholder_engagement_tmb_peg.pdf)

— Guidance for liaisons organizations — engaging stakeholders and building consensus (http://www.iso.org/iso/guidance_liaison-organizations.pdf)

— Using and referencing ISO and IEC standards to support public policy (https://www.iso.org/publication/PUB100358.html)

— ISO's global relevance policy (http://www.iso.org/iso/home/standards_development/governance_of_technical_work.htm)

— ISO Code of Conduct for the technical work and suggestions for implementation (https://www.iso.org/publication/PUB100397.html)

— Policy for the distribution of ISO publications and the protection of ISO's copyright (ISO POCOSA) (https://connect.iso.org/display/marketing/ISO+POCOSA+2017)

— How to write standards (http://www.iso.org/iso/how-to-write-standards.pdf)

— Model manuscript of a draft International Standard (known as "The Rice Model") (https://www.iso.org/publication/PUB100407.html)

— ISO Guides — adding value to International Standards (http://www.iso.org/iso/iso-iec_guides.pdf)

— Guidelines for TC/SC Chairs and Secretariats for implementation of the Agreement on technical cooperation between ISO and CEN (Vienna Agreement) (http://isotc.iso.org/livelink/livelink/open/4230689)

— Guidance on twinning in ISO standards development activities (https://www.iso.org/publication/PUB100341.html)

— Guidance for writing standards taking into account micro, small and medium-sized enterprises' needs (http://www.iso.org/iso/guidance-for-writing-standards-for-smes.pdf)

— Guide for addressing sustainability in standards ("Guide 82") (http://isotc.iso.org/livelink/livelink?func=ll&objId=8389141&objAction=browse&viewType=1)

— Guidelines on competition law (http://www.iso.org/iso/competition_law_guidelines.pdf)

— Guidelines on remote participation at committee meetings
(http://isotc.iso.org/livelink/livelink/open/17857546)

— Guidelines for the submission of text and drawings to ISO/CS
(http://isotc.iso.org/livelink/livelink/open/18862226)

— Guidance on New Work in ISO
(https://isotc.iso.org/livelink/livelink/fetch/2000/2122/15507012/20090199/Guidance_on_New_Work.pdf?nodeid=20407727&vernum=-2)